D0085584

Reinforcement Learning and Dynamic Programming Using Function Approximators

AUTOMATION AND CONTROL ENGINEERING

A Series of Reference Books and Textbooks

Series Editors

FRANK L. LEWIS, Ph.D., Fellow IEEE, Fellow IFAC
Professor
Automation and Robotics Research Institute
The University of Texas at Arlington

SHUZHI SAM GE, Ph.D., Fellow IEEE
Professor
Interactive Digital Media Institute
The National University of Singapore

Reinforcement Learning and Dynamic Programming Using Function Approximators, *Lucian Buşoniu, Robert Babuška, Bart De Schutter, and Damien Ernst*

Modeling and Control of Vibration in Mechanical Systems, *Chunling Du and Lihua Xie*

Analysis and Synthesis of Fuzzy Control Systems: A Model-Based Approach, *Gang Feng*

Lyapunov-Based Control of Robotic Systems, *Aman Behal, Warren Dixon, Darren M. Dawson, and Bin Xian*

System Modeling and Control with Resource-Oriented Petri Nets, *Naiqi Wu and MengChu Zhou*

Sliding Mode Control in Electro-Mechanical Systems, Second Edition, *Vadim Utkin, Jürgen Guldner, and Jingxin Shi*

Optimal Control: Weakly Coupled Systems and Applications, *Zoran Gajić, Myo-Taeg Lim, Dobrila Skatarić, Wu-Chung Su, and Vojislav Kecman*

Intelligent Systems: Modeling, Optimization, and Control, *Yung C. Shin and Chengying Xu*

Optimal and Robust Estimation: With an Introduction to Stochastic Control Theory, Second Edition, *Frank L. Lewis, Lihua Xie, and Dan Popa*

Feedback Control of Dynamic Bipedal Robot Locomotion, *Eric R. Westervelt, Jessy W. Grizzle, Christine Chevallereau, Jun Ho Choi, and Benjamin Morris*

Intelligent Freight Transportation, *edited by Petros A. Ioannou*

Modeling and Control of Complex Systems, *edited by Petros A. Ioannou and Andreas Pitsillides*

Wireless Ad Hoc and Sensor Networks: Protocols, Performance, and Control, *Jagannathan Sarangapani*

Stochastic Hybrid Systems, *edited by Christos G. Cassandras and John Lygeros*

Hard Disk Drive: Mechatronics and Control, *Abdullah Al Mamun, Guo Xiao Guo, and Chao Bi*

Autonomous Mobile Robots: Sensing, Control, Decision Making and Applications, *edited by Shuzhi Sam Ge and Frank L. Lewis*

Automation and Control Engineering Series

Reinforcement Learning and Dynamic Programming Using Function Approximators

Lucian Buşoniu
Delft University of Technology
Delft, The Netherlands

Robert Babuška
Delft University of Technology
Delft, The Netherlands

Bart De Schutter
Delft University of Technology
Delft, The Netherlands

Damien Ernst
University of Liège
Liège, Belgium

CRC Press is an imprint of the
Taylor & Francis Group, an **informa** business

CRC Press
Taylor & Francis Group
6000 Broken Sound Parkway NW, Suite 300
Boca Raton, FL 33487-2742

Printed in the United States of America on acid-free paper
10 9 8 7 6 5 4 3 2 1

International Standard Book Number: 978-1-4398-2108-4 (Hardback)

Visit the Taylor & Francis Web site at
http://www.taylorandfrancis.com

and the CRC Press Web site at
http://www.crcpress.com

Preface

Control systems are making a tremendous impact on our society. Though invisible to most users, they are essential for the operation of nearly all devices – from basic home appliances to aircraft and nuclear power plants. Apart from technical systems, the principles of control are routinely applied and exploited in a variety of disciplines such as economics, medicine, social sciences, and artificial intelligence.

A common denominator in the diverse applications of control is the need to influence or modify the behavior of dynamic systems to attain prespecified goals. One approach to achieve this is to assign a numerical performance index to each state trajectory of the system. The control problem is then solved by searching for a control policy that drives the system along trajectories corresponding to the best value of the performance index. This approach essentially reduces the problem of finding good control policies to the search for solutions of a mathematical optimization problem.

Early work in the field of optimal control dates back to the 1940s with the pioneering research of Pontryagin and Bellman. Dynamic programming (DP), introduced by Bellman, is still among the state-of-the-art tools commonly used to solve optimal control problems when a system model is available. The alternative idea of finding a solution *in the absence* of a model was explored as early as the 1960s. In the 1980s, a revival of interest in this model-free paradigm led to the development of the field of reinforcement learning (RL). The central theme in RL research is the design of algorithms that learn control policies solely from the knowledge of transition samples or trajectories, which are collected beforehand or by online interaction with the system. Most approaches developed to tackle the RL problem are closely related to DP algorithms.

A core obstacle in DP and RL is that solutions cannot be represented exactly for problems with large discrete state-action spaces or continuous spaces. Instead, compact representations relying on function approximators must be used. This challenge was already recognized while the first DP techniques were being developed. However, it has only been in recent years – and largely in correlation with the advance of RL – that approximation-based methods have grown in diversity, maturity, and efficiency, enabling RL and DP to scale up to realistic problems.

This book provides an accessible in-depth treatment of reinforcement learning and dynamic programming methods using function approximators. We start with a concise introduction to classical DP and RL, in order to build the foundation for the remainder of the book. Next, we present an extensive review of state-of-the-art approaches to DP and RL with approximation. Theoretical guarantees are provided on the solutions obtained, and numerical examples and comparisons are used to illustrate the properties of the individual methods. The remaining three chapters are dedicated to a detailed presentation of representative algorithms from the three major classes of techniques: value iteration, policy iteration, and policy search. The properties and the performance of these algorithms are highlighted in simulation and experimental studies on a range of control applications.

We believe that this balanced combination of practical algorithms, theoretical analysis, and comprehensive examples makes our book suitable not only for researchers, teachers, and graduate students in the fields of optimal and adaptive control, machine learning and artificial intelligence, but also for practitioners seeking novel strategies for solving challenging real-life control problems.

This book can be read in several ways. Readers unfamiliar with the field are advised to start with Chapter 1 for a gentle introduction, and continue with Chapter 2 (which discusses classical DP and RL) and Chapter 3 (which considers approximation-based methods). Those who are familiar with the basic concepts of RL and DP may consult the list of notations given at the end of the book, and then start directly with Chapter 3. This first part of the book is sufficient to get an overview of the field. Thereafter, readers can pick any combination of Chapters 4 to 6, depending on their interests: approximate value iteration (Chapter 4), approximate policy iteration and online learning (Chapter 5), or approximate policy search (Chapter 6).

Supplementary information relevant to this book, including a complete archive of the computer code used in the experimental studies, is available at the Web site:

http://www.dcsc.tudelft.nl/rlbook/

Comments, suggestions, or questions concerning the book or the Web site are welcome. Interested readers are encouraged to get in touch with the authors using the contact information on the Web site.

The authors have been inspired over the years by many scientists who undoubtedly left their mark on this book; in particular by Louis Wehenkel, Pierre Geurts, Guy-Bart Stan, Rémi Munos, Martin Riedmiller, and Michail Lagoudakis. Pierre Geurts also provided the computer program for building ensembles of regression trees, used in several examples in the book. This work would not have been possible without our colleagues, students, and the excellent professional environments at the Delft Center for Systems and Control of the Delft University of Technology, the Netherlands, the Montefiore Institute of the University of Liège, Belgium, and at Supélec Rennes, France. Among our colleagues in Delft, Justin Rice deserves special mention for carefully proofreading the manuscript. To all these people we extend our sincere thanks.

We thank Sam Ge for giving us the opportunity to publish our book with Taylor & Francis CRC Press, and the editorial and production team at Taylor & Francis for their valuable help. We gratefully acknowledge the financial support of the BSIK-ICIS project "Interactive Collaborative Information Systems" (grant no. BSIK03024) and the Dutch funding organizations NWO and STW. Damien Ernst is a Research Associate of the FRS-FNRS, the financial support of which he acknowledges. We appreciate the kind permission offered by the IEEE to reproduce material from our previous works over which they hold copyright.

Finally, we thank our families for their continual understanding, patience, and support.

Lucian Buşoniu
Robert Babuška
Bart De Schutter
Damien Ernst
November 2009

About the authors

Lucian Buşoniu is a postdoctoral fellow at the Delft Center for Systems and Control of Delft University of Technology, in the Netherlands. He received his PhD degree (*cum laude*) in 2009 from the Delft University of Technology, and his MSc degree in 2003 from the Technical University of Cluj-Napoca, Romania. His current research interests include reinforcement learning and dynamic programming with function approximation, intelligent and learning techniques for control problems, and multi-agent learning.

Robert Babuška is a full professor at the Delft Center for Systems and Control of Delft University of Technology in the Netherlands. He received his PhD degree (*cum laude*) in Control in 1997 from the Delft University of Technology, and his MSc degree (with honors) in Electrical Engineering in 1990 from Czech Technical University, Prague. His research interests include fuzzy systems modeling and identification, data-driven construction and adaptation of neuro-fuzzy systems, model-based fuzzy control and learning control. He is active in applying these techniques in robotics, mechatronics, and aerospace.

Bart De Schutter is a full professor at the Delft Center for Systems and Control and at the Marine & Transport Technology department of Delft University of Technology in the Netherlands. He received the PhD degree in Applied Sciences (*summa cum laude* with congratulations of the examination jury) in 1996 from K.U. Leuven, Belgium. His current research interests include multi-agent systems, hybrid systems control, discrete-event systems, and control of intelligent transportation systems.

Damien Ernst received the MSc and PhD degrees from the University of Liège in 1998 and 2003, respectively. He is currently a Research Associate of the Belgian FRS-FNRS and he is affiliated with the Systems and Modeling Research Unit of the University of Liège. Damien Ernst spent the period 2003–2006 with the University of Liège as a Postdoctoral Researcher of the FRS-FNRS and held during this period positions as visiting researcher at CMU, MIT and ETH. He spent the academic year 2006–2007 working at Supélec (France) as professor. His main research interests are in the fields of power system dynamics, optimal control, reinforcement learning, and design of dynamic treatment regimes.

About the authors

Contents

1

Introduction

Dynamic programming (DP) and reinforcement learning (RL) are algorithmic methods for solving problems in which actions (decisions) are applied to a system over an extended period of time, in order to achieve a desired goal. DP methods require a model of the system's behavior, whereas RL methods do not. The time variable is usually discrete and actions are taken at every discrete time step, leading to a sequential decision-making problem. The actions are taken in closed loop, which means that the outcome of earlier actions is monitored and taken into account when choosing new actions. Rewards are provided that evaluate the one-step decision-making performance, and the goal is to optimize the long-term performance, measured by the total reward accumulated over the course of interaction.

Such decision-making problems appear in a wide variety of fields, including automatic control, artificial intelligence, operations research, economics, and medicine. For instance, in automatic control, as shown in Figure 1.1(a), a controller receives output measurements from a process, and applies actions to this process in order to make its behavior satisfy certain requirements (Levine, 1996). In this context, DP and RL methods can be applied to solve optimal control problems, in which the behavior of the process is evaluated using a cost function that plays a similar role to the rewards. The decision maker is the controller, and the system is the controlled process.

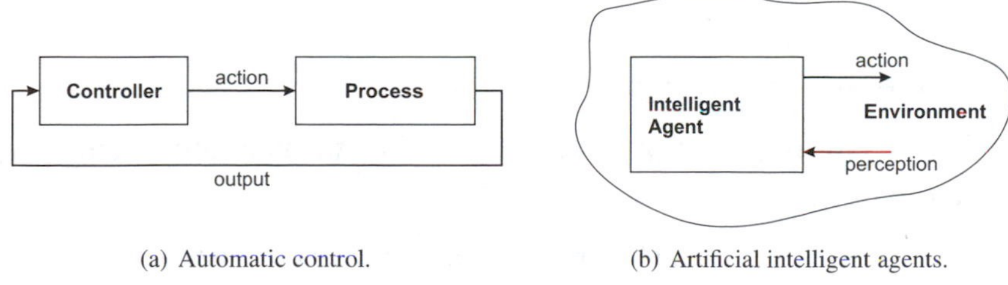

(a) Automatic control. (b) Artificial intelligent agents.

FIGURE 1.1
Two application domains for dynamic programming and reinforcement learning.

In artificial intelligence, DP and RL are useful to obtain optimal behavior for intelligent agents, which, as shown in Figure 1.1(b), monitor their environment through perceptions and influence it by applying actions (Russell and Norvig, 2003). The decision maker is now the agent, and the system is the agent's environment.

If a model of the system is available, DP methods can be applied. A key benefit of DP methods is that they make few assumptions on the system, which can generally be nonlinear and stochastic (Bertsekas, 2005a, 2007). This is in contrast to, e.g., classical techniques from automatic control, many of which require restrictive assumptions on the system, such as linearity or determinism. Moreover, many DP methods do not require an analytical expression of the model, but are able to work with a simulation model instead. Constructing a simulation model is often easier than deriving an analytical model, especially when the system behavior is stochastic.

However, sometimes a model of the system cannot be obtained at all, e.g., because the system is not fully known beforehand, is insufficiently understood, or obtaining a model is too costly. RL methods are helpful in this case, since they work using only data obtained from the system, without requiring a model of its behavior (Sutton and Barto, 1998). Offline RL methods are applicable if data can be obtained in advance. Online RL algorithms learn a solution by interacting with the system, and can therefore be applied even when data is not available in advance. For instance, intelligent agents are often placed in environments that are not fully known beforehand, which makes it impossible to obtain data in advance. Note that RL methods can, of course, also be applied when a model is available, simply by using the model instead of the real system to generate data.

In this book, we primarily adopt a control-theoretic point of view, and hence employ control-theoretical notation and terminology, and choose control systems as examples to illustrate the behavior of DP and RL algorithms. We nevertheless also exploit results from other fields, in particular the strong body of RL research from the field of artificial intelligence. Moreover, the methodology we describe is applicable to sequential decision problems in many other fields.

The remainder of this introductory chapter is organized as follows. In Section 1.1, an outline of the DP/RL problem and its solution is given. Section 1.2 then introduces the challenge of approximating the solution, which is a central topic of this book. Finally, in Section 1.3, the organization of the book is explained.

1.1 The dynamic programming and reinforcement learning problem

The main elements of the DP and RL problem, together with their flow of interaction, are represented in Figure 1.2: a controller interacts with a process by means of states and actions, and receives rewards according to a reward function. For the DP and RL algorithms considered in this book, an important requirement is the availability of a signal that completely describes the current state of the process (this requirement will be formalized in Chapter 2). This is why the process shown in Figure 1.2 outputs a state signal.

To clarify the meaning of the elements of Figure 1.2, we use a conceptual robotic navigation example. Autonomous mobile robotics is an application domain where automatic control and artificial intelligence meet in a natural way, since a mobile

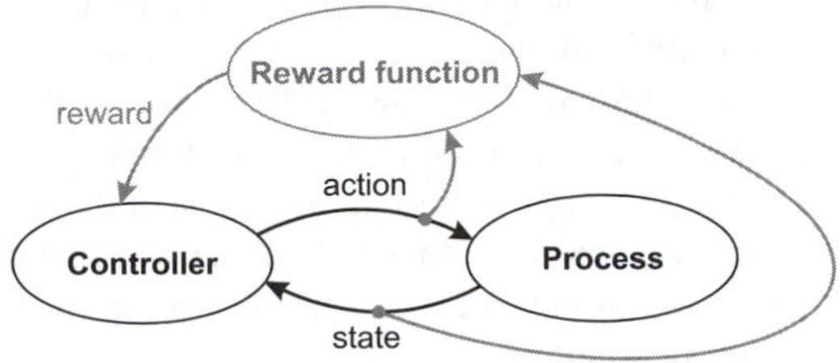

FIGURE 1.2
The elements of DP and RL and their flow of interaction. The elements related to the reward are depicted in gray.

robot and its environment comprise a process that must be controlled, while the robot is also an artificial agent that must accomplish a task in its environment. Figure 1.3 presents the navigation example, in which the robot shown in the bottom region must navigate to the goal on the top-right, while avoiding the obstacle represented by a gray block. (For instance, in the field of rescue robotics, the goal might represent the location of a victim to be rescued.) The controller is the robot's software, and the process consists of the robot's environment (the surface on which it moves, the obstacle, and the goal) together with the body of the robot itself. It should be emphasized that in DP and RL, the physical body of the decision-making entity (if it has one), its sensors and actuators, as well as any fixed lower-level controllers, are all considered to be a part of the process, whereas the controller is taken to be only the decision-making algorithm.

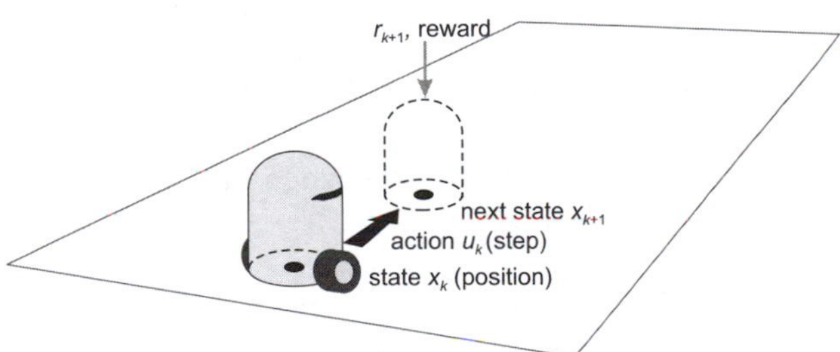

FIGURE 1.3
A robotic navigation example. An example transition is also shown, in which the current and next states are indicated by black dots, the action by a black arrow, and the reward by a gray arrow. The dotted silhouette represents the robot in the next state.

In the navigation example, the state is the position of the robot on the surface, given, e.g., in Cartesian coordinates, and the action is a step taken by the robot, similarly given in Cartesian coordinates. As a result of taking a step from the current position, the next position is obtained, according to a transition function. In this example, because both the positions and steps are represented in Cartesian coordinates, the transitions are most often additive: the next position is the sum of the current position and the step taken. More complicated transitions are obtained if the robot collides with the obstacle. Note that for simplicity, most of the dynamics of the robot, such as the motion of the wheels, have not been taken into account here. For instance, if the wheels can slip on the surface, the transitions become stochastic, in which case the next state is a random variable.

The quality of every transition is measured by a reward, generated according to the reward function. For instance, the reward could have a positive value such as 10 if the robot reaches the goal, a negative value such as -1, representing a penalty, if the robot collides with the obstacle, and a neutral value of 0 for any other transition. Alternatively, more informative rewards could be constructed, using, e.g., the distances to the goal and to the obstacle.

The behavior of the controller is dictated by its policy: a mapping from states into actions, which indicates what action (step) should be taken in each state (position).

In general, the state is denoted by x, the action by u, and the reward by r. These quantities may be subscripted by discrete time indices, where k denotes the current time index (see Figure 1.3). The transition function is denoted by f, the reward function by ρ, and the policy by h.

In DP and RL, the goal is to maximize the return, consisting of the cumulative reward over the course of interaction. We mainly consider discounted infinite-horizon returns, which accumulate rewards obtained along (possibly) infinitely long trajectories starting at the initial time step $k = 0$, and weigh the rewards by a factor that decreases exponentially as the time step increases:

$$\gamma^0 r_1 + \gamma^1 r_2 + \gamma^2 r_3 + \dots \tag{1.1}$$

The discount factor $\gamma \in [0, 1)$ gives rise to the exponential weighting, and can be seen as a measure of how "far-sighted" the controller is in considering its rewards. Figure 1.4 illustrates the computation of the discounted return for the navigation problem of Figure 1.3.

The rewards depend of course on the state-action trajectory followed, which in turn depends on the policy being used:

$$x_0,\ u_0 = h(x_0),\ x_1,\ u_1 = h(x_1),\ x_2,\ u_2 = h(x_2),\ \dots$$

In particular, each reward r_{k+1} is the result of the transition (x_k, u_k, x_{k+1}). It is convenient to consider the return separately for every initial state x_0, which means the return is a function of the initial state. Note that, if state transitions are stochastic, the goal considered in this book is to maximize the expectation of (1.1) over all the realizations of the stochastic trajectory starting from x_0.

The core challenge of DP and RL is therefore to arrive at a solution that optimizes the long-term performance given by the return, using only reward information that describes the immediate performance. Solving the DP/RL problem boils down to finding an optimal policy, denoted by h^*, that maximizes the return (1.1) for every

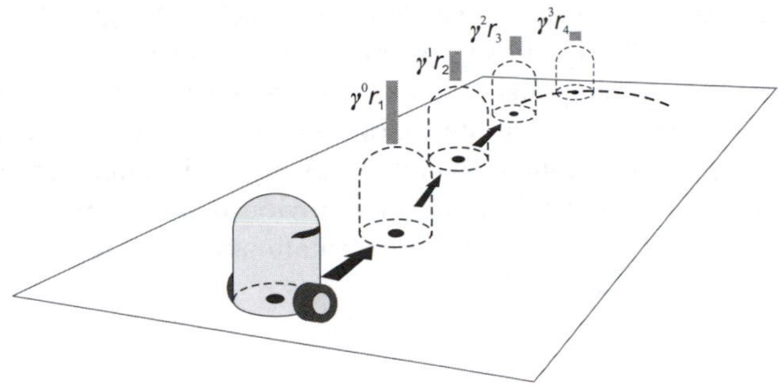

FIGURE 1.4
The discounted return along a trajectory of the robot. The decreasing heights of the gray vertical bars indicate the exponentially diminishing nature of the discounting applied to the rewards.

initial state. One way to obtain an optimal policy is to first compute the maximal returns. For example, the so-called optimal Q-function, denoted by Q^*, contains for each state-action pair (x,u) the return obtained by first taking action u in state x and then choosing optimal actions from the second step onwards:

$$Q^*(x,u) = \gamma^0 r_1 + \gamma^1 r_2 + \gamma^2 r_3 + \ldots$$
$$\text{when } x_0 = x, u_0 = u, \text{ and optimal actions are taken for } x_1, x_2, \ldots \tag{1.2}$$

If transitions are stochastic, the optimal Q-function is defined instead as the expectation of the return on the right-hand side of (1.2) over the trajectory realizations. The optimal Q-function can be found using a suitable DP or RL algorithm. Then, an optimal policy can be obtained by choosing, at each state x, an action $h^*(x)$ that maximizes the optimal Q-function for that state:

$$h^*(x) \in \arg\max_u Q^*(x,u) \tag{1.3}$$

To see that an optimal policy is obtained, recall that the optimal Q-function already contains optimal returns starting from the second step onwards; in (1.3), an action is chosen that additionally maximizes the return over the first step, therefore obtaining a return that is maximal over the entire horizon, i.e., optimal.

1.2 Approximation in dynamic programming and reinforcement learning

Consider the problem of representing a Q-function, not necessarily the optimal one. Since no prior knowledge about the Q-function is available, the only way to guarantee an exact representation is to store distinct values of the Q-function (Q-values) for every state-action pair. This is schematically depicted in Figure 1.5 for the navigation example of Section 1.1: Q-values must be stored separately for each position of the robot, and for each possible step that it might take from every such position. However, because the position and step variables are continuous, they can both take uncountably many distinct values. Therefore, even in this simple example, storing distinct Q-values for every state-action pair is obviously impossible. The only feasible way to proceed is to use a compact representation of the Q-function.

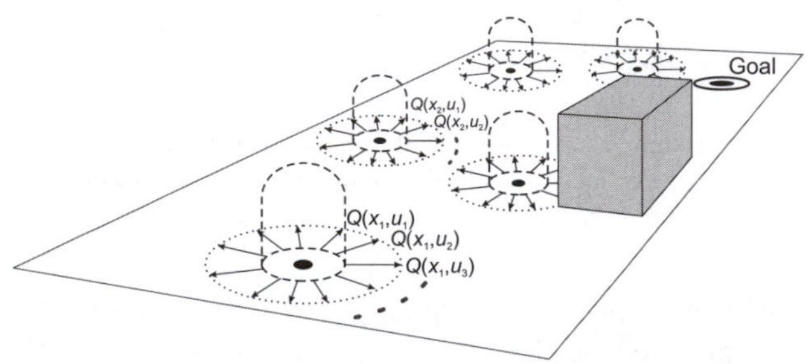

FIGURE 1.5
Illustration of an exact Q-function representation for the navigation example. For every state-action pair, there is a corresponding Q-value. The Q-values are not represented explicitly, but only shown symbolically near corresponding state-action pairs.

One type of compact Q-function representation that will often be used in the sequel relies on state-dependent basis functions (BFs) and action discretization. Such a representation is illustrated in Figure 1.6 for the navigation problem. A finite number of BFs, $\phi_1, \ldots, \phi_N$, are defined over the state space, and the action space is discretized into a finite number of actions, in this case 4: left, right, forward, and back. Instead of storing distinct Q-values for every state-action pair, such a representa-

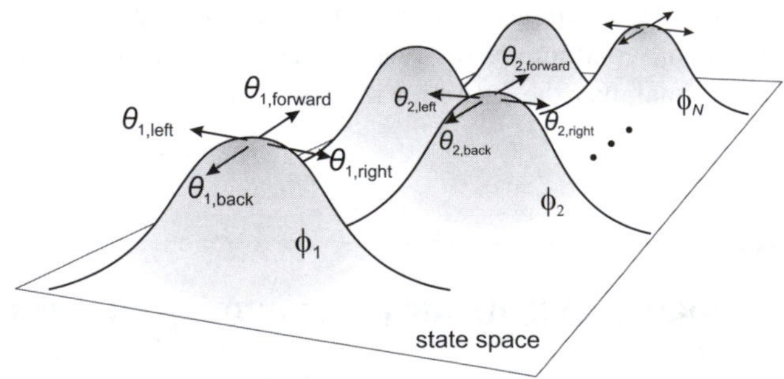

FIGURE 1.6
Illustration of a compact Q-function representation for the navigation example.

tion stores *parameters* θ, one for each combination of a BF and a discrete action. To find the Q-value of a continuous state-action pair (x,u), the action is discretized (e.g., using nearest-neighbor discretization). Assume the result of discretization is the discrete action "forward"; then, the Q-value is computed by adding the parameters $\theta_{1,\text{forward}},\ldots,\theta_{N,\text{forward}}$ corresponding to this discrete action, where the parameters are weighted by the value of their corresponding BFs at x:

$$\widehat{Q}(x,\text{forward}) = \sum_{i=1}^{N} \phi_i(x)\theta_{i,\text{forward}} \tag{1.4}$$

The DP/RL algorithm therefore only needs to remember the $4N$ parameters, which can easily be done when N is not too large. Note that this type of Q-function representation generalizes to any DP/RL problem. Even in problems with a finite number of discrete states and actions, compact representations can still be useful by reducing the number of values that must be stored.

While not all DP and RL algorithms employ Q-functions, they all generally require compact representations, so the illustration above extends to the general case. Consider, e.g., the problem of representing a policy h. An exact representation would generally require storing distinct actions for every possible state, which is impossible when the state variables are continuous. Note that continuous actions are not problematic for policy representation.

It should be emphasized at this point that, in general, a compact representation can only represent the target function up to a certain approximation error, which must be accounted for. Hence, in the sequel such representations are called "function approximators," or "approximators" for short.

Approximation in DP and RL is not only a problem of representation. Assume for instance that an approximation of the optimal Q-function is available. To obtain an approximately optimal policy, (1.3) must be applied, which requires maximizing the Q-function over the action variable. In large or continuous action spaces, this is a potentially difficult optimization problem, which can only be solved approximately in general. However, when a discrete-action Q-function of the form (1.4) is employed, it is sufficient to compute the Q-values of all the discrete actions and to find the maximum among these values using enumeration. This provides a motivation for using discretized actions. Besides approximate maximization, other approximation difficulties also arise, such as the estimation of expected values from samples. These additional challenges are outside the scope of this section, and will be discussed in detail in Chapter 3.

The classical DP and RL algorithms are only guaranteed to obtain an optimal solution if they use exact representations. Therefore, the following important questions must be kept in mind when using function approximators:

- If the algorithm is iterative, does it *converge* when approximation is employed? Or, if the algorithm is not iterative, does it obtain a meaningful solution?
- If a meaningful solution is obtained, is it *near optimal*, and more specifically, how far is it from the optimal solution?

- Is the algorithm *consistent*, i.e., does it asymptotically obtain the optimal solution as the approximation power grows?

These questions will be taken into account when discussing algorithms for approximate DP and RL.

Choosing an appropriate function approximator for a given problem is a highly nontrivial task. The complexity of the approximator must be managed, since it directly influences the memory and computational costs of the DP and RL algorithm. This is an important concern in both approximate DP and approximate RL. Equally important in approximate RL are the restrictions imposed by the limited amount of data available, since in general a more complex approximator requires more data to compute an accurate solution. If prior knowledge about the function of interest is available, it can be used in advance to design a lower-complexity, but still accurate, approximator. For instance, BFs with intuitive, relevant meanings could be defined (such as, in the navigation problem, BFs representing the distance between the robot and the goal or the obstacle). However, prior knowledge is often unavailable, especially in the model-free context of RL. In this book, we will therefore pay special attention to techniques that automatically find low-complexity approximators suited to the problem at hand, rather than relying on manual design.

1.3 About this book

This book focuses on approximate dynamic programming (DP) and reinforcement learning (RL) for control problems with continuous variables. The material is aimed at researchers, practitioners, and graduate students in the fields of systems and control (in particular optimal, adaptive, and learning control), computer science (in particular machine learning and artificial intelligence), operations research, and statistics. Although not primarily intended as a textbook, our book can nevertheless be used as support for courses that treat DP and RL methods.

Figure 1.7 presents a road map for the remaining chapters of this book, which we will detail next. Chapters 2 and 3 are prerequisite for the remainder of the book and should be read in sequence. In particular, in Chapter 2 the DP and RL problem and its solution are formalized, representative classical algorithms are introduced, and the behavior of several such algorithms is illustrated in an example with discrete states and actions. Chapter 3 gives an extensive account of DP and RL methods with function approximation, which are applicable to large and continuous-space problems. A comprehensive selection of algorithms is introduced, theoretical guarantees are provided on the approximate solutions obtained, and numerical examples involving the control of a continuous-variable system illustrate the behavior of several representative algorithms.

The material of Chapters 2 and 3 is organized along three major classes of DP and RL algorithms: value iteration, policy iteration, and policy search. In order to strengthen the understanding of these three classes of algorithms, each of the three

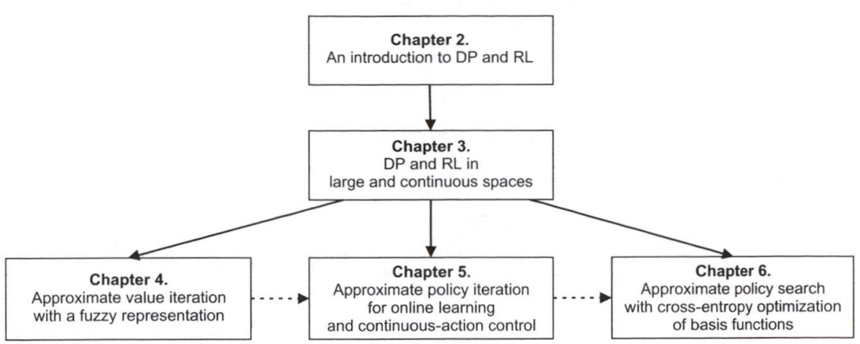

FIGURE 1.7
A road map for the remainder of this book, given in a graphical form. The full arrows indicate the recommended sequence of reading, whereas dashed arrows indicate optional ordering.

final chapters of the book considers in detail an algorithm from one of these classes. Specifically, in Chapter 4, a value iteration algorithm with fuzzy approximation is discussed, and an extensive theoretical analysis of this algorithm illustrates how convergence and consistency guarantees can be developed for approximate DP. In Chapter 5, an algorithm for approximate policy iteration is discussed. In particular, an online variant of this algorithm is developed, and some important issues that appear in online RL are emphasized along the way. In Chapter 6, a policy search approach relying on the cross-entropy method for optimization is described, which highlights one possibility to develop techniques that scale to relatively high-dimensional state spaces, by focusing the computation on important initial states. The final part of each of these three chapters contains an experimental evaluation on a representative selection of control problems.

Chapters 4, 5, and 6 can be read in any order, although, if possible, they should be read in sequence.

Two appendices are included at the end of the book (these are not shown in Figure 1.7). Appendix A outlines the so-called ensemble of extremely randomized trees, which is used as an approximator in Chapters 3 and 4. Appendix B describes the cross-entropy method for optimization, employed in Chapters 4 and 6. Reading Appendix B before these two chapters is not mandatory, since both chapters include a brief, specialized introduction to the cross-entropy method, so that they can more easily be read independently.

Additional information and material concerning this book, including the computer code used in the experimental studies, is available at the Web site:

http://www.dcsc.tudelft.nl/rlbook/

2

An introduction to dynamic programming and reinforcement learning

This chapter introduces dynamic programming and reinforcement learning techniques, and the formal model behind the problem they solve: the Markov decision process. Deterministic and stochastic Markov decision processes are discussed in turn, and their optimal solution is characterized. Three categories of dynamic programming and reinforcement learning algorithms are described: value iteration, policy iteration, and policy search.

2.1 Introduction

In dynamic programming (DP) and reinforcement learning (RL), a controller (agent, decision maker) interacts with a process (environment), by means of three signals: a state signal, which describes the state of the process, an action signal, which allows the controller to influence the process, and a scalar reward signal, which provides the controller with feedback on its immediate performance. At each discrete time step, the controller receives the state measurement and applies an action, which causes the process to transition into a new state. A reward is generated that evaluates the quality of this transition. The controller receives the new state measurement, and the whole cycle repeats. This flow of interaction is represented in Figure 2.1 (repeated from Figure 1.2).

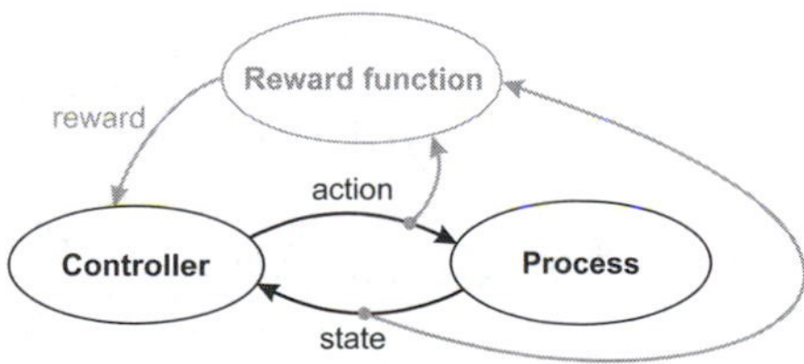

FIGURE 2.1 The flow of interaction in DP and RL.

The behavior of the controller is dictated by its policy, a function from states into actions. The behavior of the process is described by its dynamics, which determine how the state changes as a result of the controller's actions. State transitions can be deterministic or stochastic. In the deterministic case, taking a given action in a given state always results in the same next state, while in the stochastic case, the next state is a random variable. The rule according to which rewards are generated is described by the reward function. The process dynamics and the reward function, together with the set of possible states and the set of possible actions (respectively called state space and action space), constitute a so-called Markov decision process (MDP).

In the DP/RL setting, the goal is to find an optimal policy that maximizes the (expected) return, consisting of the (expected) cumulative reward over the course of interaction. In this book, we will mainly consider infinite-horizon returns, which accumulate rewards along infinitely long trajectories. This choice is made because infinite-horizon returns have useful theoretical properties. In particular, they lead to stationary optimal policies, which means that for a given state, the optimal action choices will always be the same, regardless of the time when that state is encountered.

The DP/RL framework can be used to address problems from a variety of fields, including, e.g., automatic control, artificial intelligence, operations research, and economics. Automatic control and artificial intelligence are arguably the most important fields of origin for DP and RL. In automatic control, DP can be used to solve nonlinear and stochastic optimal control problems (Bertsekas, 2007), while RL can alternatively be seen as adaptive optimal control (Sutton et al., 1992; Vrabie et al., 2009). In artificial intelligence, RL helps to build an artificial agent that learns how to survive and optimize its behavior in an unknown environment, without requiring prior knowledge (Sutton and Barto, 1998). Because of this mixed inheritance, two sets of equivalent names and notations are used in DP and RL, e.g., "controller" has the same meaning as "agent," and "process" has the same meaning as "environment." In this book, we will use the former, control-theoretical terminology and notation.

A taxonomy of DP and RL algorithms is shown in Figure 2.2 and detailed in the remainder of this section.

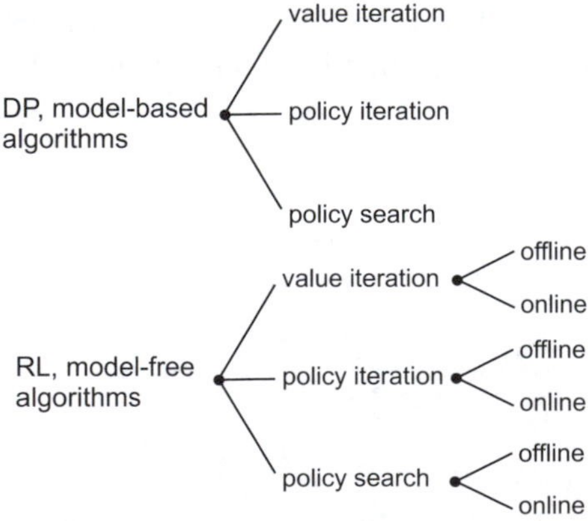

FIGURE 2.2 A taxonomy of DP and RL algorithms.

DP algorithms require a model of the MDP, including the transition dynamics and the reward function, to find an optimal policy (Bertsekas, 2007; Powell, 2007). The model DP algorithms work offline, producing a policy which is then used to control the process.[1] Usually, they do not require an analytical expression of the dynamics. Instead, given a state and an action, the model is only required to generate a next state and the corresponding reward. Constructing such a generative model is often easier than deriving an analytical expression of the dynamics, especially when the dynamics are stochastic.

RL algorithms are model-free (Bertsekas and Tsitsiklis, 1996; Sutton and Barto, 1998), which makes them useful when a model is difficult or costly to construct. RL algorithms use data obtained from the process, in the form of a set of samples, a set of process trajectories, or a single trajectory. So, RL can be seen as model-free, sample-based or trajectory-based DP, and DP can be seen as model-based RL. While DP algorithms can use the model to obtain any number of sample transitions from any state-action pair, RL algorithms must work with the limited data that can be obtained from the process – a greater challenge. Note that some RL algorithms build a model from the data; we call these algorithms "model-learning."

Both the DP and RL classes of algorithms can be broken down into three subclasses, according to the path taken to find an optimal policy. These three subclasses are value iteration, policy iteration, and policy search, and are characterized as follows.

- *Value iteration* algorithms search for the optimal value function, which consists of the maximal returns from every state or from every state-action pair. The optimal value function is used to compute an optimal policy.

- *Policy iteration* algorithms evaluate policies by constructing their value functions (instead of the optimal value function), and use these value functions to find new, improved policies.

- *Policy search* algorithms use optimization techniques to directly search for an optimal policy.

Note that, in this book, we use the name DP to refer to the class of all model-based algorithms that find solutions for MDPs, including model-based policy search. This class is larger than the category of algorithms traditionally called DP, which only includes model-based value iteration and policy iteration (Bertsekas and Tsitsiklis, 1996; Sutton and Barto, 1998; Bertsekas, 2007).

Within each of the three subclasses of RL algorithms, two categories can be further distinguished, namely offline and online algorithms. Offline RL algorithms use data collected in advance, whereas online RL algorithms learn a solution by interacting with the process. Online RL algorithms are typically not provided with any data

[1]There is also a class of model-based, DP-like *online* algorithms called model-predictive control (Maciejowski, 2002; Camacho and Bordons, 2004). In order to restrict the scope of the book, we do not discuss model-predictive control. For details about the relationship of DP/RL with model-predictive control, see, e.g., (Bertsekas, 2005b; Ernst et al., 2009).

in advance, but instead have to rely only on the data they collect while learning, and thus are useful when it is difficult or costly to obtain data in advance. Most online RL algorithms work incrementally. For instance, an incremental, online value iteration algorithm updates its estimate of the optimal value function after each collected sample. Even before this estimate becomes accurate, it is used to derive estimates of an optimal policy, which are then used to collect new data.

Online RL algorithms must balance the need to collect informative data (by *exploring* novel action choices or novel parts of the state space) with the need to control the process well (by *exploiting* the currently available knowledge). This exploration-exploitation trade-off makes online RL more challenging than offline RL. Note that, although online RL algorithms are only guaranteed (under appropriate conditions) to converge to an optimal policy when the process does not change over time, in practice they are sometimes applied also to slowly changing processes, in which case they are expected to adapt the solution so that the changes are taken into account.

The remainder of this chapter is structured as follows. Section 2.2 describes MDPs and characterizes the optimal solution for an MDP, in the deterministic as well as in the stochastic setting. The class of value iteration algorithms is introduced in Section 2.3, policy iteration in Section 2.4, and policy search in Section 2.5. When introducing value iteration and policy iteration, DP and RL algorithms are described in turn, while the introduction of policy search focuses on the model-based, DP setting. Section 2.6 concludes the chapter with a summary and discussion. Throughout the chapter, a simulation example involving a highly abstracted robotic task is employed to illustrate certain theoretical points, as well as the properties of several representative algorithms.

2.2 Markov decision processes

DP and RL problems can be formalized with the help of MDPs (Puterman, 1994). We first present the simpler case of MDPs with deterministic state transitions. Afterwards, we extend the theory to the stochastic case.

2.2.1 Deterministic setting

A deterministic MDP is defined by the state space X of the process, the action space U of the controller, the transition function f of the process (which describes how the state changes as a result of control actions), and the reward function ρ (which evaluates the immediate control performance).[2] As a result of the action u_k applied in the state x_k at the discrete time step k, the state changes to x_{k+1}, according to the

[2] As mentioned earlier, control-theoretic notation is used instead of artificial intelligence notation. For instance, in the artificial intelligence literature on DP and RL, the state space is usually denoted by S, the state by s, the action space by A, the action by a, and the policy by π.

transition function $f : X \times U \to X$:

$$x_{k+1} = f(x_k, u_k)$$

At the same time, the controller receives the scalar reward signal r_{k+1}, according to the *reward function* $\rho : X \times U \to \mathbb{R}$:

$$r_{k+1} = \rho(x_k, u_k)$$

where we assume that $\|\rho\|_\infty = \sup_{x,u} |\rho(x,u)|$ is finite.[3] The reward evaluates the immediate effect of action u_k, namely the transition from x_k to x_{k+1}, but in general does not say anything about its long-term effects.

The controller chooses actions according to its *policy* $h : X \to U$, using:

$$u_k = h(x_k)$$

Given f and ρ, the current state x_k and the current action u_k are sufficient to determine both the next state x_{k+1} and the reward r_{k+1}. This is the Markov property, which is essential in providing theoretical guarantees about DP/RL algorithms.

Some MDPs have terminal states that, once reached, can no longer be left; all the rewards received in terminal states are 0. The RL literature often uses "trials" or "episodes" to refer to trajectories starting from some initial state and ending in a terminal state.

Example 2.1 The deterministic cleaning-robot MDP. Consider the deterministic problem depicted in Figure 2.3: a cleaning robot has to collect a used can and also has to recharge its batteries.

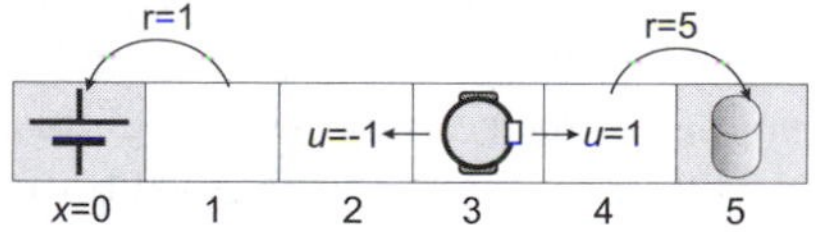

FIGURE 2.3 The cleaning-robot problem.

In this problem, the state x describes the position of the robot, and the action u describes the direction of its motion. The state space is discrete and contains six distinct states, denoted by integers 0 to 5: $X = \{0, 1, 2, 3, 4, 5\}$. The robot can move to the left ($u = -1$) or to the right ($u = 1$); the discrete action space is therefore $U = \{-1, 1\}$. States 0 and 5 are terminal, meaning that once the robot reaches either of them it can no longer leave, regardless of the action. The corresponding transition function is:

$$f(x,u) = \begin{cases} x + u & \text{if } 1 \le x \le 4 \\ x & \text{if } x = 0 \text{ or } x = 5 \text{ (regardless of } u\text{)} \end{cases}$$

[3]To simplify the notation, whenever searching for extrema, performing summations, etc., over variables whose domains are obvious from the context, we omit these domains from the formulas. For instance, in the formula $\sup_{x,u} |\rho(x,u)|$, the domains of x and u are clearly X and U, so they are omitted.

In state 5, the robot finds a can and the transition into this state is rewarded with 5. In state 0, the robot can recharge its batteries and the transition into this state is rewarded with 1. All other rewards are 0. In particular, taking any action while in a terminal state results in a reward of 0, which means that the robot will not accumulate (undeserved) rewards in the terminal states. The corresponding reward function is:

$$\rho(x,u) = \begin{cases} 5 & \text{if } x = 4 \text{ and } u = 1 \\ 1 & \text{if } x = 1 \text{ and } u = -1 \\ 0 & \text{otherwise} \end{cases}$$

□

Optimality in the deterministic setting

In DP and RL, the goal is to find an *optimal policy* that maximizes the return from any initial state x_0. The return is a cumulative aggregation of rewards along a trajectory starting at x_0. It concisely represents the reward obtained by the controller in the long run. Several types of return exist, depending on the way in which the rewards are accumulated (Bertsekas and Tsitsiklis, 1996, Section 2.1; Kaelbling et al., 1996). The *infinite-horizon discounted return* is given by:

$$R^h(x_0) = \sum_{k=0}^{\infty} \gamma^k r_{k+1} = \sum_{k=0}^{\infty} \gamma^k \rho(x_k, h(x_k)) \tag{2.1}$$

where $\gamma \in [0,1)$ is the *discount factor* and $x_{k+1} = f(x_k, h(x_k))$ for $k \geq 0$. The discount factor can be interpreted intuitively as a measure of how "far-sighted" the controller is in considering its rewards, or as a way of taking into account increasing uncertainty about future rewards. From a mathematical point of view, discounting ensures that the return will always be bounded if the rewards are bounded. The goal is therefore to maximize the long-term performance (return), while only using feedback about the immediate, one-step performance (reward). This leads to the so-called challenge of delayed rewards (Sutton and Barto, 1998): actions taken in the present affect the potential to achieve good rewards far in the future, but the immediate reward provides no information about these long-term effects.

Other types of return can also be defined. The undiscounted return, obtained by setting γ equal to 1 in (2.1), simply adds up the rewards, without discounting. Unfortunately, the infinite-horizon undiscounted return is often unbounded. An alternative is to use the infinite-horizon average return:

$$\lim_{K \to \infty} \frac{1}{K} \sum_{k=0}^{K} \rho(x_k, h(x_k))$$

which is bounded in many cases. Finite-horizon returns can be obtained by accumulating rewards along trajectories of a fixed, finite length K (the horizon), instead of along infinitely long trajectories. For instance, the finite-horizon discounted return can be defined as:

$$\sum_{k=0}^{K} \gamma^k \rho(x_k, h(x_k))$$

The undiscounted return ($\gamma = 1$) can be used more easily in the finite-horizon case, as it is bounded when the rewards are bounded.

In this book, we will mainly use the infinite-horizon discounted return (2.1), because it has useful theoretical properties. In particular, for this type of return, under certain technical assumptions, there always exists at least one *stationary*, deterministic optimal policy $h^* : X \rightarrow U$ (Bertsekas and Shreve, 1978, Chapter 9). In contrast, in the finite-horizon case, optimal policies depend in general on the time step k, i.e., they are nonstationary (Bertsekas, 2005a, Chapter 1).

While the discount factor γ can theoretically be regarded as a given part of the problem, in practice, a good value of γ has to be chosen. Choosing γ often involves a trade-off between the quality of the solution and the convergence rate of the DP/RL algorithm, for the following reasons. Some important DP/RL algorithms converge faster when γ is smaller (this is the case, e.g., for model-based value iteration, which will be introduced in Section 2.3). However, if γ is too small, the solution may be unsatisfactory because it does not sufficiently take into account rewards obtained after a large number of steps.

There is no generally valid procedure for choosing γ. Consider however, as an example, a typical stabilization problem from automatic control, where from every initial state the process should reach a steady state and remain there. In such a problem, γ should be chosen large enough that the rewards received upon reaching the steady state and remaining there have a detectable influence on the returns from every initial state. For instance, if the number of steps taken by a reasonable policy to stabilize the system from an initial state x is $K(x)$, then γ should be chosen so that $\gamma^{K_{\max}}$ is not too small, where $K_{\max} = \max_x K(x)$. However, finding $K_{\max}$ is a difficult problem in itself, which could be solved using, e.g., domain knowledge, or a suboptimal policy obtained by other means.

Value functions and the Bellman equations in the deterministic setting

A convenient way to characterize policies is by using their value functions. Two types of value functions exist: state-action value functions (Q-functions) and state value functions (V-functions). Note that the name "value function" is often used for V-functions in the literature. We will use the names "Q-function" and "V-function" to clearly differentiate between the two types of value functions, and the name "value function" to refer to Q-functions and V-functions collectively. We will first define and characterize Q-functions, and then turn our attention to V-functions.

The Q-function $Q^h : X \times U \rightarrow \mathbb{R}$ of a policy h gives the return obtained when starting from a given state, applying a given action, and following h thereafter:

$$Q^h(x,u) = \rho(x,u) + \gamma R^h(f(x,u)) \tag{2.2}$$

Here, $R^h(f(x,u))$ is the return from the next state $f(x,u)$. This concise formula can be obtained by first writing $Q^h(x,u)$ explicitly as the discounted sum of rewards obtained by taking u in x and then following h:

$$Q^h(x,u) = \sum_{k=0}^{\infty} \gamma^k \rho(x_k, u_k)$$

where $(x_0, u_0) = (x, u)$, $x_{k+1} = f(x_k, u_k)$ for $k \geq 0$, and $u_k = h(x_k)$ for $k \geq 1$. Then, the first term is separated from the sum:

$$\begin{aligned} Q^h(x,u) &= \rho(x,u) + \sum_{k=1}^{\infty} \gamma^k \rho(x_k, u_k) \\ &= \rho(x,u) + \gamma \sum_{k=1}^{\infty} \gamma^{k-1} \rho(x_k, h(x_k)) \\ &= \rho(x,u) + \gamma R^h(f(x,u)) \end{aligned} \tag{2.3}$$

where the definition (2.1) of the return was used in the last step. So, (2.2) has been obtained.

The optimal Q-function is defined as the best Q-function that can be obtained by any policy:

$$Q^*(x,u) = \max_h Q^h(x,u) \tag{2.4}$$

Any policy h^* that selects at each state an action with the largest optimal Q-value, i.e., that satisfies:

$$h^*(x) \in \arg\max_u Q^*(x,u) \tag{2.5}$$

is optimal (it maximizes the return). In general, for a given Q-function Q, a policy h that satisfies:

$$h(x) \in \arg\max_u Q(x,u) \tag{2.6}$$

is said to be *greedy* in Q. So, finding an optimal policy can be done by first finding Q^*, and then using (2.5) to compute a greedy policy in Q^*.

Note that, for simplicity of notation, we implicitly assume that the maximum in (2.4) exists, and also in similar equations in the sequel. When the maximum does not exist, the "max" operator should be replaced by the supremum operator. For the computation of greedy actions in (2.5), (2.6), and in similar equations in the sequel, the maximum must exist to ensure the existence of a greedy policy; this can be guaranteed under certain technical assumptions (Bertsekas and Shreve, 1978, Chapter 9).

The Q-functions Q^h and Q^* are recursively characterized by the *Bellman equations*, which are of central importance for value iteration and policy iteration algorithms. The Bellman equation for Q^h states that the value of taking action u in state x under the policy h equals the sum of the immediate reward and the discounted value achieved by h in the next state:

$$Q^h(x,u) = \rho(x,u) + \gamma Q^h(f(x,u), h(f(x,u))) \tag{2.7}$$

This Bellman equation can be derived from the second step in (2.3), as follows:

$$\begin{aligned} Q^h(x,u) &= \rho(x,u) + \gamma \sum_{k=1}^{\infty} \gamma^{k-1} \rho(x_k, h(x_k)) \\ &= \rho(x,u) + \gamma \left[\rho(f(x,u), h(f(x,u))) + \gamma \sum_{k=2}^{\infty} \gamma^{k-2} \rho(x_k, h(x_k)) \right] \\ &= \rho(x,u) + \gamma Q^h(f(x,u), h(f(x,u))) \end{aligned}$$

where $(x_0, u_0) = (x, u)$, $x_{k+1} = f(x_k, u_k)$ for $k \geq 0$, and $u_k = h(x_k)$ for $k \geq 1$.

The Bellman optimality equation characterizes Q^*, and states that the optimal value of action u taken in state x equals the sum of the immediate reward and the discounted optimal value obtained by the best action in the next state:

$$Q^*(x, u) = \rho(x, u) + \gamma \max_{u'} Q^*(f(x, u), u') \tag{2.8}$$

The V-function $V^h : X \to \mathbb{R}$ of a policy h is the return obtained by starting from a particular state and following h. This V-function can be computed from the Q-function of policy h:

$$V^h(x) = R^h(x) = Q^h(x, h(x)) \tag{2.9}$$

The optimal V-function is the best V-function that can be obtained by any policy, and can be computed from the optimal Q-function:

$$V^*(x) = \max_h V^h(x) = \max_u Q^*(x, u) \tag{2.10}$$

An optimal policy h^* can be computed from V^*, by using the fact that it satisfies:

$$h^*(x) \in \arg\max_u [\rho(x, u) + \gamma V^*(f(x, u))] \tag{2.11}$$

Using this formula is more difficult than using (2.5); in particular, a model of the MDP is required in the form of the dynamics f and the reward function ρ. Because the Q-function also depends on the action, it already includes information about the quality of transitions. In contrast, the V-function only describes the quality of the states; in order to infer the quality of transitions, they must be explicitly taken into account. This is what happens in (2.11), and this also explains why it is more difficult to compute policies from V-functions. Because of this difference, Q-functions will be preferred to V-functions throughout this book, even though they are more costly to represent than V-functions, as they depend both on x and u.

The V-functions V^h and V^* satisfy the following Bellman equations, which can be interpreted similarly to (2.7) and (2.8):

$$V^h(x) = \rho(x, h(x)) + \gamma V^h(f(x, h(x))) \tag{2.12}$$

$$V^*(x) = \max_u [\rho(x, u) + \gamma V^*(f(x, u))] \tag{2.13}$$

2.2.2 Stochastic setting

In a stochastic MDP, the next state is not deterministically given by the current state and action. Instead, the next state is a random variable, and the current state and action give the probability density of this random variable.

More formally, the deterministic transition function f is replaced by a transition probability function $\tilde{f} : X \times U \times X \to [0, \infty)$. After action u_k is taken in state x_k, the probability that the next state, x_{k+1}, belongs to a region $X_{k+1} \subseteq X$ is:

$$\mathrm{P}(x_{k+1} \in X_{k+1} \,|\, x_k, u_k) = \int_{X_{k+1}} \tilde{f}(x_k, u_k, x')\mathrm{d}x'$$

For any x and u, $\tilde{f}(x,u,\cdot)$ must define a valid probability density function of the argument "$\cdot$", where the dot stands for the random variable x_{k+1}. Because rewards are associated with transitions, and the transitions are no longer fully determined by the current state and action, the reward function also has to depend on the next state, $\tilde{\rho} : X \times U \times X \to \mathbb{R}$. After each transition to a state x_{k+1}, a reward r_{k+1} is received according to:

$$r_{k+1} = \tilde{\rho}(x_k, u_k, x_{k+1})$$

where we assume that $\|\tilde{\rho}\|_\infty = \sup_{x,u,x'} \tilde{\rho}(x,u,x')$ is finite. Note that $\tilde{\rho}$ is a deterministic function of the transition (x_k, u_k, x_{k+1}). This means that, once x_{k+1} has been generated, the reward r_{k+1} is fully determined. In general, the reward can also depend stochastically on the entire transition (x_k, u_k, x_{k+1}). If it does, to simplify notation, we assume that $\tilde{\rho}$ gives the *expected* reward after the transition.

When the state space is countable (e.g., discrete), the transition function can also be given as $\bar{f} : X \times U \times X \to [0,1]$, where the probability of reaching x' after taking u_k in x_k is:

$$\mathrm{P}\left(x_{k+1} = x' \,|\, x_k, u_k\right) = \bar{f}(x_k, u_k, x') \tag{2.14}$$

For any x and u, the function $\bar{f}$ must satisfy $\sum_{x'} \bar{f}(x,u,x') = 1$. The function $\tilde{f}$ is a generalization of $\bar{f}$ to uncountable (e.g., continuous) state spaces; in such spaces, the probability of ending up in a given state x' is generally 0, making a description of the form $\bar{f}$ inappropriate.

In the stochastic case, the Markov property requires that x_k and u_k fully determine the probability density of the next state.

Developing an analytical expression for the transition probability function $\tilde{f}$ is generally a difficult task. Fortunately, as previously noted in Section 2.1, most DP (model-based) algorithms can work with a generative model, which only needs to generate samples of the next state and corresponding rewards for any given pair of current state and action taken.

Example 2.2 The stochastic cleaning-robot MDP. Consider again the cleaning-robot problem of Example 2.1. Assume that, due to uncertainties in the environment, such as a slippery floor, state transitions are no longer deterministic. When trying to move in a certain direction, the robot succeeds with a probability of 0.8. With a probability of 0.15 it remains in the same state, and it may even move in the opposite direction with a probability of 0.05 (see also Figure 2.4).

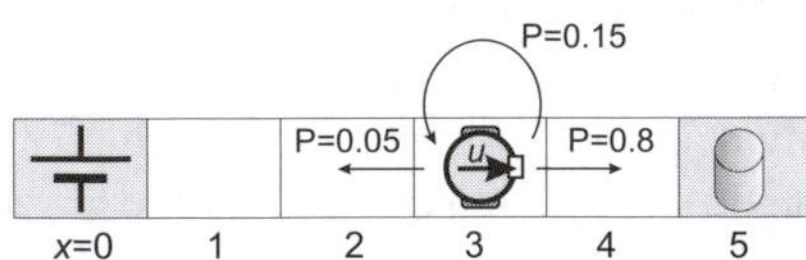

FIGURE 2.4
The stochastic cleaning-robot problem. The robot intends to move right, but it may instead end up standing still or moving left, with different probabilities.

Because the state space is discrete, a transition model of the form (2.14) is appropriate. The transition function $\bar{f}$ that models the probabilistic transitions described above is shown in Table 2.1. In this table, the rows correspond to combinations of current states and actions taken, while the columns correspond to future states. Note that the transitions from any terminal state still lead deterministically to the same terminal state, regardless of the action.

TABLE 2.1 Dynamics of the stochastic, cleaning-robot MDP.

(x,u)	$\bar{f}(x,u,0)$	$\bar{f}(x,u,1)$	$\bar{f}(x,u,2)$	$\bar{f}(x,u,3)$	$\bar{f}(x,u,4)$	$\bar{f}(x,u,5)$
$(0,-1)$	1	0	0	0	0	0
$(1,-1)$	0.8	0.15	0.05	0	0	0
$(2,-1)$	0	0.8	0.15	0.05	0	0
$(3,-1)$	0	0	0.8	0.15	0.05	0
$(4,-1)$	0	0	0	0.8	0.15	0.05
$(5,-1)$	0	0	0	0	0	1
$(0,1)$	1	0	0	0	0	0
$(1,1)$	0.05	0.15	0.8	0	0	0
$(2,1)$	0	0.05	0.15	0.8	0	0
$(3,1)$	0	0	0.05	0.15	0.8	0
$(4,1)$	0	0	0	0.05	0.15	0.8
$(5,1)$	0	0	0	0	0	1

The robot receives rewards as in the deterministic case: upon reaching state 5, it is rewarded with 5, and upon reaching state 0, it is rewarded with 1. The corresponding reward function, in the form $\tilde{\rho} : X \times U \times X \longrightarrow \mathbb{R}$, is:

$$\tilde{\rho}(x,u,x') = \begin{cases} 5 & \text{if } x \neq 5 \text{ and } x' = 5 \\ 1 & \text{if } x \neq 0 \text{ and } x' = 0 \\ 0 & \text{otherwise} \end{cases}$$

□

Optimality in the stochastic setting

The *expected* infinite-horizon discounted return of an initial state x_0 under a (deterministic) policy h is:[4]

$$\begin{aligned} R^h(x_0) &= \lim_{K\to\infty} \mathrm{E}_{x_{k+1}\sim\tilde{f}(x_k,h(x_k),\cdot)} \left\{ \sum_{k=0}^{K} \gamma^k r_{k+1} \right\} \\ &= \lim_{K\to\infty} \mathrm{E}_{x_{k+1}\sim\tilde{f}(x_k,h(x_k),\cdot)} \left\{ \sum_{k=0}^{K} \gamma^k \tilde{\rho}(x_k, h(x_k), x_{k+1}) \right\} \end{aligned} \tag{2.15}$$

[4]We assume that the MDP and the policies h have suitable properties such that the expected return and the Bellman equations in the remainder of this section are well defined. See, e.g., (Bertsekas and Shreve, 1978, Chapter 9) and (Bertsekas, 2007, Appendix A) for a discussion of these properties.

where E denotes the expectation operator, and the notation $x_{k+1} \sim \tilde{f}(x_k, h(x_k), \cdot)$ means that the random variable x_{k+1} is drawn from the density $\tilde{f}(x_k, h(x_k), \cdot)$ at each step k. The discussion of Section 2.2.1 regarding the interpretation and choice of the discount factor also applies to the stochastic case. For any stochastic or deterministic MDP, when using the infinite-horizon discounted return (2.15) or (2.1), and under certain technical assumptions on the elements of the MDP, there exists at least one stationary deterministic optimal policy (Bertsekas and Shreve, 1978, Chapter 9). Therefore, we will mainly consider stationary deterministic policies in the sequel.

Expected undiscounted, average, and finite-horizon returns (see Section 2.2.1) can be defined analogously to (2.15).

Value functions and the Bellman equations in the stochastic setting

To obtain the Q-function of a policy h, the definition (2.2) is generalized to the stochastic case, as follows. The Q-function is the *expected* return under the stochastic transitions, when starting in a particular state, applying a particular action, and following the policy h thereafter:

$$Q^h(x,u) = \mathrm{E}_{x' \sim \tilde{f}(x,u,\cdot)} \left\{ \tilde{\rho}(x,u,x') + \gamma R^h(x') \right\} \tag{2.16}$$

The definition of the optimal Q-function Q^* remains unchanged from the deterministic case (2.4), and is repeated here for easy reference:

$$Q^*(x,u) = \max_h Q^h(x,u)$$

Similarly, optimal policies can still be computed from Q^* as in the deterministic case, because they satisfy (2.5), also repeated here:

$$h^*(x) \in \arg\max_u Q^*(x,u)$$

The Bellman equations for Q^h and Q^* are given in terms of expectations over the one-step stochastic transitions:

$$Q^h(x,u) = \mathrm{E}_{x' \sim \tilde{f}(x,u,\cdot)} \left\{ \tilde{\rho}(x,u,x') + \gamma Q^h(x', h(x')) \right\} \tag{2.17}$$

$$Q^*(x,u) = \mathrm{E}_{x' \sim \tilde{f}(x,u,\cdot)} \left\{ \tilde{\rho}(x,u,x') + \gamma \max_{u'} Q^*(x',u') \right\} \tag{2.18}$$

The definition of the V-function V^h of a policy h, as well as of the optimal V-function V^*, are the same as for the deterministic case (2.9), (2.10):

$$V^h(x) = R^h(x)$$
$$V^*(x) = \max_h V^h(x)$$

However, the computation of optimal policies from V^* becomes more difficult, involving an expectation that did not appear in the deterministic case:

$$h^*(x) \in \arg\max_u \mathrm{E}_{x' \sim \tilde{f}(x,u,\cdot)} \left\{ \tilde{\rho}(x,u,x') + \gamma V^*(x') \right\} \tag{2.19}$$

In contrast, computing an optimal policy from Q^* is as simple as in the deterministic case, which is yet another reason for using Q-functions in practice.

The Bellman equations for V^h and V^* are obtained from (2.12) and (2.13), by considering expectations over the one-step stochastic transitions:

$$V^h(x) = \mathrm{E}_{x' \sim \tilde{f}(x,h(x),\cdot)} \left\{ \tilde{\rho}(x,h(x),x') + \gamma V^h(x') \right\} \tag{2.20}$$

$$V^*(x) = \max_u \mathrm{E}_{x' \sim \tilde{f}(x,u,\cdot)} \left\{ \tilde{\rho}(x,u,x') + \gamma V^*(x') \right\} \tag{2.21}$$

Note that in the Bellman equation for V^* (2.21), the maximization is outside the expectation operator, whereas in the Bellman equation for Q^* (2.18), the order of the expectation and maximization is reversed.

Clearly, all the equations for deterministic MDPs are a special case of the equations for stochastic MDPs. The deterministic case is obtained by using a degenerate density $\tilde{f}(x,u,\cdot)$ that assigns all the probability mass to $f(x,u)$. The deterministic reward function is obtained as $\rho(x,u) = \tilde{\rho}(x,u,f(x,u))$.

The entire class of value iteration algorithms, introduced in Section 2.3, revolves around solving the Bellman optimality equations (2.18) or (2.21) to find, respectively, the optimal Q-function or the optimal V-function (in the deterministic case, (2.8) or (2.13) are solved instead). Similarly, policy evaluation, which is a core component of the policy iteration algorithms introduced in Section 2.4, revolves around solving (2.17) or (2.20) to find, respectively, Q^h or V^h (in the deterministic case (2.7) or (2.12) are solved instead).

2.3 Value iteration

Value iteration techniques use the Bellman optimality equation to iteratively compute an optimal value function, from which an optimal policy is derived. We first present DP (model-based) algorithms for value iteration, followed by RL (model-free) algorithms. DP algorithms like V-iteration (Bertsekas, 2007, Section 1.3) solve the Bellman optimality equation by using knowledge of the transition and reward functions. RL techniques either learn a model, e.g., Dyna (Sutton, 1990), or do not use an explicit model at all, e.g., Q-learning (Watkins and Dayan, 1992).

2.3.1 Model-based value iteration

We will next introduce the model-based *Q-iteration* algorithm, as an illustrative example from the class of model-based value iteration algorithms. Let the set of all the Q-functions be denoted by $\mathcal{Q}$. Then, the Q-iteration mapping $T : \mathcal{Q} \to \mathcal{Q}$, computes the right-hand side of the Bellman optimality equation (2.8) or (2.18) for any

Q-function.[5] In the deterministic case, this mapping is:

$$[T(Q)](x,u) = \rho(x,u) + \gamma \max_{u'} Q(f(x,u),u') \tag{2.22}$$

and in the stochastic case, it is:

$$[T(Q)](x,u) = \mathrm{E}_{x' \sim \tilde{f}(x,u,\cdot)} \left\{ \tilde{\rho}(x,u,x') + \gamma \max_{u'} Q(x',u') \right\} \tag{2.23}$$

Note that if the state space is countable (e.g., finite), a transition model of the form (2.14) is appropriate, and the Q-iteration mapping for the stochastic case (2.23) can be written as the simpler summation:

$$[T(Q)](x,u) = \sum_{x'} \bar{f}(x,u,x') \left[\tilde{\rho}(x,u,x') + \gamma \max_{u'} Q(x',u') \right] \tag{2.24}$$

The same notation is used for the Q-iteration mapping both in the deterministic case and in the stochastic case, because the analysis given below applies to both cases, and the definition (2.22) of T is a special case of (2.23) (or of (2.24) for countable state spaces).

The Q-iteration algorithm starts from an arbitrary Q-function Q_0 and at each iteration ℓ updates the Q-function using:

$$Q_{\ell+1} = T(Q_\ell) \tag{2.25}$$

It can be shown that T is a contraction with factor $\gamma < 1$ in the infinity norm, i.e., for any pair of functions Q and Q', it is true that:

$$\|T(Q) - T(Q')\|_\infty \leq \gamma \|Q - Q'\|_\infty$$

Because T is a contraction, it has a unique fixed point (Istratescu, 2002). Additionally, when rewritten using the Q-iteration mapping, the Bellman optimality equation (2.8) or (2.18) states that Q^* is a fixed point of T, i.e.:

$$Q^* = T(Q^*) \tag{2.26}$$

Hence, the unique fixed point of T is actually Q^*, and Q-iteration asymptotically converges to Q^* as $\ell \to \infty$. Moreover, Q-iteration converges to Q^* at a rate of γ, in the sense that $\|Q_{\ell+1} - Q^*\|_\infty \leq \gamma \|Q_\ell - Q^*\|_\infty$. An optimal policy can be computed from Q^* with (2.5).

Algorithm 2.1 presents Q-iteration for deterministic MDPs in an explicit, procedural form, wherein T is computed using (2.22). Similarly, Algorithm 2.2 presents Q-iteration for stochastic MDPs with countable state spaces, using the expression (2.24) for T.

[5] The term "mapping" is used to refer to functions that work with other functions as inputs and/or outputs; as well as to compositions of such functions. The term is used to differentiate mappings from ordinary functions, which only have numerical scalars, vectors, or matrices as inputs and/or outputs.

ALGORITHM 2.1 Q-iteration for deterministic MDPs.

Input: dynamics f, reward function ρ, discount factor γ
initialize Q-function, e.g., $Q_0 \leftarrow 0$
repeat at every iteration $\ell = 0, 1, 2, \ldots$
    for every (x, u) **do**
        $Q_{\ell+1}(x, u) \leftarrow \rho(x, u) + \gamma \max_{u'} Q_\ell(f(x, u), u')$
    end for
until $Q_{\ell+1} = Q_\ell$
Output: $Q^* = Q_\ell$

ALGORITHM 2.2 Q-iteration for stochastic MDPs with countable state spaces.

Input: dynamics $\bar{f}$, reward function $\tilde{\rho}$, discount factor γ
initialize Q-function, e.g., $Q_0 \leftarrow 0$
repeat at every iteration $\ell = 0, 1, 2, \ldots$
    for every (x, u) **do**
        $Q_{\ell+1}(x, u) \leftarrow \sum_{x'} \bar{f}(x, u, x') \left[\tilde{\rho}(x, u, x') + \gamma \max_{u'} Q_\ell(x', u')\right]$
    end for
until $Q_{\ell+1} = Q_\ell$
Output: $Q^* = Q_\ell$

The results given above only guarantee the asymptotic convergence of Q-iteration, hence the stopping criterion of Algorithms 2.1 and 2.2 may only be satisfied asymptotically. In practice, it is also important to guarantee the performance of Q-iteration when the algorithm is stopped after a finite number of iterations. The following result holds both in the deterministic case and in the stochastic case. Given a suboptimality bound $\varsigma_{\mathrm{QI}} > 0$, where the subscript "QI" stands for "Q-iteration," a finite number L of iterations can be (conservatively) chosen with:

$$L = \left\lceil \log_\gamma \frac{\varsigma_{\mathrm{QI}}(1-\gamma)^2}{2\|\rho\|_\infty} \right\rceil \tag{2.27}$$

so that the suboptimality of a policy h_L that is greedy in Q_L is guaranteed to be at most ς_{QI}, in the sense that $\|V^{h_L} - V^*\|_\infty \leq \varsigma_{\mathrm{QI}}$. Here, $\lceil \cdot \rceil$ is the smallest integer larger than or equal to the argument (ceiling). Equation (2.27) follows from the bound (Ernst et al., 2005):

$$\|V^{h_L} - V^*\|_\infty \leq 2\frac{\gamma^L \|\rho\|_\infty}{(1-\gamma)^2}$$

on the suboptimality of h_L, by requiring that $2\frac{\gamma^L\|\rho\|_\infty}{(1-\gamma)^2} \leq \varsigma_{\mathrm{QI}}$.

Alternatively, Q-iteration could be stopped when the difference between two consecutive Q-functions decreases below a given threshold $\varepsilon_{\mathrm{QI}} > 0$, i.e., when $\|Q_{\ell+1} - Q_\ell\|_\infty \leq \varepsilon_{\mathrm{QI}}$. This can also be guaranteed to happen after a finite number of iterations, due to the contracting nature of the Q-iteration updates.

A V-iteration algorithm that computes the optimal V-function can be developed along similar lines, using the Bellman optimality equation (2.13) in the deterministic case, or (2.21) in the stochastic case. Note that the name "value iteration" is typically used for the V-iteration algorithm in the literature, whereas we use it to refer more generally to the entire class of algorithms that use the Bellman optimality equations to compute optimal value functions. (Recall that we similarly use "value function" to refer to Q-functions and V-functions collectively.)

Computational cost of Q-iteration for finite MDPs

Next, we investigate the computational cost of Q-iteration when applied to an MDP with a finite number of states and actions. Denote by $|\cdot|$ the cardinality of the argument set "$\cdot$", so that $|X|$ denotes the finite number of states and $|U|$ denotes the finite number of actions.

Consider first the deterministic case, for which Algorithm 2.1 can be used. Assume that, when updating the Q-value for a given state-action pair (x,u), the maximization over the action space U is solved by enumeration over its $|U|$ elements, and $f(x,u)$ is computed once and then stored and reused. Updating the Q-value then requires $2+|U|$ function evaluations, where the functions being evaluated are f, ρ, and the current Q-function Q_ℓ. Since at every iteration, the Q-values of $|X|\,|U|$ state-action pairs have to be updated, the cost per iteration is $|X|\,|U|\,(2+|U|)$. So, the total cost of L Q-iterations for a deterministic, finite MDP is:

$$L\,|X|\,|U|\,(2+|U|) \tag{2.28}$$

The number L of iterations can be chosen, e.g., by imposing a suboptimality bound ς_{QI} and using (2.27).

In the stochastic case, because the state space is finite, Algorithm 2.2 can be used. Assuming that the maximization over u' is implemented using enumeration, the cost of updating the Q-value for a given pair (x,u) is $|X|\,(2+|U|)$, where the functions being evaluated are $\bar{f}$, $\tilde{\rho}$, and Q_ℓ. The cost per iteration is $|X|^2\,|U|\,(2+|U|)$, and the total cost of L Q-iterations for a stochastic, finite MDP is thus:

$$L\,|X|^2\,|U|\,(2+|U|) \tag{2.29}$$

which is larger by a factor $|X|$ than the cost (2.28) for the deterministic case.

Example 2.3 Q-iteration for the cleaning robot. In this example, we apply Q-iteration to the cleaning-robot problem of Examples 2.1 and 2.2. The discount factor γ is set to 0.5.

Consider first the deterministic variant of Example 2.1. For this variant, Q-iteration is implemented as Algorithm 2.1. Starting from an identically zero initial Q-function, $Q_0 = 0$, this algorithm produces the sequence of Q-functions given in the first part of Table 2.2 (above the dashed line), where each cell shows the Q-values of the two actions in a certain state, separated by a semicolon. For instance:

$$Q_3(2,1) = \rho(2,1) + \gamma \max_u Q_2(f(2,1),u) = 0 + 0.5 \max_u Q_2(3,u) = 0 + 0.5\cdot 2.5 = 1.25$$

TABLE 2.2
Q-iteration results for the deterministic cleaning-robot problem.

	$x=0$	$x=1$	$x=2$	$x=3$	$x=4$	$x=5$
Q_0	0; 0	0; 0	0; 0	0; 0	0; 0	0; 0
Q_1	0; 0	1; 0	0.5; 0	0.25; 0	0.125; 5	0; 0
Q_2	0; 0	1; 0.25	0.5; 0.125	0.25; 2.5	1.25; 5	0; 0
Q_3	0; 0	1; 0.25	0.5; 1.25	0.625; 2.5	1.25; 5	0; 0
Q_4	0; 0	1; 0.625	0.5; 1.25	0.625; 2.5	1.25; 5	0; 0
Q_5	0; 0	1; 0.625	0.5; 1.25	0.625; 2.5	1.25; 5	0; 0
h^*	$*$	-1	1	1	1	$*$
V^*	0	1	1.25	2.5	5	0

The algorithm converges after 5 iterations; $Q_5 = Q_4 = Q^*$. The last two rows of the table (below the dashed line) also give the optimal policies, computed from Q^* with (2.5), and the optimal V-function V^*, computed from Q^* with (2.10). In the policy representation, the symbol "$*$" means that any action can be taken in that state without changing the quality of the policy. The total number of function evaluations required by the algorithm in the deterministic case is:

$$5\,|X|\,|U|\,(2+|U|) = 5 \cdot 6 \cdot 2 \cdot 4 = 240$$

Consider next the stochastic variant of the cleaning-robot problem, introduced in Example 2.2. For this variant, Q-iteration is implemented by using Algorithm 2.2, and produces the sequence of Q-functions illustrated in the first part of Table 2.3 (not all the iterations are shown). The algorithm fully converges after 22 iterations.

TABLE 2.3
Q-iteration results for the stochastic cleaning-robot problem. Q-function and V-function values are rounded to 3 decimal places.

	$x=0$	$x=1$	$x=2$	$x=3$	$x=4$	$x=5$
Q_0	0; 0	0; 0	0; 0	0; 0	0; 0	0; 0
Q_1	0; 0	0.800; 0.110	0.320; 0.044	0.128; 0.018	0.301; 4.026	0; 0
Q_2	0; 0	0.868; 0.243	0.374; 0.101	0.260; 1.639	1.208; 4.343	0; 0
Q_3	0; 0	0.874; 0.265	0.419; 0.709	0.515; 1.878	1.327; 4.373	0; 0
Q_4	0; 0	0.883; 0.400	0.453; 0.826	0.581; 1.911	1.342; 4.376	0; 0
...	...	...	...	...	...	...
Q_{12}	0; 0	0.888; 0.458	0.467; 0.852	0.594; 1.915	1.344; 4.376	0; 0
...	...	...	...	...	...	...
Q_{22}	0; 0	0.888; 0.458	0.467; 0.852	0.594; 1.915	1.344; 4.376	0; 0
h^*	$*$	-1	1	1	1	$*$
V^*	0	0.888	0.852	1.915	4.376	0

The optimal policies and the optimal V-function obtained are also shown in Table 2.3 (below the dashed line). While the optimal Q-function and the optimal V-function are different from those obtained in the deterministic case, the optimal policies remain the same. The total number of function evaluations required by the algorithm in the stochastic case is:

$$22|X|^2|U|(2+|U|) = 22\cdot 6^2\cdot 2\cdot 4 = 6336$$

which is considerably greater than in the deterministic case.

If we impose a suboptimality bound $\varsigma_{\text{QI}} = 0.01$ and apply (2.27), we find that Q-iteration should run for $L = 12$ iterations in order to guarantee this bound, where the maximum absolute reward $\|\rho\|_\infty = 5$ and the discount factor $\gamma = 0.5$ have also been used. So, in the deterministic case, the algorithm fully converges to its fixed point in fewer iterations than the conservative number given by (2.27). In the stochastic case, even though the algorithm does not fully converge after 12 iterations (instead requiring 22 iterations), the Q-function at iteration 12 (shown in Table 2.3) is already very accurate, and a policy that is greedy in this Q-function is fully optimal. The suboptimality of such a policy is 0, which is smaller than the imposed bound ς_{QI}. In fact, for any iteration $\ell \geq 3$, the Q-function Q_ℓ would produce an optimal policy, which means that $L = 12$ is also conservative in the stochastic case. □

2.3.2 Model-free value iteration and the need for exploration

We have discussed until now model-based value iteration. We next consider RL, model-free value iteration algorithms, and discuss *Q-learning*, the most widely used algorithm from this class. Q-learning starts from an arbitrary initial Q-function Q_0 and updates it without requiring a model, using instead observed state transitions and rewards, i.e., data tuples of the form $(x_k, u_k, x_{k+1}, r_{k+1})$ (Watkins, 1989; Watkins and Dayan, 1992). After each transition, the Q-function is updated using such a data tuple, as follows:

$$Q_{k+1}(x_k, u_k) = Q_k(x_k, u_k) + \alpha_k[r_{k+1} + \gamma \max_{u'} Q_k(x_{k+1}, u') - Q_k(x_k, u_k)] \tag{2.30}$$

where $\alpha_k \in (0, 1]$ is the learning rate. The term between square brackets is the temporal difference, i.e., the difference between the updated estimate $r_{k+1} + \gamma \max_{u'} Q_k(x_{k+1}, u')$ of the optimal Q-value of (x_k, u_k), and the current estimate $Q_k(x_k, u_k)$. In the deterministic case, the new estimate is actually the Q-iteration mapping (2.22) applied to Q_k in the state-action pair (x_k, u_k), where $\rho(x_k, u_k)$ has been replaced by the observed reward r_{k+1}, and $f(x_k, u_k)$ by the observed next-state x_{k+1}. In the stochastic case, these replacements provide a single sample of the random quantity whose expectation is computed by the Q-iteration mapping (2.23), and thus Q-learning can be seen as a sample-based, stochastic approximation procedure based on this mapping (Singh et al., 1995; Bertsekas and Tsitsiklis, 1996, Section 5.6).

As the number of transitions k approaches infinity, Q-learning asymptotically converges to Q^* if the state and action spaces are discrete and finite, and under the following conditions (Watkins and Dayan, 1992; Tsitsiklis, 1994; Jaakkola et al., 1994):

- The sum $\sum_{k=0}^{\infty} \alpha_k^2$ produces a finite value, whereas the sum $\sum_{k=0}^{\infty} \alpha_k$ produces an infinite value.
- All the state-action pairs are (asymptotically) visited infinitely often.

The first condition is not difficult to satisfy. For instance, a satisfactory standard choice is:

$$\alpha_k = \frac{1}{k} \tag{2.31}$$

In practice, the learning rate schedule may require tuning, because it influences the number of transitions required by Q-learning to obtain a good solution. A good choice for the learning rate schedule depends on the problem at hand.

The second condition can be satisfied if, among other things, the controller has a nonzero probability of selecting any action in every encountered state; this is called exploration. The controller also has to exploit its current knowledge in order to obtain good performance, e.g., by selecting greedy actions in the current Q-function. This is a typical illustration of the exploration-exploitation trade-off in online RL. A classical way to balance exploration with exploitation in Q-learning is ε-greedy exploration (Sutton and Barto, 1998, Section 2.2), which selects actions according to:

$$u_k = \begin{cases} u \in \arg\max_{\bar{u}} Q_k(x_k, \bar{u}) & \text{with probability } 1 - \varepsilon_k \\ \text{a uniformly random action in } U & \text{with probability } \varepsilon_k \end{cases} \tag{2.32}$$

where $\varepsilon_k \in (0, 1)$ is the exploration probability at step k. Another option is to use Boltzmann exploration (Sutton and Barto, 1998, Section 2.3), which at step k selects an action u with probability:

$$\mathrm{P}(u \,|\, x_k) = \frac{e^{Q_k(x_k,u)/\tau_k}}{\sum_{\bar{u}} e^{Q_k(x_k,\bar{u})/\tau_k}} \tag{2.33}$$

where the temperature $\tau_k \geq 0$ controls the randomness of the exploration. When $\tau_k \to 0$, (2.33) is equivalent to greedy action selection, while for $\tau_k \to \infty$, action selection is uniformly random. For nonzero, finite values of τ_k, higher-valued actions have a greater chance of being selected than lower-valued ones.

Usually, the exploration diminishes over time, so that the policy used asymptotically becomes greedy and therefore (as $Q_k \to Q^*$) optimal. This can be achieved by making ε_k or τ_k approach 0 as k grows. For instance, an ε-greedy exploration schedule of the form $\varepsilon_k = 1/k$ diminishes to 0 as $k \to \infty$, while still satisfying the second convergence condition of Q-learning, i.e., while allowing infinitely many visits to all the state-action pairs (Singh et al., 2000). Notice the similarity of this exploration schedule with the learning rate schedule (2.31). For a schedule of the Boltzmann exploration temperature τ_k that decreases to 0 while satisfying the convergence conditions, see (Singh et al., 2000). Like the learning rate schedule, the exploration schedule has a significant effect on the performance of Q-learning.

Algorithm 2.3 presents Q-learning with ε-greedy exploration. Note that an idealized, infinite-time online setting is considered for this algorithm, in which no termination condition is specified and no explicit output is produced. Instead, the result of the algorithm is the improvement of the control performance achieved while interacting with the process. A similar setting will be considered for other online learning algorithms described in this book, with the implicit understanding that, in practice, the algorithms will of course be stopped after a finite number of steps. When Q-learning is stopped, the resulting Q-function and the corresponding greedy policy can be interpreted as outputs and reused.

ALGORITHM 2.3 Q-learning with ε-greedy exploration.

Input: discount factor γ,
exploration schedule $\{\varepsilon_k\}_{k=0}^{\infty}$, learning rate schedule $\{\alpha_k\}_{k=0}^{\infty}$

initialize Q-function, e.g., $Q_0 \leftarrow 0$

measure initial state x_0

for every time step $k = 0, 1, 2, \ldots$ **do**

$$u_k \leftarrow \begin{cases} u \in \arg\max_{\bar{u}} Q_k(x_k, \bar{u}) & \text{with probability } 1 - \varepsilon_k \text{ (exploit)} \\ \text{a uniformly random action in } U & \text{with probability } \varepsilon_k \text{ (explore)} \end{cases}$$

  apply u_k, measure next state x_{k+1} and reward r_{k+1}

  $Q_{k+1}(x_k, u_k) \leftarrow Q_k(x_k, u_k) + \alpha_k[r_{k+1} + \gamma \max_{u'} Q_k(x_{k+1}, u') - Q_k(x_k, u_k)]$

end for

Note that this discussion has not been all-encompassing; the ε-greedy and Boltzmann exploration procedures can also be used in other online RL algorithms besides Q-learning, and a variety of other exploration procedures exist. For instance, the policy can be biased towards actions that have not recently been taken, or that may lead the system towards rarely visited areas of the state space (Thrun, 1992). The value function can also be initialized to be larger than the true returns, in a method known as "optimism in the face of uncertainty" (Sutton and Barto, 1998, Section 2.7). Because the return estimates have been adjusted downwards for any actions already taken, greedy action selection leads to exploring novel actions. Confidence intervals for the returns can be estimated, and the action with largest upper confidence bound, i.e., with the best potential for good returns, can be chosen (Kaelbling, 1993). Many authors have also studied the exploration-exploitation trade-off for specific types of problems, such as problems with linear transition dynamics (Feldbaum, 1961), and problems without any dynamics, for which the state space reduces to a single element (Auer et al., 2002; Audibert et al., 2007).

2.4 Policy iteration

Having introduced value iteration in Section 2.3, we now consider *policy iteration*, the second major class of DP/RL algorithms. Policy iteration algorithms evaluate policies by constructing their value functions, and use these value functions to find new, improved policies (Bertsekas, 2007, Section 1.3). As a representative example of policy iteration, consider an offline algorithm that evaluates policies using their Q-functions. This algorithm starts with an arbitrary policy h_0. At every iteration ℓ, the Q-function Q^{h_ℓ} of the current policy h_ℓ is determined; this step is called *policy evaluation*. Policy evaluation is performed by solving the Bellman equation (2.7) in the deterministic case, or (2.17) in the stochastic case. When policy evaluation is complete, a new policy $h_{\ell+1}$ that is greedy in Q^h is found:

$$h_{\ell+1}(x) \in \arg\max_{u} Q^{h_\ell}(x,u) \tag{2.34}$$

This step is called *policy improvement*. The entire procedure is summarized in Algorithm 2.4. The sequence of Q-functions produced by policy iteration asymptotically converges to Q^* as $\ell \to \infty$. Simultaneously, an optimal policy h^* is obtained.

ALGORITHM 2.4 Policy iteration with Q-functions.

initialize policy h_0
repeat at every iteration $\ell = 0, 1, 2, \ldots$
  find Q^{h_ℓ}, the Q-function of h_ℓ ▷ policy evaluation
  $h_{\ell+1}(x) \in \arg\max_u Q^{h_\ell}(x,u)$ ▷ policy improvement
until $h_{\ell+1} = h_\ell$
Output: $h^* = h_\ell$, $Q^* = Q^{h_\ell}$

The crucial component of policy iteration is policy evaluation. Policy improvement can be performed by solving static optimization problems, e.g., of the form (2.34) when Q-functions are used – often an easier challenge.

In the remainder of this section, we first discuss DP (model-based) policy iteration, followed by RL (model-free) policy iteration. We pay special attention to the policy evaluation component.

2.4.1 Model-based policy iteration

In the model-based setting, the policy evaluation step employs knowledge of the transition and reward functions. A model-based iterative algorithm for policy evaluation can be given that is similar to Q-iteration, which will be called *policy evaluation for Q-functions*. Analogously to the Q-iteration mapping T (2.22), a policy evaluation mapping $T^h : \mathcal{Q} \to \mathcal{Q}$ is defined, which computes the right-hand side of the Bellman equation for an arbitrary Q-function. In the deterministic case, this mapping is:

$$[T^h(Q)](x,u) = \rho(x,u) + \gamma Q(f(x,u), h(f(x,u))) \tag{2.35}$$

and in the stochastic case, it is:

$$[T^h(Q)](x,u) = \mathrm{E}_{x' \sim \tilde{f}(x,u,\cdot)} \left\{ \tilde{\rho}(x,u,x') + \gamma Q(x', h(x')) \right\} \tag{2.36}$$

Note that when the state space is countable, the transition model (2.14) is appropriate, and the policy evaluation mapping for the stochastic case (2.36) can be written as the simpler summation:

$$[T^h(Q)](x,u) = \sum_{x'} \bar{f}(x,u,x') \left[\tilde{\rho}(x,u,x') + \gamma Q(x',h(x'))\right] \tag{2.37}$$

Policy evaluation for Q-functions starts from an arbitrary Q-function Q_0^h and at each iteration τ updates the Q-function using:[6]

$$Q_{\tau+1}^h = T^h(Q_\tau^h) \tag{2.38}$$

Like the Q-iteration mapping T, the policy evaluation mapping T^h is a contraction with a factor $\gamma < 1$ in the infinity norm, i.e., for any pair of functions Q and Q':

$$\|T^h(Q) - T^h(Q')\|_\infty \leq \gamma \|Q - Q'\|_\infty$$

So, T^h has a unique fixed point. Written in terms of the mapping T^h, the Bellman equation (2.7) or (2.17) states that this unique fixed point is actually Q^h:

$$Q^h = T^h(Q^h) \tag{2.39}$$

Therefore, policy evaluation for Q-functions (2.38) asymptotically converges to Q^h. Moreover, also because T^h is a contraction with factor γ, this variant of policy evaluation converges to Q^h at a rate of γ, in the sense that $\|Q_{\tau+1}^h - Q^h\|_\infty \leq \gamma \|Q_\tau^h - Q^h\|_\infty$.

Algorithm 2.5 presents policy evaluation for Q-functions in deterministic MDPs, while Algorithm 2.6 is used for stochastic MDPs with countable state spaces. In Algorithm 2.5, T^h is computed with (2.35), while in Algorithm 2.6, (2.37) is employed. Since the convergence condition of these algorithms is only guaranteed to be satisfied asymptotically, in practice they can be stopped, e.g., when the difference between consecutive Q-functions decreases below a given threshold, i.e., when $\|Q_{\tau+1}^h - Q_\tau^h\|_\infty \leq \varepsilon_{\text{PE}}$, where $\varepsilon_{\text{PE}} > 0$. Here, the subscript "PE" stands for "policy evaluation."

ALGORITHM 2.5 Policy evaluation for Q-functions in deterministic MDPs.

Input: policy h to be evaluated, dynamics f, reward function ρ, discount factor γ

initialize Q-function, e.g., $Q_0^h \leftarrow 0$
repeat at every iteration $\tau = 0, 1, 2, \ldots$
  for every (x,u) **do**
    $Q_{\tau+1}^h(x,u) \leftarrow \rho(x,u) + \gamma Q_\tau^h(f(x,u), h(f(x,u)))$
  end for
until $Q_{\tau+1}^h = Q_\tau^h$

Output: $Q^h = Q_\tau^h$

[6] A different iteration index τ is used for policy evaluation, because it runs in the inner loop of every (offline) policy iteration ℓ.

ALGORITHM 2.6
Policy evaluation for Q-functions in stochastic MDPs with countable state spaces.

Input: policy h to be evaluated, dynamics $\bar{f}$, reward function $\tilde{\rho}$, discount factor γ
initialize Q-function, e.g., $Q_0^h \leftarrow 0$
repeat at every iteration $\tau = 0, 1, 2, \ldots$
  for every (x,u) **do**
    $Q_{\tau+1}^h(x,u) \leftarrow \sum_{x'} \bar{f}(x,u,x') \left[\tilde{\rho}(x,u,x') + \gamma Q_\tau^h(x', h(x'))\right]$
  end for
until $Q_{\tau+1}^h = Q_\tau^h$
Output: $Q^h = Q_\tau^h$

There are also other ways to compute Q^h. For example, in the deterministic case, the mapping T^h (2.35) and equivalently the Bellman equation (2.7) are obviously linear in the Q-values. In the stochastic case, because X has a finite cardinality, the policy evaluation mapping T^h and equivalently the Bellman equation (2.39) can be written by using the summation (2.37), and are therefore also linear. Hence, when the state and action spaces are finite and the cardinality of $X \times U$ is not too large (e.g., up to several thousands), Q^h can be found by directly solving the linear system of equations given by the Bellman equation.

The entire derivation can be repeated and similar algorithms can be given for V-functions instead of Q-functions. Such algorithms are more popular in the literature, see, e.g., (Sutton and Barto, 1998, Section 4.1) and (Bertsekas, 2007, Section 1.3). Recall however, that policy improvements are more problematic when V-functions are used, because a model is required to find greedy policies, as seen in (2.11). Additionally, in the stochastic case, expectations over the one-step stochastic transitions have to be computed to find greedy policies, as seen in (2.19).

An important advantage of policy iteration over value iteration stems from the linearity of the Bellman equation for Q^h in the Q-values. In contrast, the Bellman optimality equation (for Q^*) is highly nonlinear due to the maximization at the right-hand side. This makes policy evaluation generally easier to solve than value iteration. Moreover, in practice, offline policy iteration algorithms often converge in a small number of iterations (Madani, 2002; Sutton and Barto, 1998, Section 4.3), possibly smaller than the number of iterations taken by offline value iteration algorithms. However, this does not mean that policy iteration is computationally less costly than value iteration. For instance, even though policy evaluation using Q-functions is generally less costly than Q-iteration, every single policy iteration requires a complete policy evaluation.

Computational cost of policy evaluation for Q-functions in finite MDPs

We next investigate the computational cost of policy evaluation for Q-functions (2.38) for an MDP with a finite number of states and actions. We also provide a comparison with the computational cost of Q-iteration. Note again that policy evaluation is only one component of policy iteration; for an illustration of the computational cost of the entire policy iteration algorithm, see the upcoming Example 2.4.

In the deterministic case, policy evaluation for Q-functions can be implemented as in Algorithm 2.5. The computational cost of one iteration of this algorithm, measured by the number of function evaluations, is:

$$4\,|X|\,|U|$$

where the functions being evaluated are ρ, f, h, and the current Q-function Q^h_τ. In the stochastic case, Algorithm 2.6 can be used, which requires at each iteration the following number of function evaluations:

$$4\,|X|^2\,|U|$$

where the functions being evaluated are $\tilde{\rho}, \bar{f}, h$, and Q^h_τ. The cost in the stochastic case is thus larger by a factor $|X|$ than the cost in the deterministic case.

Table 2.4 collects the computational cost of policy evaluation for Q-functions, and compares it with the computational cost of Q-iteration (Section 2.3.1). A single Q-iteration requires $|X|\,|U|\,(2+|U|)$ function evaluations in the deterministic case (2.28), and $|X|^2\,|U|\,(2+|U|)$ function evaluations in the stochastic case (2.29). Whenever $|U|>2$, the cost of a single Q-iteration is therefore larger than the cost of a policy evaluation iteration.

TABLE 2.4
Computational cost of policy evaluation for Q-functions and of Q-iteration, measured by the number of function evaluations. The cost for a single iteration is shown.

	Deterministic case	Stochastic case
Policy evaluation	$4\,|X|\,|U|$	$4\,|X|^2\,|U|$
Q-iteration	$|X|\,|U|\,(2+|U|)$	$|X|^2\,|U|\,(2+|U|)$

Note also that evaluating the policy by directly solving the linear system given by the Bellman equation typically requires $\mathrm{O}(|X|^3\,|U|^3)$ computation. This is an asymptotic measure of computational complexity (Knuth, 1976), and is no longer directly related to the number of function evaluations. By comparison, the complexity of the complete iterative policy evaluation algorithm is $\mathrm{O}(L\,|X|\,|U|)$ in the deterministic case and $\mathrm{O}(L\,|X|^2\,|U|)$ in the stochastic case, where L is the number of iterations.

Example 2.4 Model-based policy iteration for the cleaning robot. In this example, we apply a policy iteration algorithm to the cleaning-robot problem introduced in Examples 2.1 and 2.2. Recall that *every* single policy iteration requires a *complete execution* of policy evaluation for the current policy, together with a policy improvement. The (model-based) policy evaluation for Q-functions (2.38) is employed, starting from identically zero Q-functions. Each policy evaluation is run until the

Q-function fully converges. The same discount factor is used as for Q-iteration in Example 2.3, namely $\gamma = 0.5$.

Consider first the deterministic variant of Example 2.1, in which policy evaluation for Q-functions takes the form shown in Algorithm 2.5. Starting from a policy that always moves right ($h_0(x) = 1$ for all x), policy iteration produces the sequence of Q-functions and policies given in Table 2.5. In this table, the sequence of Q-functions produced by a given execution of policy evaluation is separated by dashed lines from the policy being evaluated (shown above the sequence of Q-functions) and from the improved policy (shown below the sequence). The policy iteration algorithm converges after 2 iterations. In fact, the policy is already optimal after the first policy improvement: $h_2 = h_1 = h^*$.

TABLE 2.5
Policy iteration results for the deterministic cleaning-robot problem. Q-values are rounded to 3 decimal places.

	$x = 0$	$x = 1$	$x = 2$	$x = 3$	$x = 4$	$x = 5$
h_0	$*$	1	1	1	1	$*$
Q_0	0; 0	0; 0	0; 0	0; 0	0; 0	0; 0
Q_1	0; 0	1; 0	0; 0	0; 0	0; 5	0; 0
Q_2	0; 0	1; 0	0; 0	0; 2.5	1.25; 5	0; 0
Q_3	0; 0	1; 0	0; 1.25	0.625; 2.5	1.25; 5	0; 0
Q_4	0; 0	1; 0.625	0.313; 1.25	0.625; 2.5	1.25; 5	0; 0
Q_5	0; 0	1; 0.625	0.313; 1.25	0.625; 2.5	1.25; 5	0; 0
h_1	$*$	-1	1	1	1	$*$
Q_0	0; 0	0; 0	0; 0	0; 0	0; 0	0; 0
Q_1	0; 0	1; 0	0.5; 0	0; 0	0; 5	0; 0
Q_2	0; 0	1; 0	0.5; 0	0; 2.5	1.25; 5	0; 0
Q_3	0; 0	1; 0	0.5; 1.25	0.625; 2.5	1.25; 5	0; 0
Q_4	0; 0	1; 0.625	0.5; 1.25	0.625; 2.5	1.25; 5	0; 0
Q_5	0; 0	1; 0.625	0.5; 1.25	0.625; 2.5	1.25; 5	0; 0
h_2	$*$	-1	1	1	1	$*$

Five iterations of the policy evaluation algorithm are required for the first policy, and the same number of iterations are required for the second policy. Recall that the computational cost of every iteration of the policy evaluation algorithm, measured by the number of function evaluations, is $4\,|X|\,|U|$, leading to a total cost of $5 \cdot 4 \cdot |X|\,|U|$ for each of the two policy evaluations. Assuming that the maximization over U in the policy improvement is solved by enumeration, the computational cost of every policy improvement is $|X|\,|U|$. Each of the two policy iterations consists of a policy evaluation and a policy improvement, requiring:

$$5 \cdot 4 \cdot |X|\,|U| + |X|\,|U| = 21\,|X|\,|U|$$

function evaluations, and thus the entire policy iteration algorithm has a cost of:

$$2 \cdot 21 \cdot |X|\,|U| = 2 \cdot 21 \cdot 6 \cdot 2 = 504$$

Compared to the cost 240 of Q-iteration in Example 2.3, policy iteration is in this case more computationally expensive. This is true even though the cost of any single policy evaluation, $5 \cdot 4 \cdot |X|\,|U| = 240$, is the same as the cost of Q-iteration. The latter fact is expected from the theory (Table 2.4), which indicated that policy evaluation for Q-functions and Q-iteration have similar costs when $|U| = 2$, as is the case here.

Consider now the stochastic case of Example 2.2. For this case, policy evaluation for Q-functions takes the form shown in Algorithm 2.6. Starting from the same policy as in the deterministic case (always going right), policy iteration produces the sequence of Q-functions and policies illustrated in Table 2.6 (not all Q-functions are shown). Although the Q-functions are different from those in the deterministic case, the same sequence of policies is produced.

TABLE 2.6

Policy iteration results for the stochastic cleaning-robot problem. Q-values are rounded to 3 decimal places.

	$x=0$	$x=1$	$x=2$	$x=3$	$x=4$	$x=5$
h_0	$*$	1	1	1	1	$*$
Q_0	0; 0	0; 0	0; 0	0; 0	0; 0	0; 0
Q_1	0; 0	0.800; 0.050	0.020; 0.001	0.001; 0	0.250; 4	0; 0
Q_2	0; 0	0.804; 0.054	0.022; 0.001	0.101; 1.600	1.190; 4.340	0; 0
Q_3	0; 0	0.804; 0.055	0.062; 0.641	0.485; 1.872	1.324; 4.372	0; 0
Q_4	0; 0	0.820; 0.311	0.219; 0.805	0.572; 1.909	1.342; 4.376	0; 0
...	...	...	...	...	...	...
Q_{24}	0; 0	0.852; 0.417	0.278; 0.839	0.589; 1.915	1.344; 4.376	0; 0
h_1	$*$	−1	1	1	1	$*$
Q_0	0; 0	0; 0	0; 0	0; 0	0; 0	0; 0
Q_1	0; 0	0.800; 0.110	0.320; 0.020	0.008; 0.001	0.250; 4	0; 0
Q_2	0; 0	0.861; 0.123	0.346; 0.023	0.109; 1.601	1.190; 4.340	0; 0
Q_3	0; 0	0.865; 0.124	0.388; 0.664	0.494; 1.873	1.325; 4.372	0; 0
Q_4	0; 0	0.881; 0.382	0.449; 0.821	0.578; 1.910	1.342; 4.376	0; 0
...	...	...	...	...	...	...
Q_{22}	0; 0	0.888; 0.458	0.467; 0.852	0.594; 1.915	1.344; 4.376	0; 0
h_2	$*$	−1	1	1	1	$*$

Twenty-four iterations of the policy evaluation algorithm are required to evaluate the first policy, and 22 iterations are required for the second. Recall that the cost of every iteration of the policy evaluation algorithm, measured by the number of function evaluations, is $4\,|X|^2\,|U|$ in the stochastic case, while the cost for policy improvement is the same as in the deterministic case: $|X|\,|U|$. So, the first policy iteration requires $24 \cdot 4 \cdot |X|^2 |U| + |X| |U|$ function evaluations, and the second requires $22 \cdot 4 \cdot |X|^2 |U| + |X| |U|$ function evaluations. The total cost of policy iteration is obtained by adding these two costs:

$$46 \cdot 4 \cdot |X|^2 |U| + 2|X| |U| = 46 \cdot 4 \cdot 6^2 \cdot 2 + 2 \cdot 6 \cdot 2 = 13272$$

Comparing this to the 6336 function evaluations necessary for Q-iteration in the stochastic problem (see Example 2.3), it appears that policy iteration is also more computationally expensive in the stochastic case. Moreover, policy iteration is more computationally costly in the stochastic case than in the deterministic case; in the latter case, policy iteration required only 504 function evaluations. □

2.4.2 Model-free policy iteration

After having discussed above model-based policy iteration, we now turn our attention to the class of class of RL, model-free policy iteration algorithms, and within this class, we focus on *SARSA*, an online algorithm proposed by Rummery and Niranjan (1994) as an alternative to the value-iteration based Q-learning. The name SARSA is obtained by joining together the initials of every element in the data tuples employed by the algorithm, namely: state, action, reward, (next) state, (next) action. Formally, such a tuple is denoted by $(x_k, u_k, r_{k+1}, x_{k+1}, u_{k+1})$. SARSA starts with an arbitrary initial Q-function Q_0 and updates it at each step using tuples of this form, as follows:

$$Q_{k+1}(x_k, u_k) = Q_k(x_k, u_k) + \alpha_k [r_{k+1} + \gamma Q_k(x_{k+1}, u_{k+1}) - Q_k(x_k, u_k)] \tag{2.40}$$

where $\alpha_k \in (0, 1]$ is the learning rate. The term between square brackets is the temporal difference, obtained as the difference between the updated estimate $r_{k+1} + \gamma Q_k(x_{k+1}, u_{k+1})$ of the Q-value for (x_k, u_k), and the current estimate $Q_k(x_k, u_k)$. This is not the same as the temporal difference used in Q-learning (2.30). While the Q-learning temporal difference includes the maximal Q-value in the next state, the SARSA temporal difference includes the Q-value of the action actually taken in this next state. This means that SARSA performs online, model-free policy evaluation of the policy that is currently being followed. In the deterministic case, the new estimate $r_{k+1} + \gamma Q_k(x_{k+1}, u_{k+1})$ of the Q-value for (x_k, u_k) is actually the policy evaluation mapping (2.35) applied to Q_k in the state-action pair (x_k, u_k). Here, $\rho(x_k, u_k)$ has been replaced by the observed reward r_{k+1}, and $f(x_k, u_k)$ by the observed next state x_{k+1}. In the stochastic case, these replacements provide a single sample of the random quantity whose expectation is found by the policy evaluation mapping (2.36).

Next, the policy employed by SARSA is considered. Unlike offline policy iteration, SARSA cannot afford to wait until the Q-function has (almost) converged before it improves the policy. This is because convergence may take a long time, during which the unimproved (and possibly bad) policy would be used. Instead, to select actions, SARSA combines a greedy policy in the current Q-function with exploration, using, e.g., ε-greedy (2.32) or Boltzmann (2.33) exploration. Because of the greedy component, SARSA implicitly performs a policy improvement at every time step, and is therefore a type of online policy iteration. Such a policy iteration algorithm, which improves the policy after every sample, is sometimes called fully optimistic (Bertsekas and Tsitsiklis, 1996, Section 6.4).

Algorithm 2.3 presents SARSA with ε-greedy exploration. In this algorithm, because the update at step k involves the action u_{k+1}, this action has to be chosen prior to updating the Q-function.

ALGORITHM 2.7 SARSA with ε-greedy exploration.

Input: discount factor γ,
exploration schedule $\{\varepsilon_k\}_{k=0}^{\infty}$, learning rate schedule $\{\alpha_k\}_{k=0}^{\infty}$

initialize Q-function, e.g., $Q_0 \leftarrow 0$

measure initial state x_0

$$u_0 \leftarrow \begin{cases} u \in \arg\max_{\bar{u}} Q_0(x_0, \bar{u}) & \text{with probability } 1 - \varepsilon_0 \text{ (exploit)} \\ \text{a uniformly random action in } U & \text{with probability } \varepsilon_0 \text{ (explore)} \end{cases}$$

for every time step $k = 0, 1, 2, \ldots$ **do**

  apply u_k, measure next state x_{k+1} and reward r_{k+1}

$$u_{k+1} \leftarrow \begin{cases} u \in \arg\max_{\bar{u}} Q_k(x_{k+1}, \bar{u}) & \text{with probability } 1 - \varepsilon_{k+1} \\ \text{a uniformly random action in } U & \text{with probability } \varepsilon_{k+1} \end{cases}$$

$$Q_{k+1}(x_k, u_k) \leftarrow Q_k(x_k, u_k) + \alpha_k[r_{k+1} + \gamma Q_k(x_{k+1}, u_{k+1}) - Q_k(x_k, u_k)]$$

end for

In order to converge to the optimal Q-function Q^*, SARSA requires conditions similar to those of Q-learning, which demand exploration, and *additionally* that the exploratory policy being followed asymptotically becomes greedy (Singh et al., 2000). Such a policy can be obtained by using, e.g., ε-greedy (2.32) exploration with an exploration probability ε_k that asymptotically decreases to 0, or Boltzmann (2.33) exploration with an exploration temperature τ_k that asymptotically decreases to 0. Note that, as already explained in Section 2.3.2, the exploratory policy used by Q-learning can also be made greedy asymptotically, even though the convergence of Q-learning does not rely on this condition.

Algorithms like SARSA, which evaluate the policy they are currently using to control the process, are also called "on-policy" in the RL literature (Sutton and Barto, 1998). In contrast, algorithms like Q-learning, which act on the process using one policy and evaluate another policy, are called "off-policy." In Q-learning, the policy used to control the system typically includes exploration, whereas the algorithm implicitly evaluates a policy that is greedy in the current Q-function, since maximal Q-values are used in the Q-function updates (2.30).

2.5 Policy search

The previous two sections have introduced value iteration and policy iteration. In this section, we consider the third major class of DP/RL methods, namely *policy search* algorithms. These algorithms use optimization techniques to directly search for an optimal policy, which maximizes the return from every initial state. The optimization criterion should therefore be a combination (e.g., average) of the returns from every initial state. In principle, any optimization technique can be used to search for an optimal policy. For a general problem, however, the optimization criterion may be a nondifferentiable function with multiple local optima. This means that global, gradient-free optimization techniques are more appropriate than local, gradient-based techniques. Particular examples of global, gradient-free techniques include genetic algorithms (Goldberg, 1989), tabu search (Glover and Laguna, 1997), pattern search (Torczon, 1997; Lewis and Torczon, 2000), cross-entropy optimization (Rubinstein and Kroese, 2004), etc.

Consider the return estimation procedure of a model-based policy search algorithm. The returns are infinite sums of discounted rewards (2.1), (2.15). However, in practice, the returns have to be estimated in a finite time. To this end, the infinite sum in the return can be approximated with a finite sum over the first K steps. To guarantee that the approximation obtained in this way is within a bound $\varepsilon_{\text{MC}} > 0$ of the infinite sum, K can be chosen with (e.g., Mannor et al., 2003):

$$K = \left\lceil \log_\gamma \frac{\varepsilon_{\text{MC}}(1-\gamma)}{\|\rho\|_\infty} \right\rceil \tag{2.41}$$

Note that, in the stochastic case, usually many sample trajectories need to be simulated to obtain an accurate estimate of the expected return.

Evaluating the optimization criterion of policy search requires the accurate estimation of returns from all the initial states. This procedure is likely to be computationally expensive, especially in the stochastic case. Since optimization algorithms typically require many evaluations of the criterion, policy search algorithms are therefore computationally expensive, usually more so than value iteration and policy iteration.

Computational cost of exhaustive policy search for finite MDPs

We next investigate the computational cost of a policy search algorithm for deterministic MDPs with a finite number of states and actions. Since the state and action spaces are finite and therefore discrete, any combinatorial optimization technique could be used to look for an optimal policy. However, for simplicity, we consider an algorithm that exhaustively searches the entire policy space.

In the deterministic case, a single trajectory consisting of K simulation steps suffices to estimate the return from a given initial state. The number of possible policies is $|U|^{|X|}$ and the return has to be evaluated for all the $|X|$ initial states. It follows that the total number of simulation steps that have to be performed to find an optimal policy is at most $K\,|U|^{|X|}\,|X|$. Since f, ρ, and h are each evaluated once at every simulation step, the computational cost, measured by the number of function evaluations, is:

$$3K\,|U|^{|X|}\,|X|$$

Compared to the cost $L|X||U|(2+|U|)$ of Q-iteration for deterministic systems (2.28), this implementation of policy search is, in most cases, clearly more costly.

In the stochastic case, when computing the expected return from a given initial state x_0, the exhaustive search algorithm considers all the possible realizations of a trajectory of length K. Starting from initial state x_0 and taking action $h(x_0)$, there are $|X|$ possible values of x_1, the state at step 1 of the trajectory. The algorithm considers all these possible values, together with their respective probabilities of being reached, namely $\bar{f}(x_0,h(x_0),x_1)$. Then, for each of these values of x_1, given the respective actions $h(x_1)$, there are again $|X|$ possible values of x_2, each reachable with a certain probability, and so on until K steps have been considered. With a recursive implementation, a total number of $|X|+|X|^2+\cdots+|X|^K$ steps have to be considered. Each such step requires 3 function evaluations, where the functions being evaluated are $\bar{f}$, $\tilde{\rho}$, and h. Moreover, $|U|^{|X|}$ policies have to be evaluated for $|X|$ initial states, so the total cost of exhaustive policy search in the stochastic case is:

$$3\left(\sum_{k=1}^{K}|X|^k\right)|U|^{|X|}|X| = 3\frac{|X|^{K+1}-|X|}{|X|-1}|U|^{|X|}|X|$$

Unsurprisingly, this cost grows roughly exponentially with K, rather than linearly as in the deterministic case, so exhaustive policy search is more computationally expensive in the stochastic case than in the deterministic case. In most problems, the cost of exhaustive policy search in the stochastic case is also greater than the cost $L|X|^2|U|(2+|U|)$ of Q-iteration (2.29).

Of course, much more efficient optimization techniques than exhaustive search are available, and the estimation of the expected returns can also be accelerated. For instance, after the return of a state has been estimated, this estimate can be reused at every occurrence of that state along subsequent trajectories, thereby reducing the computational cost. Nevertheless, the costs derived above can be seen as worst-case values that illustrate the inherently large complexity of policy search.

Example 2.5 Exhaustive policy search for the cleaning robot. Consider again the cleaning-robot problem introduced in Examples 2.1 and 2.2, and assume that the exhaustive policy search described above is applied. Take the approximation tolerance in the evaluation of the return to be $\varepsilon_{\mathrm{MC}} = 0.01$, which is equal to the suboptimality bound ς_{QI} for Q-iteration in Example 2.3. Using ς_{QI}, maximum absolute reward $\|\rho\|_\infty = 5$, and discount factor $\gamma = 0.5$ in (2.41), a time horizon of $K = 10$ steps is obtained. Therefore, in the deterministic case, the computational cost of the algorithm, measured by the number of function evaluations, is:

$$3K|U|^{|X|}|X| = 3\cdot 10\cdot 2^6\cdot 6 = 11520$$

whereas in the stochastic case, it is:

$$3\frac{|X|^{K+1}-|X|}{|X|-1}|U|^{|X|}|X| = 3\cdot\frac{6^{11}-6}{6-1}\cdot 2^6\cdot 6 \approx 8\cdot 10^{10}$$

By observing that it is unnecessary to look for optimal actions and to evaluate returns in the terminal states, the cost can further be reduced to $3 \cdot 10 \cdot 2^4 \cdot 4 = 1920$ in the deterministic case, and to $3 \cdot \frac{6^{11}-6}{6-1} \cdot 2^4 \cdot 4 \approx 1 \cdot 10^{10}$ in the stochastic case. Additional reductions in cost can be obtained by stopping the simulation of trajectories as soon as they reach a terminal state, which will often happen in fewer than 10 steps.

Table 2.7 compares the computational cost of exhaustive policy search with the cost of Q-iteration from Example 2.3 and of policy iteration from Example 2.4. For the cleaning-robot problem, the exhaustive implementation of direct policy search is very likely to be more expensive than both Q-iteration and policy iteration. □

TABLE 2.7

Computational cost of exhaustive policy search for the cleaning robot, compared with the cost of Q-iteration and of policy iteration. The cost is measured by the number of function evaluations.

	Deterministic case	Stochastic case
Exhaustive policy search	11520	$8 \cdot 10^{10}$
Exhaustive policy search, no optimization in terminal states	1920	$1 \cdot 10^{10}$
Q-iteration	240	6336
Policy iteration	504	13272

2.6 Summary and discussion

In this chapter, deterministic and stochastic MDPs have been introduced, and their optimal solution has been characterized. Three classes of DP and RL algorithms have been described: value iteration, policy iteration, and direct search for control policies. This presentation provides the necessary background for the remainder of this book, but is by no means exhaustive. For the reader interested in more details about classical DP and RL, we recommend the textbook of Bertsekas (2007) on DP, and that of Sutton and Barto (1998) on RL.

A central challenge in the DP and RL fields is that, in their original form, DP and RL algorithms cannot be implemented for general problems. They can only be implemented when the state and action spaces consist of a finite number of discrete elements, because (among other reasons) they require the exact representation of value functions or policies, which is generally impossible for state spaces with an infinite number of elements. In the case of Q-functions, an infinite number of actions also prohibits an exact representation. For instance, most problems in automatic control have continuous states and actions, which can take infinitely many distinct values. Even when the states and actions take finitely many values, the cost of representing value functions and policies grows exponentially with the number of state variables (and action variables, for Q-functions). This problem is called the curse of dimensionality, and makes the classical DP and RL algorithms impractical when there are many state and action variables.

To cope with these problems, versions of the classical algorithms that *approximately* represent value functions and/or policies must be used. Such algorithms for approximate DP and RL form the subject of the remainder of this book.

In practice, it is essential to provide more comprehensive performance guarantees than simply the asymptotical maximization of the return. For instance, online RL algorithms should guarantee an increase in performance over time. Note that the performance cannot increase monotonically, since exploration is necessary, which can cause temporary degradations of the performance. In order to use DP and RL algorithms in industrial applications of automatic control, it should be guaranteed that they can never destabilize the process. For instance, Perkins and Barto (2002); Balakrishnan et al. (2008) discuss DP and RL approaches that guarantee stability using the Lyapunov framework (Khalil, 2002, Chapters 4 and 14).

Designing a good reward function is an important and nontrivial step of applying DP and RL. Classical texts on RL recommend making the reward function as simple as possible; it should only reward the achievement of the final goal (Sutton and Barto, 1998). However, a simple reward function often makes online RL slow, and including more information may be required for successful learning. Moreover, other high-level requirements on the behavior of the controller often have to be considered in addition to achieving the final goal. For instance, in automatic control the controlled state trajectories often have to satisfy requirements on overshoot and the rate of convergence to an equilibrium, etc. Translating such requirements into the language of rewards can be very challenging.

DP and RL algorithms can also be greatly helped by domain knowledge. Although RL is usually envisioned as purely model-free, it can be very beneficial to use prior knowledge about the problem, if such knowledge is available. If a partial model is available, a DP algorithm can be run with this partial model, in order to obtain a rough initial solution for the RL algorithm. Prior knowledge about the policy can also be used to restrict the class of policies considered by (model-based or model-free) policy iteration and policy search. A good way to provide domain knowledge to any DP or RL algorithm is to encode it in the reward function (Dorigo and Colombetti, 1994; Matarić, 1997; Randløv and Alstrøm, 1998; Ng et al., 1999). This procedure is related to the problem of reward function design discussed above. For instance, prior knowledge about promising control actions can be exploited by associating these actions with high rewards. Encoding prior knowledge in the reward function should be done with care, because doing so incorrectly can lead to unexpected and possibly undesirable behavior.

Other work aiming to expand the boundaries of DP and RL includes: problems in which the state is not fully measurable, called partially observable MDPs (Lovejoy, 1991; Kaelbling et al., 1998; Singh et al., 2004; Pineau et al., 2006; Porta et al., 2006), exploiting modular and hierarchical task decompositions (Dietterich, 2000; Hengst, 2002; Russell and Zimdars, 2003; Barto and Mahadevan, 2003; Ghavamzadeh and Mahadevan, 2007), and applying DP and RL to distributed, multi-agent problems (Panait and Luke, 2005; Shoham et al., 2007; Buşoniu et al., 2008a).

3

Dynamic programming and reinforcement learning in large and continuous spaces

This chapter describes dynamic programming and reinforcement learning for large and continuous-space problems. In such problems, exact solutions cannot be found in general, and approximation is necessary. The algorithms of the previous chapter can therefore no longer be applied in their original form. Instead, approximate versions of value iteration, policy iteration, and policy search are introduced. Theoretical guarantees are provided on the performance of the algorithms, and numerical examples are used to illustrate their behavior. Techniques to automatically find value function approximators are reviewed, and the three categories of algorithms are compared.

3.1 Introduction

The classical dynamic programming (DP) and reinforcement learning (RL) algorithms introduced in Chapter 2 require exact representations of the value functions and policies. In general, an exact value function representation can only be achieved by storing distinct estimates of the return for every state-action pair (when Q-functions are used) or for every state (in the case of V-functions). Similarly, to represent policies exactly, distinct actions have to be stored for every state. When some of the variables have a very large or infinite number of possible values (e.g., when they are continuous), such exact representations are no longer possible, and value functions and policies need to be represented approximately. Since most problems of practical interest have large or continuous state and action spaces, approximation is essential in DP and RL.

Approximators can be separated into two main types: parametric and nonparametric. *Parametric* approximators are mappings from a parameter space into the space of functions they aim to represent. The form of the mapping and the number of parameters are given *a priori*, while the parameters themselves are tuned using data about the target function. A representative example is a weighted linear combination of a fixed set of basis functions, in which the weights are the parameters. In contrast, the structure of a *nonparametric* approximator is derived from the data. Despite its name, a nonparametric approximator typically still has parameters, but unlike in the parametric case, the number of parameters (as well as their values) is determined from the data. For instance, kernel-based approximators considered in this book define one kernel per data point, and represent the target function as a weighted linear combination of these kernels, where again the weights are the parameters.

This chapter provides an extensive, in-depth review of approximate DP and RL in large and continuous-space problems. The three basic classes of DP and RL algorithms discussed in Chapter 2, namely value iteration, policy iteration, and policy search, are all extended to use approximation, resulting in *approximate value iteration*, *approximate policy iteration*, and *approximate policy search*. Algorithm derivations are complemented by theoretical guarantees on their performance, by numerical examples illustrating their behavior, and by comparisons of the different approaches. Several other important topics in value function and policy approximation are also treated. To help in navigating this large body of material, Figure 3.1 presents a road map of the chapter in graphical form, and the remainder of this section details this road map.

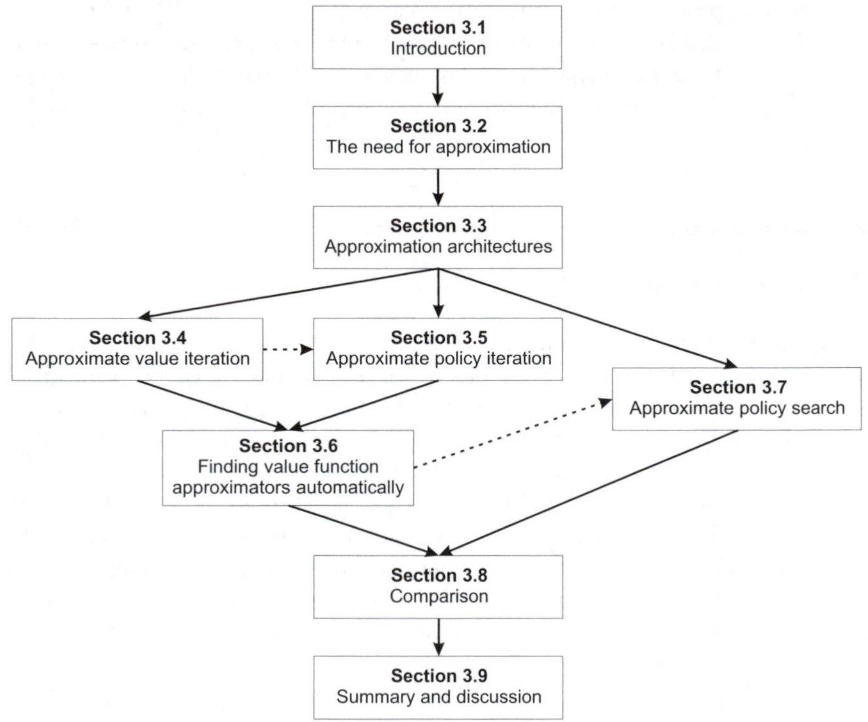

FIGURE 3.1
A road map of this chapter. The arrows indicate the recommended sequence of reading. Dashed arrows indicate optional ordering.

In Section 3.2, the need for approximation in DP and RL for large and continuous spaces is explained. Approximation is not only a problem of compact representation, but also plays a role in several other parts of DP and RL algorithms. In Section 3.3, parametric and nonparametric approximation architectures are introduced and compared.

This introduction is followed by an in-depth discussion of approximate value iteration in Section 3.4, and of approximate policy iteration in Section 3.5. Techniques to automatically derive value function approximators, useful in approximate value iteration and policy iteration, are reviewed right after these two classes of algorithms, in Section 3.6. Approximate policy search is discussed in detail in Section 3.7. Representative algorithms from each of the three classes are applied to a numerical example involving the optimal control of a DC motor.

In closing the chapter, approximate value iteration, policy iteration, and policy search are compared in Section 3.8, while Section 3.9 provides a summary and discussion.

In order to reasonably restrict the scope of this chapter, several choices are made regarding the material that will be presented:

- In the context of value function approximation, we focus on Q-function approximation and Q-function based algorithms, because a significant portion of the remainder of this book concerns such algorithms. Nevertheless, a majority of the concepts and algorithms introduced extend in a straightforward manner to the case of V-function approximation.
- We mainly consider parametric approximation, because the remainder of the book relies on this type of approximation, but we also review nonparametric approaches to approximate value iteration and policy iteration.
- When discussing parametric approximation, whenever appropriate, we consider general (possibly nonlinear) parametrizations. Sometimes, however, we consider linear parametrizations in more detail, e.g., because they allow the derivation of better theoretical guarantees on the resulting approximate solutions.

Next, we give some additional details about the organization of the core material of this chapter, which consists of approximate value iteration (Section 3.4), approximate policy iteration (Section 3.5), and approximate policy search (Section 3.7). To this end, Figure 3.2 shows how the algorithms selected for presentation are organized, using a graphical tree format. This organization will be explained below. All the terminal (right-most) nodes in the trees correspond to subsections in Sections 3.4, 3.5, and 3.7. Note that Figure 3.2 does not contain an exhaustive taxonomy of all the approaches.

Within the context of approximate value iteration, algorithms employing parametric approximation are presented first, separating model-based from model-free approaches. Then, value iteration with nonparametric approximation is reviewed.

Approximate policy iteration consists of two distinct problems: approximate policy evaluation, i.e., finding an approximate value function for a given policy, and policy improvement. Out of these two problems, approximate policy evaluation poses more interesting theoretical questions, because, like approximate value iteration, it

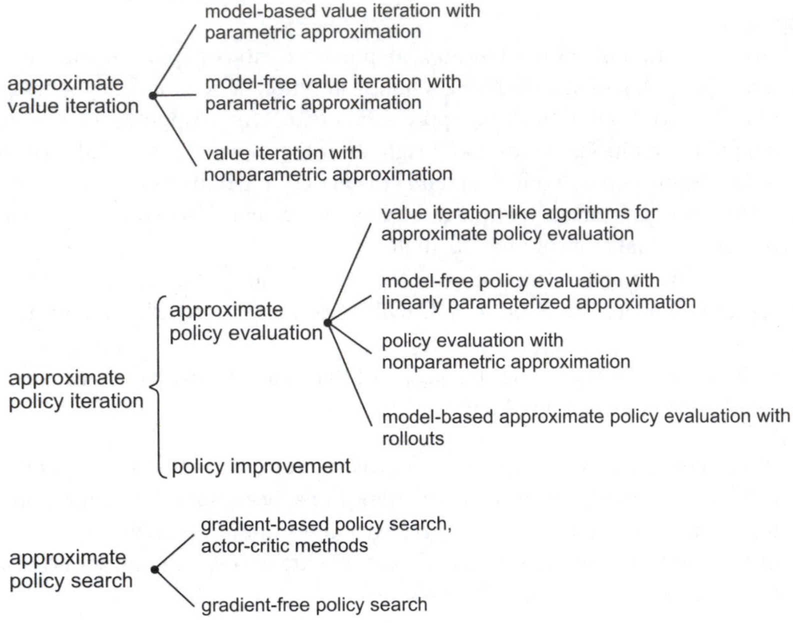

FIGURE 3.2
The organization of the algorithms for approximate value iteration, policy iteration, and policy search presented in this chapter.

involves finding an approximate solution to a Bellman equation. Special requirements have to be imposed to ensure that a meaningful approximate solution exists and can be found by appropriate algorithms. In contrast, policy improvement relies on solving maximization problems over the action variables, which involve fewer technical difficulties (although they may still be hard to solve when the action space is large). Therefore, we pay special attention to approximate policy evaluation in our presentation. We first describe a class of algorithms for policy evaluation that are derived along the same lines as approximate value iteration. Then, we introduce model-free policy evaluation with linearly parameterized approximation, and briefly review nonparametric approaches to approximate policy evaluation. Additionally, a model-based, direct simulation approach for policy evaluation is discussed that employs Monte Carlo estimates called "rollouts."

From the class of approximate policy search methods (Section 3.7), gradient-based and gradient-free methods for policy optimization are discussed in turn. In the context of gradient-based methods, special attention is paid to the important category of actor-critic techniques.

3.2 The need for approximation in large and continuous spaces

The algorithms for exact value iteration (Section 2.3) require the storage of distinct return estimates for every state (if V-functions are used) or for every state-action pair (in the case of Q-functions). When some of the state variables have a very large or infinite number of possible values (e.g., when they are continuous), exact storage is no longer possible, and the value functions must be represented approximately. Large or continuous action spaces make the representation of Q-functions additionally challenging. In policy iteration (Section 2.4), value functions and sometimes policies also need to be represented approximately in general. Similarly, in policy search (Section 2.5), policies must be represented approximately when the state space is large or continuous.

Approximation in DP/RL is not only a problem of representation. Two additional types of approximation are needed. First, sample-based approximation is necessary in any DP/RL algorithm. Second, value iteration and policy iteration must repeatedly solve potentially difficult nonconcave maximization problems over the action variables, whereas policy search must find optimal policy parameters, which involves similar difficulties. In general, these optimization problems can only be solved approximately. These two types of approximation are detailed below.

Sample-based approximation is required for two distinct purposes in value function estimation. Consider first, as an example, the Q-iteration algorithm for deterministic problems, namely Algorithm 2.1. Every iteration of this algorithm would have to be implemented as follows:

$$\textbf{for every } (x,u) \textbf{ do: } Q_{\ell+1}(x,u) = \rho(x,u) + \gamma \max_{u'} Q_\ell(f(x,u),u') \tag{3.1}$$

When the state-action space contains an infinite number of elements, it is impossible to loop over all the state-action pairs in finite time. Instead, a sample-based, approximate update has to be used that only considers a finite number of state-action samples.

Such sample-based updates are also necessary in stochastic problems. Moreover, in the stochastic case, sample-based approximation is required for a second, distinct purpose. Consider, e.g., the Q-iteration algorithm for general stochastic problems, which for every state-action pair (x,u) considered would have to be implemented as follows:

$$Q_{\ell+1}(x,u) = \mathrm{E}_{x' \sim \tilde{f}(x,u,\cdot)} \left\{ \tilde{\rho}(x,u,x') + \gamma \max_{u'} Q_\ell(x',u') \right\} \tag{3.2}$$

Clearly, the expectation on the right-hand side of (3.2) cannot be computed exactly in general, and must be estimated from a finite number of samples, e.g., by using Monte Carlo methods. Note that, in many RL algorithms, the estimation of the expectation does not appear explicitly, but is performed implicitly while processing samples. For instance, Q-learning (Algorithm 2.3) is such an algorithm, in which stochastic approximation is employed to estimate the expectation.

The maximization over the action variable in (3.1) or (3.2) (as well as in other value iteration algorithms) has to be solved for every sample considered. In large or continuous action spaces, this maximization is a potentially difficult nonconcave optimization problem, which can only be solved approximately in general. To simplify this problem, many algorithms discretize the action space in a small number of values, compute the value function for all the discrete actions, and find the maximum among these values using enumeration.

In policy iteration, sample-based approximation is required at the policy evaluation step, for reasons similar to those explained above. The maximization issues affect the policy improvement step, which in the case of Q-functions computes a policy $h_{\ell+1}$ using (2.34), repeated here for easy reference:

$$h_{\ell+1}(x) \in \arg\max_{u} Q^{h_\ell}(x,u)$$

Note that these sampling and maximization issues also affect algorithms that employ V-functions.

In policy search, some methods (e.g., actor-critic algorithms) estimate value functions and are therefore affected by the sampling issues mentioned above. Even methods that do not employ value functions must estimate returns in order to evaluate the policies, and return estimation requires sample-based approximation, as described next. In principle, a policy that maximizes the return from every initial state should be found. However, the return can only be estimated for a finite subset of initial states (samples) from the possibly infinite state space. Additionally, in stochastic problems, for every initial state considered, the expected return (2.15) must be evaluated using a finite set of sampled trajectories, e.g., by using Monte Carlo methods.

Besides these sampling problems, policy search methods must of course find the best policy within the class of policies considered. This is a difficult optimization problem, which can only be solved approximately in general. However, it only needs to be solved once, unlike the maximization over actions in value iteration and policy iteration, which needs to be solved for every sample considered. In this sense, policy search methods are less affected from the maximization difficulties than value iteration or policy iteration.

A different view on the benefits of approximation can be taken in the model-free, RL setting. Consider a value iteration algorithm that estimates Q-functions, such as Q-learning (Algorithm 2.3). Without approximation, the Q-value of every state-action pair must be estimated separately (assuming it is possible to do so). If little or no data is available for some states, their Q-values are poorly estimated, and the algorithm makes poor control decisions in those states. However, when approximation is used, the approximator can be designed so that the Q-values of each state influence the Q-values of other, usually nearby, states (this requires the assumption of a certain degree of smoothness for the Q-function). Then, if good estimates of the Q-values of a certain state are available, the algorithm can also make reasonable control decisions in nearby states. This is called *generalization* in the RL literature, and can help algorithms work well despite using only a limited number of samples.

3.3 Approximation architectures

Two major classes of approximators can be identified, namely parametric and non-parametric approximators. We introduce parametric approximators in Section 3.3.1, nonparametric approximators in Section 3.3.2, and compare the two classes in Section 3.3.3. Section 3.3.4 contains some additional remarks.

3.3.1 Parametric approximation

Parametric approximators are mappings from a parameter space into the space of functions they aim to represent (in DP/RL, value functions or policies). The functional form of the mapping and the number of parameters are typically established in advance and do not depend on the data. The parameters of the approximator are tuned using data about the target function.

Consider a Q-function approximator parameterized by an n-dimensional vector[1] θ. The approximator is denoted by an *approximation mapping* $F : \mathbb{R}^n \to \mathscr{Q}$, where $\mathbb{R}^n$ is the parameter space and $\mathscr{Q}$ is the space of Q-functions. Every parameter vector θ provides a compact representation of a corresponding approximate Q-function:

$$\widehat{Q} = F(\theta)$$

or equivalently, element-wise:

$$\widehat{Q}(x,u) = [F(\theta)](x,u)$$

where $[F(\theta)](x,u)$ denotes the Q-function $F(\theta)$ evaluated at the state-action pair (x,u). So, instead of storing distinct Q-values for every pair (x,u), which would be impractical in many cases, it is only necessary to store n parameters. When the state-action space is discrete, n is usually much smaller than $|X| \cdot |U|$, thereby providing a compact representation (recall that, when applied to sets, the notation $|\cdot|$ stands for cardinality). However, since the set of Q-functions representable by F is only a subset of $\mathscr{Q}$, an arbitrary Q-function can generally only be represented up to a certain approximation error, which must be accounted for.

In general, the mapping F can be nonlinear in the parameters. A typical example of a nonlinearly parameterized approximator is a feed-forward neural network (Hassoun, 1995; Bertsekas and Tsitsiklis, 1996, Chapter 3). However, linearly parameterized approximators are often preferred in DP and RL, because they make it easier to analyze the theoretical properties of the resulting DP/RL algorithms. A linearly parameterized Q-function approximator employs n basis functions (BFs) $\phi_1, \dots, \phi_n : X \times U \to \mathbb{R}$ and an n-dimensional parameter vector θ. Approximate Q-values are computed with:

$$[F(\theta)](x,u) = \sum_{l=1}^{n} \phi_l(x,u)\theta_l = \phi^{\mathrm{T}}(x,u)\theta \tag{3.3}$$

[1] All the vectors used in this book are column vectors.

where $\phi(x,u) = [\phi_1(x,u), \dots, \phi_n(x,u)]^{\mathrm{T}}$ is the vector of BFs. In the literature, the BFs are also called features (Bertsekas and Tsitsiklis, 1996).

Example 3.1 Approximating Q-functions with state-dependent BFs and discrete actions. As explained in Section 3.2, in order to simplify the maximization over actions, in many DP/RL algorithms the action space is discretized into a small number of values. In this example we consider such a discrete-action approximator, which additionally employs state-dependent BFs to approximate over the state space.

A discrete, finite set of actions $u_1, \dots, u_M$ is chosen from the original action space U. The resulting discretized action space is denoted by $U_{\mathrm{d}} = \{u_1, \dots, u_M\}$. A number of N state-dependent BFs $\bar{\phi}_1, \dots, \bar{\phi}_N : X \to \mathbb{R}$ are defined and replicated for each discrete action in U_{d}. Approximate Q-values can be computed for any state-discrete action pair with:

$$[F(\theta)](x,u_j) = \phi^{\mathrm{T}}(x,u_j)\,\theta, \tag{3.4}$$

where, in the state-action BF vector $\phi^{\mathrm{T}}(x,u_j)$, all the BFs that do not correspond to the current discrete action are taken to be equal to 0:

$$\phi(x,u_j) = [\underbrace{0,\dots,0}_{u_1},\dots,0,\underbrace{\bar{\phi}_1(x),\dots,\bar{\phi}_N(x)}_{u_j},0,\dots,\underbrace{0,\dots,0}_{u_M}]^{\mathrm{T}} \in \mathbb{R}^{NM} \tag{3.5}$$

The parameter vector θ therefore has NM elements. This type of approximator can be seen as representing M distinct state-dependent slices through the Q-function, one slice for each of the M discrete actions. Note that it is only meaningful to use such an approximator for the discrete actions in U_{d}; for any other actions, the approximator outputs 0. For this reason, only the discrete actions are considered in (3.4) and (3.5).

In this book, we will often use such discrete-action approximators. For instance, consider normalized (elliptical) Gaussian radial basis functions (RBFs). This type of RBF can be defined as follows:

$$\bar{\phi}_i(x) = \frac{\phi_i'(x)}{\sum_{i'=1}^{N} \phi_{i'}'(x)}, \quad \phi_i'(x) = \exp\left(-\frac{1}{2}[x-c_i]^{\mathrm{T}} B_i^{-1} [x-c_i]\right) \tag{3.6}$$

Here, ϕ_i' are the nonnormalized RBFs, the vector $c_i = [c_{i,1}, \dots, c_{i,D}]^{\mathrm{T}} \in \mathbb{R}^D$ is the center of the ith RBF, and the symmetric positive-definite matrix $B_i \in \mathbb{R}^{D \times D}$ is its width. Depending on the structure of the width matrix, RBFs of various shapes can be obtained. For a general width matrix, the RBFs are elliptical, while axis-aligned RBFs are obtained if the width matrix is diagonal, i.e., if $B_i = \mathrm{diag}(b_{i,1}, \dots, b_{i,D})$. In this case, the width of an RBF can also be expressed using a vector $b_i = [b_{i,1}, \dots, b_{i,D}]^{\mathrm{T}}$. Furthermore, spherical RBFs are obtained if, in addition, $b_{i,1} = \dots = b_{i,D}$.

Another class of discrete-action approximators uses *state aggregation* (Bertsekas and Tsitsiklis, 1996, Section 6.7). For state aggregation, the state space is partitioned into N disjoint subsets. Let X_i be the ith subset in this partition, for $i = 1, \dots, N$. For a given action, the approximator assigns the same Q-values for all the states in X_i. This corresponds to a BF vector of the form (3.5), with binary-valued (0 or 1) state-dependent BFs:

$$\bar{\phi}_i(x) = \begin{cases} 1 & \text{if } x \in X_i \\ 0 & \text{otherwise} \end{cases} \tag{3.7}$$

Because the subsets X_i are disjoint, exactly one BF is active at any point in the state space. All the individual states belonging to X_i can thus be seen as a single, larger *aggregate* (or quantized) state; hence the name "state aggregation" (or state quantization). By additionally identifying each subset X_i with a prototype state $x_i \in X_i$, state aggregation can also be seen as *state discretization*, where the discretized state space is $X_\mathrm{d} = \{x_1, \ldots, x_N\}$. The prototype state can be, e.g., the geometrical center of X_i (assuming this center belongs to X_i), or some other representative state.

Using the definition (3.7) of the state-dependent BFs and the expression (3.5) for the state-action BFs, the state-action BFs can be written compactly as follows:

$$\phi_{[i,j]}(x,u) = \begin{cases} 1 & \text{if } x \in X_i \text{ and } u = u_j \\ 0 & \text{otherwise} \end{cases} \tag{3.8}$$

The notation $[i, j]$ represents the scalar index corresponding to i and j, which can be computed as $[i, j] = i + (j-1)N$. If the n elements of the BF vector were arranged into an $N \times M$ matrix, by first filling in the first column with the first N elements, then the second column with the subsequent N elements, etc., then the element at index $[i, j]$ of the vector would be placed at row i and column j of the matrix. Note that exactly one state-action BF (3.8) is active at any point of $X \times U_\mathrm{d}$, and no BF is active if $u \notin U_\mathrm{d}$. □

Other types of linearly parameterized approximators used in the literature include tile coding (Watkins, 1989; Sherstov and Stone, 2005), multilinear interpolation (Davies, 1997), and Kuhn triangulation (Munos and Moore, 2002).

3.3.2 Nonparametric approximation

Nonparametric approximators, despite their name, still have parameters. However, unlike in the parametric case, the number of parameters, as well as the form of the nonparametric approximator, are derived from the available data.

Kernel-based approximators are typical representatives of the nonparametric class. Consider a kernel-based approximator of the Q-function. In this case, the *kernel function* is a function defined over two state-action pairs, $\kappa : X \times U \times X \times U \to \mathbb{R}$:

$$(x, u, x', u') \mapsto \kappa((x,u),(x',u')) \tag{3.9}$$

that must also satisfy certain additional conditions (see, e.g., Smola and Schölkopf, 2004). Under these conditions, the function κ can be interpreted as an inner product between feature vectors of its two arguments (the two state-action pairs) in a high-dimensional feature space. Using this property, a powerful approximator can be obtained by only computing the kernels, without ever working explicitly in the feature space. Note that in (3.9), as well as in the sequel, the state-action pairs are grouped together for clarity.

A widely used type of kernel is the Gaussian kernel, which for the problem of approximating the Q-function is given by:

$$\kappa((x,u),(x',u')) = \exp\left(-\frac{1}{2}\begin{bmatrix}x-x'\\u-u'\end{bmatrix}^{\mathrm{T}} B^{-1}\begin{bmatrix}x-x'\\u-u'\end{bmatrix}\right) \tag{3.10}$$

where the kernel width matrix $B \in \mathbb{R}^{(D+C)\times(D+C)}$ must be symmetric and positive definite. Here, D denotes the number of state variables and C denotes the number of action variables. For instance, a diagonal matrix $B = \operatorname{diag}(b_1,\ldots,b_{D+C})$ can be used. Note that, when the pair (x',u') is fixed, the kernel (3.10) has the same shape as a Gaussian state-action RBF centered on (x',u').

Assume that a set of state-action samples is available: $\{(x_{l_\mathrm{s}},u_{l_\mathrm{s}}) \,|\, l_\mathrm{s} = 1,\ldots,n_\mathrm{s}\}$. For this set of samples, the kernel-based approximator takes the form:

$$\widehat{Q}(x,u) = \sum_{l_\mathrm{s}=1}^{n_\mathrm{s}} \kappa((x,u),(x_{l_\mathrm{s}},u_{l_\mathrm{s}}))\theta_{l_\mathrm{s}} \tag{3.11}$$

where $\theta_1,\ldots,\theta_{n_\mathrm{s}}$ are the parameters. This form is superficially similar to the linearly parameterized approximator (3.3). However, there is a crucial difference between these two approximators. In the parametric case, the number and form of the BFs were defined in advance, and therefore led to a fixed functional form F of the approximator. In contrast, in the nonparametric case, the number of kernels and their form, and thus also the number of parameters and the functional form of the approximator, are determined from the samples.

One situation in which the kernel-based approximator can be seen as a parametric approximator is when the set of samples is selected in advance. Then, the resulting kernels can be identified with predefined BFs:

$$\phi_{l_\mathrm{s}}(x,u) = \kappa((x,u),(x_{l_\mathrm{s}},u_{l_\mathrm{s}})), \quad l_\mathrm{s} = 1,\ldots,n_\mathrm{s}$$

and the kernel-based approximator (3.11) is equivalent to a linearly parameterized approximator (3.3). However, in many cases, such as in online RL, the samples are not available in advance.

Important classes of nonparametric approximators that have been used in DP and RL include kernel-based methods (Shawe-Taylor and Cristianini, 2004), among which support vector machines are the most popular (Schölkopf et al., 1999; Cristianini and Shawe-Taylor, 2000; Smola and Schölkopf, 2004), Gaussian processes, which also employ kernels (Rasmussen and Williams, 2006), and regression trees (Breiman et al., 1984; Breiman, 2001). For instance, kernel-based and related approximators have been applied to value iteration (Ormoneit and Sen, 2002; Deisenroth et al., 2009; Farahmand et al., 2009a) and to policy evaluation and policy iteration (Lagoudakis and Parr, 2003b; Engel et al., 2003, 2005; Xu et al., 2007; Jung and Polani, 2007a; Bethke et al., 2008; Farahmand et al., 2009b). Ensembles of regression trees have been used with value iteration by Ernst et al. (2005, 2006a) and with policy iteration by Jodogne et al. (2006).

Note that nonparametric approximators are themselves driven by certain meta-parameters, such as the width B of the Gaussian kernel (3.10). These meta-parameters influence the accuracy of the approximator and may require tuning, which can be difficult to perform manually. However, there also exist methods for automating the tuning process (Deisenroth et al., 2009; Jung and Stone, 2009).

3.3.3 Comparison of parametric and nonparametric approximation

Because they are designed in advance, parametric approximators have to be flexible enough to accurately model the target functions solely by tuning the parameters. Highly flexible, nonlinearly parameterized approximators are available, such as neural networks. However, when used to approximate value functions in DP and RL, general nonlinear approximators make it difficult to guarantee the convergence of the resulting algorithms, and indeed can sometimes lead to divergence. Often, linearly parameterized approximators (3.3) must be used to guarantee convergence. Such approximators are specified by their BFs. When prior knowledge is not available to guide the selection of BFs (as is usually the case), a large number of BFs must be defined to evenly cover the state-action space. This is impractical in high-dimensional problems. To address this issue, methods have been proposed to automatically derive a small number of good BFs from data. We review these methods in Section 3.6. Because they derive BFs from data, such methods can be seen as residing between parametric and nonparametric approximation.

Nonparametric approximators are highly flexible. However, because their shape depends on the data, it may change while the DP/RL algorithm is running, which makes it difficult to provide convergence guarantees. A nonparametric approximator adapts its complexity to the amount of available data. This is beneficial in situations where data is costly or difficult to obtain. It can, however, become a disadvantage when a large amount of data is used, because the computational and memory demands of the approximator usually grow with the number of samples. For instance, the kernel-based approximator (3.11) has a number of parameters equal to the number of samples n_s used. This is especially problematic in online RL algorithms, which keep receiving new samples for their entire lifetime. There exist approaches to mitigate this problem. For instance, in kernel-based methods, the number of samples used to derive the approximator can be limited by only employing a subset of samples that contribute significantly to the accuracy of the approximation, and discarding the rest. Various measures can be used for the contribution of a given sample to the approximation accuracy. Such kernel sparsification methods were employed by Xu et al. (2007); Engel et al. (2003, 2005), and a related, so-called subset of regressors method was applied by Jung and Polani (2007a). Ernst (2005) proposed the selection of informative samples for an offline RL algorithm by iteratively choosing those samples for which the error in the Bellman equation is maximal under the current value function.

3.3.4 Remarks

The approximation architectures introduced above for Q-functions can be extended in a straightforward manner to V-function and policy approximation. For instance, a linearly parameterized policy approximator can be described as follows. A set of state-dependent BFs $\varphi_1, \ldots, \varphi_{\mathcal{N}} : X \to \mathbb{R}$ are defined, and given a parameter vector $\vartheta \in \mathbb{R}^{\mathcal{N}}$, the approximate policy is:

$$\widehat{h}(x) = \sum_{i=1}^{\mathcal{N}} \varphi_i(x) \vartheta_i = \varphi^{\mathrm{T}}(x) \vartheta \tag{3.12}$$

where $\varphi(x) = [\varphi_1(x), \ldots, \varphi_{\mathcal{N}}(x)]^{\mathrm{T}}$. For simplicity, the parametrization (3.12) is only given for scalar actions, but it can easily be extended to the case of multiple action variables. Note that we use calligraphic notation to differentiate variables related to policy approximation from variables related to value function approximation. So, the policy parameter is ϑ and the policy BFs are denoted by φ, whereas the value function parameter is θ and the value function BFs are denoted by ϕ. Furthermore, the number of policy parameters and BFs is $\mathcal{N}$. When samples are used to approximate the policy, their number is denoted by $\mathcal{N}_{\mathrm{s}}$.

In the parametric case, whenever we wish to explicitly highlight the dependence of an approximate policy $\widehat{h}$ on the parameter vector ϑ, we will use the notation $\widehat{h}(x; \vartheta)$. Similarly, when the dependence of a value function on the parameters needs to be made explicit without using the mapping F, we will use $\widehat{Q}(x, u; \theta)$ and $\widehat{V}(x; \theta)$ to denote Q-functions and V-functions, respectively.

Throughout the remainder of this chapter, we will mainly focus on DP and RL with parametric approximation, because the remainder of the book relies on this type of approximation, but we will also overview nonparametric approaches to value iteration and policy iteration.

3.4 Approximate value iteration

In order to apply value iteration to large or continuous-space problems, the value function must be approximated. Figure 3.3 (repeated from the relevant part of Figure 3.2) shows how our presentation of the algorithms for approximate value iteration is organized. First, we describe value iteration with parametric approximation in detail. Specifically, in Section 3.4.1 we present model-based algorithms from this class, and in Section 3.4.2 we describe offline and online model-free algorithms. Then, in Section 3.4.3, we briefly review value iteration with nonparametric approximation.

Having completed our review of the algorithms for approximate value iteration, we then provide convergence guarantees for these algorithms, in Section 3.4.4. Finally, in Section 3.4.5, we apply two representative algorithms for approximate value iteration to a DC motor control problem.

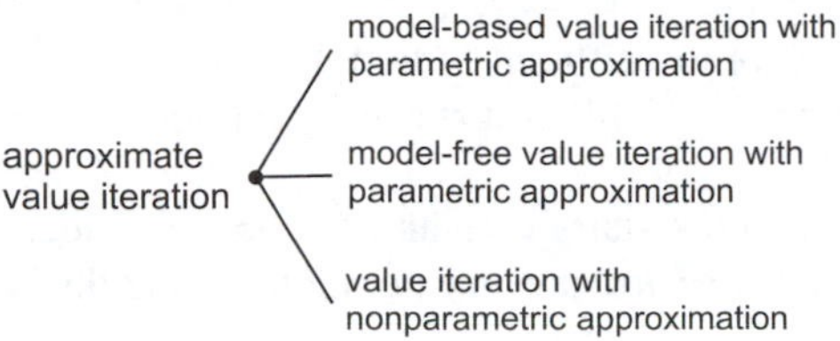

FIGURE 3.3
The organization of the algorithms for approximate value iteration presented in this section.

3.4.1 Model-based value iteration with parametric approximation

This section considers Q-iteration with a general parametric approximator, which is a representative model-based algorithm for approximate value iteration.

Approximate Q-iteration is an extension of the exact Q-iteration algorithm introduced in Section 2.3.1. Recall that exact Q-iteration starts from an arbitrary Q-function Q_0 and at each iteration ℓ updates the Q-function using the rule (2.25), repeated here for easy reference:

$$Q_{\ell+1} = T(Q_\ell)$$

where T is the Q-iteration mapping (2.22) or (2.23). In approximate Q-iteration, the Q-function Q_ℓ cannot be represented exactly. Instead, an approximate version is compactly represented by a parameter vector $\theta_\ell \in \mathbb{R}^n$, using a suitable approximation mapping $F : \mathbb{R}^n \to \mathscr{Q}$ (see Section 3.3):

$$\widehat{Q}_\ell = F(\theta_\ell)$$

This approximate Q-function is provided, instead of Q_ℓ, as an input to the Q-iteration mapping T. So, the Q-iteration update would become:

$$Q^{\ddagger}_{\ell+1} = (T \circ F)(\theta_\ell) \tag{3.13}$$

However, in general, the newly found Q-function $Q^{\ddagger}_{\ell+1}$ cannot be explicitly stored, either. Instead, it must also be represented approximately, using a new parameter vector $\theta_{\ell+1}$. This parameter vector is obtained by a *projection mapping* $P : \mathscr{Q} \to \mathbb{R}^n$:

$$\theta_{\ell+1} = P(Q^{\ddagger}_{\ell+1})$$

which ensures that $\widehat{Q}_{\ell+1} = F(\theta_{\ell+1})$ is as close as possible to $Q^{\ddagger}_{\ell+1}$. A natural choice for P is least-squares regression, which, given a Q-function Q, produces:[2]

$$P(Q) = \theta^{\ddagger}, \text{ where } \theta^{\ddagger} \in \arg\min_{\theta} \sum_{l_s=1}^{n_s} \left(Q(x_{l_s}, u_{l_s}) - [F(\theta)](x_{l_s}, u_{l_s})\right)^2 \tag{3.14}$$

[2] In the absence of additional restrictions, the use of least-squares projections can cause convergence problems, as we will discuss in Section 3.4.4.

for some set of state-action samples $\{(x_{l_s}, u_{l_s}) \,|\, l_s = 1, \dots, n_s\}$. Some care is required to ensure that $\theta^\ddagger$ exists and that it is not too difficult to find. For instance, when the approximator F is linearly parameterized, (3.14) is a convex quadratic optimization problem.

To summarize, approximate Q-iteration starts with an arbitrary (e.g., identically 0) parameter vector θ_0, and updates this vector at every iteration ℓ using the composition of mappings P, T, and F:

$$\theta_{\ell+1} = (P \circ T \circ F)(\theta_\ell) \tag{3.15}$$

Of course, in practice, the intermediate results of F and T cannot be fully computed and stored. Instead, $P \circ T \circ F$ can be implemented as a single mapping, or T and F can be sampled at a finite number of points. The algorithm is stopped once a satisfactory parameter vector $\widehat{\theta}^*$ has been found (see below for examples of stopping criteria). Ideally, $\widehat{\theta}^*$ is near to a fixed point θ^* of $P \circ T \circ F$. In Section 3.4.4, we will give conditions under which a unique fixed point exists and is obtained asymptotically as $\ell \to \infty$.

Given $\widehat{\theta}^*$, a greedy policy in $F(\widehat{\theta}^*)$ can be found, i.e., a policy h that satisfies:

$$h(x) \in \arg\max_u [F(\widehat{\theta}^*)](x,u) \tag{3.16}$$

Here, as well as in the sequel, we assume that the Q-function approximator is structured in a way that guarantees the existence of at least one maximizing action for any state. Because the approximator is under the control of the designer, ensuring this property should not be too difficult.

Figure 3.4 illustrates approximate Q-iteration and the relations between the various mappings, parameter vectors, and Q-functions considered by the algorithm.

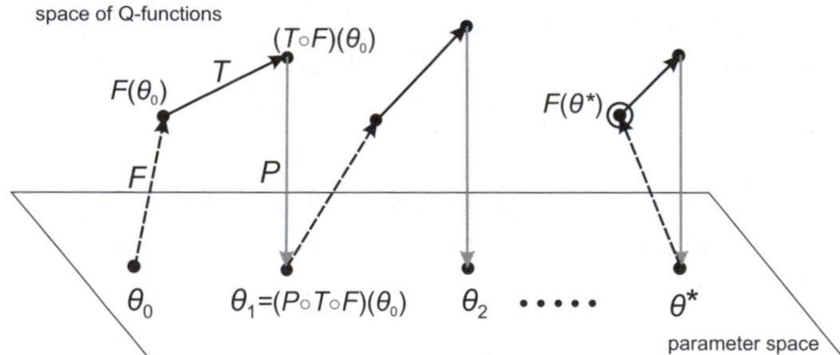

FIGURE 3.4
A conceptual illustration of approximate Q-iteration. At every iteration, the approximation mapping F is applied to the current parameter vector to obtain an approximate Q-function, which is then passed through the Q-iteration mapping T. The result of T is then projected back onto the parameter space with the projection mapping P. Ideally, the algorithm asymptotically converges to a fixed point θ^*, which leads back to itself when passed through $P \circ T \circ F$. The asymptotically obtained solution of approximate Q-iteration is then the Q-function $F(\theta^*)$.

Algorithm 3.1 presents an example of approximate Q-iteration for a deterministic Markov decision process (MDP), using the least-squares projection (3.14). At line 4 of this algorithm, $Q^{\ddagger}_{\ell+1}(x_{l_s}, u_{l_s})$ has been computed according to (3.13), in which the definition (2.22) of the Q-iteration mapping has been substituted.

ALGORITHM 3.1 Least-squares approximate Q-iteration for deterministic MDPs.

Input: dynamics f, reward function ρ, discount factor γ,
approximation mapping F, samples $\{(x_{l_s}, u_{l_s}) \,|\, l_s = 1, \ldots, n_s\}$
1: initialize parameter vector, e.g., $\theta_0 \leftarrow 0$
2: **repeat** at every iteration $\ell = 0, 1, 2, \ldots$
3: **for** $l_s = 1, \ldots, n_s$ **do**
4: $Q^{\ddagger}_{\ell+1}(x_{l_s}, u_{l_s}) \leftarrow \rho(x_{l_s}, u_{l_s}) + \gamma \max_{u'} [F(\theta_\ell)](f(x_{l_s}, u_{l_s}), u')$
5: **end for**
6: $\theta_{\ell+1} \leftarrow \theta^{\ddagger}$, where $\theta^{\ddagger} \in \arg\min_\theta \sum_{l_s=1}^{n_s} \left(Q^{\ddagger}_{\ell+1}(x_{l_s}, u_{l_s}) - [F(\theta)](x_{l_s}, u_{l_s}) \right)^2$
7: **until** $\theta_{\ell+1}$ is satisfactory
Output: $\widehat{\theta}^* = \theta_{\ell+1}$

There still remains the question of when to stop approximate Q-iteration, i.e., when to consider the parameter vector satisfactory. One possibility is to stop after a predetermined number of iterations L. Under the (reasonable) assumption that, at every iteration ℓ, the approximate Q-function $\widehat{Q}_\ell = F(\theta_\ell)$ is close to the Q-function Q_ℓ that would have been obtained by exact Q-iteration, the number L of iterations can be chosen with Equation (2.27) of Section 2.3.1, repeated here:

$$L = \left\lceil \log_\gamma \frac{\varsigma_{\mathrm{QI}}(1-\gamma)^2}{2\|\rho\|_\infty} \right\rceil$$

where $\varsigma_{\mathrm{QI}} > 0$ is a desired bound on the suboptimality of a policy greedy in the Q-function obtained at iteration L. Of course, because $F(\theta_\ell)$ is not identical to Q_ℓ, it cannot be guaranteed that this bound is achieved. Nevertheless, L is still useful as an initial guess for the number of iterations needed to achieve a good performance.

Another possibility is to stop the algorithm when the distance between $\theta_{\ell+1}$ and θ_ℓ decreases below a certain threshold $\varepsilon_{\mathrm{QI}} > 0$. This criterion is only useful if approximate Q-iteration is convergent to a fixed point (see Section 3.4.4 for convergence conditions). When convergence is not guaranteed, this criterion should be combined with a maximum number of iterations, to ensure that the algorithm stops in finite time.

Note that we have not explicitly considered the maximization issues or the estimation of expected values in the stochastic case. As explained in Section 3.2, one way to address the maximization difficulties is to discretize the action space. The expected values in the Q-iteration mapping for the stochastic case (2.23) need to be estimated from samples. For additional insight into this problem, see the fitted Q-iteration algorithm introduced in the next section.

A similar derivation can be given for approximate V-iteration, which is more popular in the literature (Gonzalez and Rofman, 1985; Chow and Tsitsiklis, 1991; Gordon, 1995; Tsitsiklis and Van Roy, 1996; Munos and Moore, 2002; Grüne, 2004). Many results from the literature deal with the discretization of continuous-variable problems (Gonzalez and Rofman, 1985; Chow and Tsitsiklis, 1991; Munos and Moore, 2002; Grüne, 2004). Such discretization procedures sometimes use interpolation, which leads to linearly parameterized approximators similar to (3.3).

3.4.2 Model-free value iteration with parametric approximation

From the class of model-free algorithms for approximate value iteration, we first discuss offline, batch algorithms, followed by online algorithms. Online algorithms, mainly approximate versions of Q-learning, have been studied since the beginning of the nineties (Lin, 1992; Singh et al., 1995; Horiuchi et al., 1996; Jouffe, 1998; Glorennec, 2000; Tuyls et al., 2002; Szepesvári and Smart, 2004; Murphy, 2005; Sherstov and Stone, 2005; Melo et al., 2008). A strong research thread in offline model-free value iteration emerged later (Ormoneit and Sen, 2002; Ernst et al., 2005; Riedmiller, 2005; Szepesvári and Munos, 2005; Ernst et al., 2006b; Antos et al., 2008a; Munos and Szepesvári, 2008; Farahmand et al., 2009a).

Offline model-free approximate value iteration

In the offline model-free case, the transition dynamics f and the reward function ρ are unknown.[3] Instead, only a batch of transition samples is available:

$$\{(x_{l_s}, u_{l_s}, x'_{l_s}, r_{l_s}) \,|\, l_s = 1, \ldots, n_s\}$$

where for every l_s, the next state x'_{l_s} and the reward r_{l_s} have been obtained as a result of taking action u_{l_s} in the state x_{l_s}. The transition samples may be independent, they may belong to a set of trajectories, or to a single trajectory. For instance, when the samples come from a single trajectory, they are typically ordered so that $x_{l_s+1} = x'_{l_s}$ for all $l_s < n_s$.

In this section, we present *fitted Q-iteration* (Ernst et al., 2005), a model-free version of approximate Q-iteration (3.15) that employs such a batch of samples. To obtain this version, two changes are made in the original, model-based algorithm. First, one has to use a sample-based projection mapping that considers only the samples (x_{l_s}, u_{l_s}), such as the least-squares regression (3.14). Second, because f and ρ are not available, the updated Q-function $Q^{\ddagger}_{\ell+1} = (T \circ F)(\theta_\ell)$ (3.13) at a given iteration ℓ cannot be computed directly. Instead, the Q-values $Q^{\ddagger}_{\ell+1}(x_{l_s}, u_{l_s})$ are replaced by quantities derived from the available data.

To understand how this is done, consider first the deterministic case. In this case,

[3]We take the point of view prevalent in the RL literature, which considers that the learning controller has no prior information about the problem to be solved. This means the reward function is unknown. In practice, of course, the reward function is almost always designed by the experimenter and is therefore known.

the updated Q-values are:

$$Q^{\ddagger}_{\ell+1}(x_{l_s},u_{l_s}) = \rho(x_{l_s},u_{l_s}) + \gamma \max_{u'} [F(\theta_\ell)](f(x_{l_s},u_{l_s}),u') \tag{3.17}$$

where the Q-iteration mapping (2.22) has been used. Recall that $\rho(x_{l_s},u_{l_s}) = r_{l_s}$ and that $f(x_{l_s},u_{l_s}) = x'_{l_s}$. By performing these substitutions in (3.17), we get:

$$Q^{\ddagger}_{\ell+1}(x_{l_s},u_{l_s}) = r_{l_s} + \gamma \max_{u'} [F(\theta_\ell)](x'_{l_s},u') \tag{3.18}$$

and hence the updated Q-value can be computed exactly from the transition sample $(x_{l_s},u_{l_s},x'_{l_s},r_{l_s})$, without using f or ρ.

Fitted Q-iteration works in deterministic and stochastic problems, and replaces each Q-value $Q^{\ddagger}_{\ell+1}(x_{l_s},u_{l_s})$ by the quantity:

$$Q^{\ddagger}_{\ell+1,l_s} = r_{l_s} + \gamma \max_{u'} [F(\theta_\ell)](x'_{l_s},u') \tag{3.19}$$

identical to the right-hand side of (3.18). As already discussed, in the deterministic case, this replacement is exact. In the stochastic case, the updated Q-value is the expectation of a random variable, of which $Q^{\ddagger}_{\ell+1,l_s}$ is only a *sample*. This updated Q-value is:

$$Q^{\ddagger}_{\ell+1}(x_{l_s},u_{l_s}) = \mathrm{E}_{x'\sim\tilde{f}(x_{l_s},u_{l_s},\cdot)} \left\{ \tilde{\rho}(x_{l_s},u_{l_s},x') + \gamma \max_{u'} [F(\theta_\ell)](x',u') \right\}$$

where the Q-iteration mapping (2.23) has been used (note that $Q^{\ddagger}_{\ell+1}(x_{l_s},u_{l_s})$ is the true Q-value and not a data point, so it is no longer subscripted by the sample index l_s). Nevertheless, most projection algorithms, including the least-squares regression (3.14), seek to approximate the expected value of their output variable conditioned by the input. In fitted Q-iteration, this means that the projection actually looks for θ_ℓ such that $F(\theta_\ell) \approx Q^{\ddagger}_{\ell+1}$, even though only samples of the form (3.19) are used. Therefore, the algorithm remains valid in the stochastic case.

Algorithm 3.2 presents fitted Q-iteration using least-squares projection (3.14). Note that, in the deterministic case, fitted Q-iteration is identical to model-based approximate Q-iteration (e.g., Algorithm 3.1), whenever both algorithms use the same approximator F, the same projection P, and the same state-action samples (x_{l_s},u_{l_s}).

The considerations of Section 3.4.1 about the stopping criteria of approximate Q-iteration also apply to fitted Q-iteration, so they will not be repeated here. Moreover, once fitted Q-iteration has found a satisfactory parameter vector, a policy can be derived with (3.16).

We have introduced fitted Q-iteration in the parametric case, to clearly establish its link with model-based approximate Q-iteration. Neural networks are one class of parametric approximators that have been combined with fitted Q-iteration, leading to the so-called "neural fitted Q-iteration" (Riedmiller, 2005). However, fitted Q-iteration is more popular in combination with nonparametric approximators, so we will revisit it in the nonparametric context in Section 3.4.3.

ALGORITHM 3.2 Least-squares fitted Q-iteration with parametric approximation.

Input: discount factor γ,
approximation mapping F, samples $\{(x_{l_s}, u_{l_s}, x'_{l_s}, r_{l_s}) \,|\, l_s = 1, \dots, n_s\}$
initialize parameter vector, e.g., $\theta_0 \leftarrow 0$
repeat at every iteration $\ell = 0, 1, 2, \dots$
  for $l_s = 1, \dots, n_s$ **do**
    $Q^{\ddagger}_{\ell+1, l_s} \leftarrow r_{l_s} + \gamma \max_{u'} [F(\theta_\ell)](x'_{l_s}, u')$
  end for
  $\theta_{\ell+1} \leftarrow \theta^{\ddagger}$, where $\theta^{\ddagger} \in \arg\min_\theta \sum_{l_s=1}^{n_s} \left(Q^{\ddagger}_{\ell+1, l_s} - [F(\theta)](x_{l_s}, u_{l_s}) \right)^2$
until $\theta_{\ell+1}$ is satisfactory
Output: $\widehat{\theta}^* = \theta_{\ell+1}$

Although we have assumed that the batch of samples is given in advance, fitted Q-iteration, together with other offline RL algorithms, can also be modified to use different batches of samples at different iterations. This property can be exploited, e.g., to add new, more informative samples in-between iterations. Ernst et al. (2006b) proposed a different, but related approach that integrates fitted Q-iteration into a larger iterative process. At every larger iteration, the entire fitted Q-iteration algorithm is run on the current batch of samples. Then, the solution obtained by fitted Q-iteration is used to generate new samples, e.g., by using an ε-greedy policy in the obtained Q-function. The entire cycle is then repeated.

Online model-free approximate value iteration

From the class of online algorithms for approximate value iteration, approximate versions of Q-learning are the most popular (Lin, 1992; Singh et al., 1995; Horiuchi et al., 1996; Jouffe, 1998; Glorennec, 2000; Tuyls et al., 2002; Szepesvári and Smart, 2004; Murphy, 2005; Sherstov and Stone, 2005; Melo et al., 2008). Recall from Section 2.3.2 that the original Q-learning updates the Q-function with (2.30):

$$Q_{k+1}(x_k, u_k) = Q_k(x_k, u_k) + \alpha_k [r_{k+1} + \gamma \max_{u'} Q_k(x_{k+1}, u') - Q_k(x_k, u_k)]$$

after observing the next state x_{k+1} and reward r_{k+1}, as a result of taking action u_k in state x_k. A straightforward way to integrate approximation in Q-learning is by using gradient descent. We next explain how gradient-based Q-learning is obtained, following Sutton and Barto (1998, Chapter 8). We require that the approximation mapping F is differentiable in the parameters.

To simplify the formulas below, we denote the approximate Q-function at time k by $\widehat{Q}_k(x_k, u_k) = [F(\theta_k)](x_k, u_k)$, leaving the dependence on the parameter vector implicit. In order to derive gradient-based Q-learning, assume for now that after taking action u_k in state x_k, the algorithm is provided with the true optimal Q-value of the current state action pair, $Q^*(x_k, u_k)$, in addition to the next state x_{k+1} and reward r_{k+1}. Under these circumstances, the algorithm could aim to minimize the squared error between this optimal value and the current Q-value:

$$\theta_{k+1} = \theta_k - \frac{1}{2}\alpha_k \frac{\partial}{\partial \theta_k}\left[Q^*(x_k,u_k) - \widehat{Q}_k(x_k,u_k)\right]^2$$
$$= \theta_k + \alpha_k \left[Q^*(x_k,u_k) - \widehat{Q}_k(x_k,u_k)\right] \frac{\partial}{\partial \theta_k}\widehat{Q}_k(x_k,u_k)$$

Of course, $Q^*(x_k,u_k)$ is not available, but it can be replaced by an estimate derived from the Q-iteration mapping (2.22) or (2.23):

$$r_{k+1} + \gamma \max_{u'} \widehat{Q}_k(x_{k+1},u')$$

Note the similarity with the Q-function samples (3.19) used in fitted Q-iteration. The substitution leads to the approximate Q-learning update:

$$\theta_{k+1} = \theta_k + \alpha_k \left[r_{k+1} + \gamma \max_{u'} \widehat{Q}_k(x_{k+1},u') - \widehat{Q}_k(x_k,u_k)\right] \frac{\partial}{\partial \theta_k}\widehat{Q}_k(x_k,u_k) \tag{3.20}$$

We have actually obtained, in the square brackets, an approximation of the temporal difference. With a linearly parameterized approximator (3.3), the update (3.20) simplifies to:

$$\theta_{k+1} = \theta_k + \alpha_k \left[r_{k+1} + \gamma \max_{u'}\left(\phi^{\mathrm{T}}(x_{k+1},u')\theta_k\right) - \phi^{\mathrm{T}}(x_k,u_k)\theta_k\right] \phi(x_k,u_k) \tag{3.21}$$

Note that, like the original Q-learning algorithm of Section 2.3.2, approximate Q-learning requires exploration. As an example, Algorithm 3.3 presents gradient-based Q-learning with a linear parametrization and ε-greedy exploration. For an explanation and examples of the learning rate and exploration schedules used in this algorithm, see Section 2.3.2.

ALGORITHM 3.3 Q-learning with a linear parametrization and ε-greedy exploration.

Input: discount factor γ,
BFs $\phi_1, \ldots, \phi_n : X \times U \to \mathbb{R}$,
exploration schedule $\{\varepsilon_k\}_{k=0}^{\infty}$, learning rate schedule $\{\alpha_k\}_{k=0}^{\infty}$

initialize parameter vector, e.g., $\theta_0 \leftarrow 0$
measure initial state x_0
for every time step $k = 0, 1, 2, \ldots$ **do**

$$u_k \leftarrow \begin{cases} u \in \arg\max_{\bar{u}}\left(\phi^{\mathrm{T}}(x_k,\bar{u})\theta_k\right) & \text{with probability } 1-\varepsilon_k \text{ (exploit)} \\ \text{a uniform random action in } U & \text{with probability } \varepsilon_k \text{ (explore)} \end{cases}$$

apply u_k, measure next state x_{k+1} and reward r_{k+1}
$\theta_{k+1} \leftarrow \theta_k + \alpha_k\left[r_{k+1} + \gamma \max_{u'}\left(\phi^{\mathrm{T}}(x_{k+1},u')\theta_k\right) - \phi^{\mathrm{T}}(x_k,u_k)\theta_k\right]\phi(x_k,u_k)$
end for

In the literature, Q-learning has been combined with a variety of approximators, for example:

- linearly parameterized approximators, including tile coding (Watkins, 1989; Sherstov and Stone, 2005), as well as so-called interpolative representations (Szepesvári and Smart, 2004) and "soft" state aggregation (Singh et al., 1995).
- fuzzy rule-bases (Horiuchi et al., 1996; Jouffe, 1998; Glorennec, 2000), which can also be linear in the parameters.
- neural networks (Lin, 1992; Touzet, 1997).

While approximate Q-learning is easy to use, it typically requires many transition samples (i.e., many steps, k) before it can obtain a good approximation of the optimal Q-function. One possible approach to alleviate this problem is to store transition samples in a database and reuse them multiple times, similarly to how the batch algorithms of the previous section work. This procedure is known as experience replay (Lin, 1992; Kalyanakrishnan and Stone, 2007). Another option is to employ so-called eligibility traces, which allow the parameter updates at the current step to also incorporate information about recently observed transitions (e.g., Singh and Sutton, 1996). This mechanism makes use of the fact that the latest transition is the causal result of an entire trajectory.

3.4.3 Value iteration with nonparametric approximation

In this section, we first describe fitted Q-iteration with nonparametric approximation. We then point out some other algorithms that combine value iteration with nonparametric approximators.

The fitted Q-iteration algorithm was introduced in a parametric context in Section 3.4.2, see Algorithm 3.2. In the nonparametric case, fitted Q-iteration can no longer be described using approximation and projection mappings that remain unchanged from one iteration to the next. Instead, fitted Q-iteration can be regarded as generating an entirely new, nonparametric approximator at every iteration. Algorithm 3.4 outlines a general template for fitted Q-iteration with nonparametric approximation. The nonparametric regression at line 6 is responsible for generating a new approximator $\widehat{Q}_{\ell+1}$ that accurately represents the updated Q-function $Q^{\ddagger}_{\ell+1}$, using the information provided by the available samples $Q^{\ddagger}_{\ell+1,l_s}$, $l_s = 1, \dots n_s$.

Fitted Q-iteration has been combined with several types of nonparametric approximators, including kernel-based approximators (Farahmand et al., 2009a) and ensembles of regression trees (Ernst et al., 2005, 2006b); see Appendix A for a description of such an ensemble.

Of course, other DP and RL algorithms besides fitted Q-iteration can also be combined with nonparametric approximation. For instance, Deisenroth et al. (2009) employed Gaussian processes in approximate value iteration. They proposed two algorithms: one that assumes that a model of the (deterministic) dynamics is known, and another that estimates a Gaussian-process approximation of the dynamics from

ALGORITHM 3.4 Fitted Q-iteration with nonparametric approximation.

Input: discount factor γ,
samples $\{(x_{l_s}, u_{l_s}, x'_{l_s}, r_{l_s}) \mid l_s = 1, \dots, n_s\}$

initialize Q-function approximator, e.g., $\widehat{Q}_0 \leftarrow 0$
repeat at every iteration $\ell = 0, 1, 2, \dots$
  for $l_s = 1, \dots, n_s$ **do**
    $Q^{\ddagger}_{\ell+1, l_s} \leftarrow r_{l_s} + \gamma \max_{u'} \widehat{Q}_\ell(x'_{l_s}, u')$
  end for
  find $\widehat{Q}_{\ell+1}$ using
    nonparametric regression on $\{((x_{l_s}, u_{l_s}), Q^{\ddagger}_{\ell+1, l_s}) \mid l_s = 1, \dots, n_s\}$
until $\widehat{Q}_{\ell+1}$ is satisfactory

Output: $\widehat{Q}^* = \widehat{Q}_{\ell+1}$

transition data. Ormoneit and Sen (2002) employed kernel-based approximation in model-free approximate value iteration for discrete-action problems.

3.4.4 Convergence and the role of nonexpansive approximation

An important question in approximate DP/RL is whether the approximate solution computed by the algorithm converges, and, if it does converge, how far the convergence point is from the optimal solution. Convergence is important because a convergent algorithm is more amenable to analysis and meaningful performance guarantees.

Convergence of model-based approximate value iteration

The convergence proofs for approximate value iteration often rely on contraction mapping arguments. Consider for instance approximate Q-iteration (3.15). The Q-iteration mapping T is a contraction in the infinity norm with factor $\gamma < 1$, as already explained in Section 2.3.1. If the composite mapping $P \circ T \circ F$ of approximate Q-iteration is also a contraction, i.e., if for any pair of parameter vectors θ, θ' and for some $\gamma' < 1$:

$$\|(P \circ T \circ F)(\theta) - (P \circ T \circ F)(\theta')\|_\infty \leq \gamma' \|\theta - \theta'\|_\infty$$

then approximate Q-iteration asymptotically converges to a unique fixed point, which we denote by θ^*.

One way to ensure that $P \circ T \circ F$ is a contraction is to require F and P to be nonexpansions, i.e.:

$$\begin{aligned} \|F(\theta) - F(\theta')\|_\infty &\leq \|\theta - \theta'\|_\infty \quad \text{for all pairs } \theta, \theta' \\ \|P(Q) - P(Q')\|_\infty &\leq \|Q - Q'\|_\infty \quad \text{for all pairs } Q, Q' \end{aligned}$$

Note that in this case the contraction factor of $P \circ T \circ F$ is the same as that of T: $\gamma' = \gamma < 1$. Under these conditions, as we will describe next, suboptimality bounds can be derived on the approximate Q-function $F(\theta^*)$ and on any policy $\widehat{h}^*$ that is greedy in this Q-function, i.e., that satisfies:

$$\widehat{h}^*(x) \in \arg\max_u [F(\theta^*)](x,u) \tag{3.22}$$

Denote by $\mathscr{F}_{F \circ P} \subset \mathscr{Q}$ the set of fixed points of the composite mapping $F \circ P$, which is assumed nonempty. Define the minimum distance between Q^* and any fixed point of $F \circ P$:[4]

$$\varsigma^*_{\mathrm{QI}} = \min_{Q' \in \mathscr{F}_{F \circ P}} \|Q^* - Q'\|_\infty$$

This distance characterizes the representation power of the approximator: the better the representation power, the closer the nearest fixed point of $F \circ P$ will be to Q^*, and the smaller $\varsigma^*_{\mathrm{QI}}$ will be. Using this distance, the convergence point θ^* of approximate Q-iteration satisfies the following suboptimality bounds:

$$\|Q^* - F(\theta^*)\|_\infty \leq \frac{2\varsigma^*_{\mathrm{QI}}}{1-\gamma} \tag{3.23}$$

$$\|Q^* - Q^{\widehat{h}^*}\|_\infty \leq \frac{4\gamma\varsigma^*_{\mathrm{QI}}}{(1-\gamma)^2} \tag{3.24}$$

where $Q^{\widehat{h}^*}$ is the Q-function of the near-optimal policy $\widehat{h}^*$ (3.22). These bounds can be derived similarly to those for approximate V-iteration found by Gordon (1995); Tsitsiklis and Van Roy (1996). Equation (3.23) gives the suboptimality bound of the approximately optimal Q-function, whereas (3.24) gives the suboptimality bound of the resulting approximately optimal policy, and may be more relevant in practice. The following relationship between the policy suboptimality and the Q-function suboptimality was used to obtain (3.24), and is also valid in general:

$$\|Q^* - Q^h\|_\infty \leq \frac{2\gamma}{(1-\gamma)} \|Q^* - Q\|_\infty \tag{3.25}$$

where the policy h is greedy in the (arbitrary) Q-function Q.

Ideally, the optimal Q-function Q^* is a fixed point of $F \circ P$, in which case $\varsigma^*_{\mathrm{QI}} = 0$, and approximate Q-iteration asymptotically converges to Q^*. For instance, when Q^* happens to be exactly representable by F, a well-chosen tandem of approximation and projection mappings should ensure that Q^* is in fact a fixed point of $F \circ P$. In practice, of course, $\varsigma^*_{\mathrm{QI}}$ will rarely be 0, and only near-optimal solutions can be obtained.

In order to take advantage of these theoretical guarantees, F and P should be nonexpansions. When F is linearly parameterized (3.3), it is fairly easy to ensure its nonexpansiveness by normalizing the BFs ϕ_l, so that for every x and u, we have:

$$\sum_{l=1}^{n} \phi_l(x,u) = 1$$

[4]For simplicity, we assume that the minimum in this equation exists. If the minimum does not exist, then $\varsigma^*_{\mathrm{QI}}$ should be taken as small as possible so that there still exists a $Q' \in \mathscr{F}_{F \circ P}$ with $\|Q' - Q^*\|_\infty \leq \varsigma^*_{\mathrm{QI}}$.

Ensuring that P is nonexpansive is more difficult. For instance, the least-squares projection (3.14) can in general be an expansion, and examples of divergence when using it have been given (Tsitsiklis and Van Roy, 1996; Wiering, 2004). One way to make least-squares projection nonexpansive is to choose exactly $n_s = n$ state-action samples (x_l, u_l), $l = 1, \dots, n$, and require that:

$$\phi_l(x_l, u_l) = 1, \quad \phi_{l'}(x_l, u_l) = 0 \ \forall l' \neq l$$

These samples could be, e.g., the centers of the BFs. Then, the projection (3.14) simplifies to an assignment that associates each parameter with the Q-value of the corresponding sample:

$$[P(Q)]_l = Q(x_l, u_l) \tag{3.26}$$

where the notation $[P(Q)]_l$ refers to the lth component in the parameter vector $P(Q)$. This mapping is clearly nonexpansive. More general, but still restrictive conditions on the BFs under which convergence and near optimality are guaranteed are given in (Tsitsiklis and Van Roy, 1996).

Convergence of model-free approximate value iteration

Like in the model-based case, convergence guarantees for offline, batch model-free value iteration typically rely on nonexpansive approximation. In fitted Q-iteration with parametric approximation (Algorithm 3.2), care must be taken when selecting F and P, to prevent possible expansion and divergence. Similarly, in fitted Q-iteration with nonparametric approximation (Algorithm 3.4), the nonparametric regression algorithm should have nonexpansive properties. Certain types of kernel-based approximators satisfy this condition (Ernst et al., 2005). The convergence of the kernel-based V-iteration algorithm of Ormoneit and Sen (2002) is also guaranteed under nonexpansiveness assumptions.

More recently, a different class of theoretical results for batch value iteration have been developed, which do not rely on nonexpansion properties and do not concern the asymptotic case. Instead, these results provide probabilistic bounds on the suboptimality of the policy obtained by using a finite number of samples, after a finite number of iterations. Besides the number of samples and iterations, such *finite-sample* bounds typically depend on the representation power of the approximator and on certain properties of the MDP. For instance, Munos and Szepesvári (2008) provided finite-sample bounds for approximate V-iteration in discrete-action MDPs, while Farahmand et al. (2009a) focused on fitted Q-iteration in the same type of MDPs. Antos et al. (2008a) gave finite-sample bounds for fitted Q-iteration in the more difficult case of continuous-action MDPs.

In the area of online approximate value iteration, as already discussed in Section 3.4.2, the main representative is approximate Q-learning. Many variants of approximate Q-learning are heuristic and do not guarantee convergence (Horiuchi et al., 1996; Touzet, 1997; Jouffe, 1998; Glorennec, 2000; Millán et al., 2002). Convergence of approximate Q-learning has been proven for linearly parameterized approximators, under the requirement that the policy followed by Q-learning remains *unchanged* during the learning process (Singh et al., 1995; Szepesvári and Smart,

2004; Melo et al., 2008). This requirement is restrictive, because it does not allow the controller to improve its performance, even if it has gathered knowledge that would enable it to do so. Among these results, Singh et al. (1995) and Szepesvári and Smart (2004) proved the convergence of approximate Q-learning with nonexpansive, linearly parameterized approximation. Melo et al. (2008) showed that gradient-based Q-learning (3.21) converges without requiring nonexpansive approximation, but at the cost of other restrictive assumptions.

Consistency of approximate value iteration

Besides convergence, another important theoretical property of algorithms for approximate DP and RL is consistency. In model-based value iteration, and more generally in DP, an algorithm is said to be consistent if the approximate value function converges to the optimal one as the approximation accuracy increases (e.g., Gonzalez and Rofman, 1985; Chow and Tsitsiklis, 1991; Santos and Vigo-Aguiar, 1998). In model-free value iteration, and more generally in RL, consistency is sometimes understood as the convergence to a well-defined solution as the number of samples increases. The stronger result of convergence to an optimal solution as the approximation accuracy also increases was proven in (Ormoneit and Sen, 2002; Szepesvári and Smart, 2004).

3.4.5 Example: Approximate Q-iteration for a DC motor

In closing the discussion on approximate value iteration, we provide a numerical example involving a DC motor control problem. This example shows how approximate value iteration algorithms can be used in practice. The first part of the example concerns a basic version of approximate Q-iteration that relies on a gridding of the state space and on a discretization of the action space, while the second part employs the state-of-the-art, fitted Q-iteration algorithm with nonparametric approximation (Algorithm 3.4).

Consider a second-order discrete-time model of an electrical DC (direct current) motor:

$$\begin{aligned} x_{k+1} &= f(x_k, u_k) = Ax_k + Bu_k \\ A &= \begin{bmatrix} 1 & 0.0049 \\ 0 & 0.9540 \end{bmatrix}, \quad B = \begin{bmatrix} 0.0021 \\ 0.8505 \end{bmatrix} \end{aligned} \tag{3.27}$$

This model was obtained by discretizing a continuous-time model of the DC motor, which was developed by first-principles modeling (e.g., Khalil, 2002, Chapter 1) of a real DC motor. The discretization was performed with the zero-order-hold method (Franklin et al., 1998), using a sampling time of $T_s = 0.005$ s. Using saturation, the shaft angle $x_{1,k} = \alpha$ is bounded to $[-\pi, \pi]$ rad, the angular velocity $x_{2,k} = \dot{\alpha}$ to $[-16\pi, 16\pi]$ rad/s, and the control input u_k to $[-10, 10]$ V.

The control goal is to stabilize the DC motor in the zero equilibrium ($x = 0$). The following quadratic reward function is chosen to express this goal:

$$
\begin{aligned}
r_{k+1} &= \rho(x_k, u_k) = -x_k^{\mathrm{T}} Q_{\mathrm{rew}} x_k - R_{\mathrm{rew}} u_k^2 \\
Q_{\mathrm{rew}} &= \begin{bmatrix} 5 & 0 \\ 0 & 0.01 \end{bmatrix}, \quad R_{\mathrm{rew}} = 0.01
\end{aligned}
\tag{3.28}
$$

This reward function leads to a discounted quadratic regulation problem. A (near-)optimal policy will drive the state (close) to 0, while also minimizing the magnitude of the states along the trajectory and the control effort. The discount factor was chosen to be $\gamma = 0.95$, which is sufficiently large to lead to an optimal policy that produces a good stabilizing control behavior.[5]

Figure 3.5 presents a near-optimal solution to this problem, including a representative state-dependent slice through the Q-function (obtained by setting the action

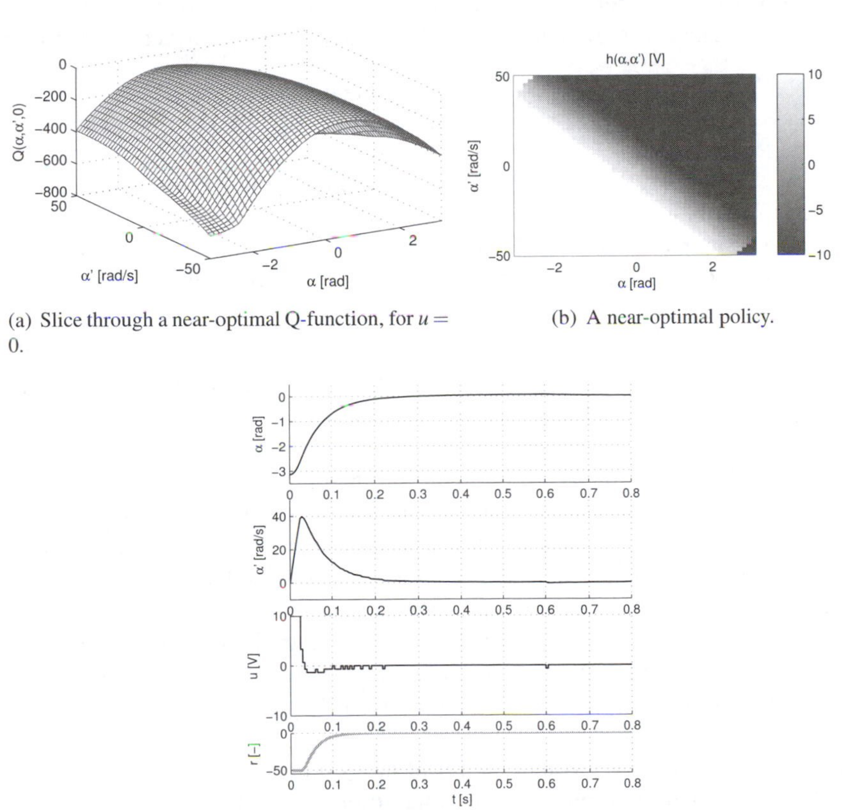

(a) Slice through a near-optimal Q-function, for $u = 0$.

(b) A near-optimal policy.

(c) Controlled trajectory from $x_0 = [-\pi, 0]^{\mathrm{T}}$.

FIGURE 3.5 A near-optimal solution for the DC motor.

[5]Note that a distinction is made between the optimality under the chosen reward function and discount factor, and the actual (albeit subjective) quality of the control behavior.

argument u to 0), a greedy policy in this Q-function, and a representative trajectory that is controlled by this policy. To find the near-optimal solution, the convergent and consistent fuzzy Q-iteration algorithm (which will be discussed in detail in Chapter 4) was applied. An accurate approximator over the state space was used, together with a fine discretization of the action space, which contains 31 equidistant actions.

Grid Q-iteration

As an example of approximate value iteration, we apply a Q-iteration algorithm that relies on state aggregation and action discretization, a type of approximator introduced in Example 3.1. The state space is partitioned into N disjoint rectangles. Denote by X_i the ith rectangle in the state space partition. For this problem, the following three discrete actions suffice to produce an acceptable stabilizing control behavior: $u_1 = -10$, $u_2 = 0$, $u_3 = 10$ (i.e., applying maximum torque in either direction, and no torque at all). So, the discrete action space is $U_\mathrm{d} = \{-10, 0, 10\}$. Recall from Example 3.1 that the state-action BFs are given by (3.8), repeated here for easy reference:

$$\phi_{[i,j]}(x,u) = \begin{cases} 1 & \text{if } x \in X_i \text{ and } u = u_j \\ 0 & \text{otherwise} \end{cases} \tag{3.29}$$

where $[i, j] = i + (j-1)N$. To derive the projection mapping P, the least-squares projection (3.14) is used, taking the cross-product of the sets $\{x_1, \dots, x_N\}$ and U_d as state-action samples, where x_i denotes the center of the ith rectangle X_i. These samples satisfy the conditions to simplify P to an assignment of the form (3.26), namely:

$$[P(Q)]_{[i,j]} = Q(x_i, u_j) \tag{3.30}$$

Using a linearly parameterized approximator with the BFs (3.29) and the projection (3.30) yields the grid Q-iteration algorithm. Because F and P are nonexpansions, the algorithm is convergent.

To apply grid Q-iteration to the DC motor problem, two different grids over the state space are used: a coarse grid, with 20 equidistant bins on each axis (leading to $20^2 = 400$ rectangles); and a fine grid, with 400 equidistant bins on each axis (leading to $400^2 = 160\,000$ rectangles). The algorithm is considered to have converged when the maximum amount by which any parameter changes between two consecutive iterations does not exceed $\varepsilon_\mathrm{QI} = 0.001$. For the coarse grid, convergence occurred after 160 iterations, and for the fine grid, after 123. This shows that the number of iterations required for convergence does not necessarily increase with the number of parameters.

Figure 3.6 shows slices through the resulting Q-functions, together with corresponding policies and representative controlled trajectories. The accuracy in representing the Q-function and policy is better for the fine grid (Figures 3.6(b) and 3.6(d)) than for the coarse grid (Figures 3.6(a) and 3.6(c)). Axis-oriented policy artifacts appear for both grid sizes, due to the limitations of the chosen type of approximator. For instance, the piecewise-constant nature of the approximator is clearly visible in Figure 3.6(a). Compared to the near-optimal trajectory of Figure 3.5(c), the grid Q-iteration trajectories in Figures 3.6(e) and 3.6(f) do not reach the goal state $x = 0$ with the same accuracy. With the coarse-grid policy, there is a large steady-state error of the angle α, while the fine-grid policy leads to chattering of the control action.

The execution time of grid Q-iteration was 0.06 s for the coarse grid, and 7.80 s

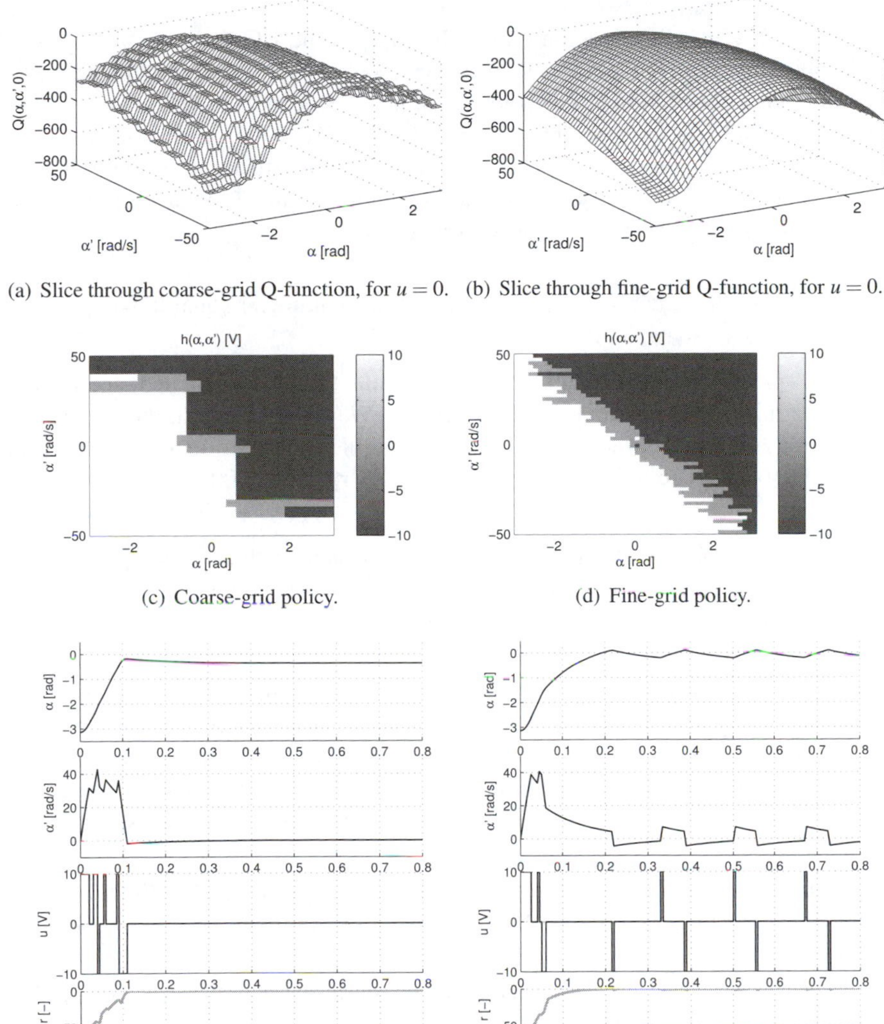

(a) Slice through coarse-grid Q-function, for $u = 0$. (b) Slice through fine-grid Q-function, for $u = 0$.

(c) Coarse-grid policy. (d) Fine-grid policy.

(e) Trajectory from $x_0 = [-\pi, 0]^{\mathrm{T}}$, controlled by the coarse-grid policy.

(f) Trajectory from $x_0 = [-\pi, 0]^{\mathrm{T}}$, controlled by the fine-grid policy.

FIGURE 3.6
Grid Q-iteration solutions for the DC motor. The results obtained with the coarse grid are shown on the left-hand side of the figure, and those obtained with the fine grid on the right-hand side.

for the fine grid.[6] The fine grid is significantly more computationally expensive to use, because it has a much larger number of parameters to update (480000, versus 1200 for the coarse grid).

Fitted Q-iteration

Next, we apply fitted Q-iteration (Algorithm 3.4) to the DC motor problem, using ensembles of extremely randomized trees (Geurts et al., 2006) to approximate the Q-function. For a description of this approximator, see Appendix A. The same discrete actions are employed as for grid Q-iteration: $U_{\rm d} = \{-10, 0, 10\}$. A distinct ensemble of regression trees is used to approximate the Q-function for each of these discrete actions – in analogy to the discrete-action grid approximator. The construction of the tree ensembles is driven by three meta-parameters:

- Each ensemble contains $N_{\rm tr}$ trees. We set this parameter equal to 50.
- To split a node, $K_{\rm tr}$ randomly chosen cut directions are evaluated, and the one that maximizes a certain score is selected. We set $K_{\rm tr}$ equal to the dimensionality 2 of the input to the regression trees (the 2-dimensional state variable), which is its recommended default value (Geurts et al., 2006).
- A node is only split further when it is associated with at least $n_{\rm tr}^{\rm min}$ samples. Otherwise, it remains a leaf node. We set $n_{\rm tr}^{\rm min}$ to its default value of 2, which means that the trees are fully developed.

Fitted Q-iteration is supplied with a set of samples consisting of the cross-product between a regular grid of 100×100 points in the state space, and the 3 discrete actions. This ensures the meaningfulness of the comparison with grid Q-iteration, which employed similarly placed samples. Fitted Q-iteration is run for a predefined number of 100 iterations, and the Q-function found after the 100th iteration is considered satisfactory.

Figure 3.7 shows the solution obtained. This is similar in quality to the solution obtained by grid Q-iteration with the fine grid, and better than the solution obtained with the coarse grid (Figure 3.6).

The execution time of fitted Q-iteration was approximately 2151 s, several orders of magnitude larger than the execution time of grid Q-iteration (recall that the latter was 0.06 s for the coarse grid, and 7.80 s for the fine grid). Clearly, finding a more powerful nonparametric approximator is much more computationally intensive than updating the parameters of the simple, grid-based approximator.

---

[6] All the execution times reported in this chapter were recorded while running the algorithms in MATLAB® 7 on a PC with an Intel Core 2 Duo T9550 2.66 GHz CPU and with 3 GB RAM. For value iteration and policy iteration, the reported execution times do not include the time required to simulate the system for every state-action sample in order to obtain the next state and reward.

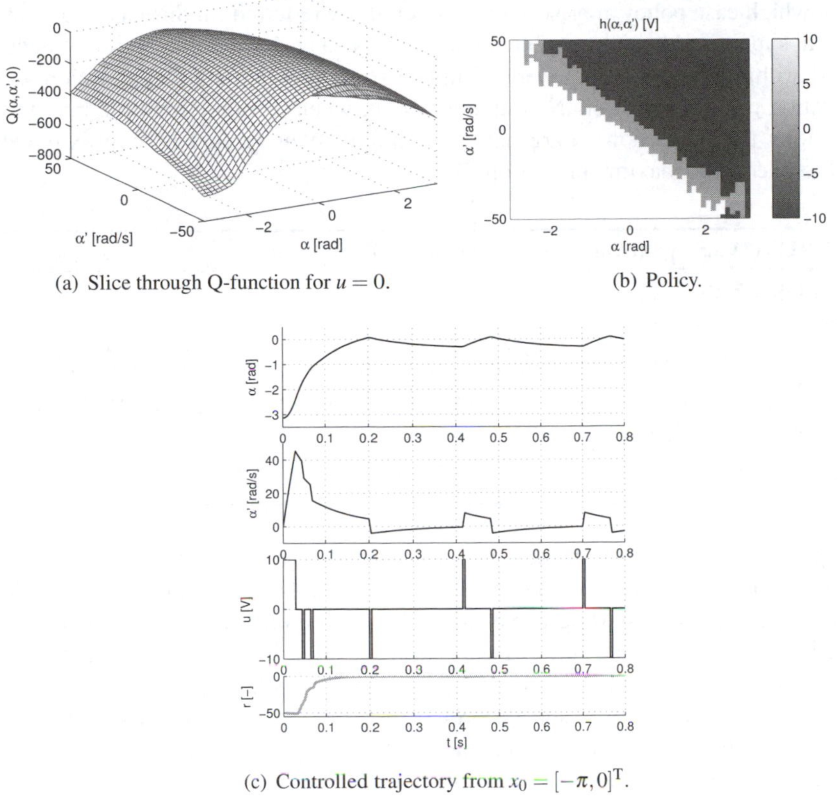

(a) Slice through Q-function for $u = 0$.

(b) Policy.

(c) Controlled trajectory from $x_0 = [-\pi, 0]^T$.

FIGURE 3.7 Fitted Q-iteration solution for the DC motor.

3.5 Approximate policy iteration

Policy iteration algorithms evaluate policies by constructing their value functions, and use these value functions to find new, improved policies. They were introduced in Section 2.4. In large or continuous spaces, policy evaluation cannot be solved exactly, and the value function has to be approximated. *Approximate policy evaluation* is a difficult problem, because, like approximate value iteration, it involves finding an approximate solution to a Bellman equation. Special requirements must be imposed to ensure that a meaningful approximate solution exists and can be found by appropriate algorithms. *Policy improvement* relies on solving maximization problems over the action variables, which involve fewer technical difficulties (although they may still be hard to solve when the action space is large). Often, an explicit representation of the policy can be avoided, by computing improved actions on demand from the current value function. Alternatively, the policy can be represented explicitly, in which case policy approximation is generally required. In this case, solving a classical supervised learning problem is necessary to perform policy improvement.

Algorithm 3.5 outlines a general template for approximate policy iteration with Q-function policy evaluation. Note that at line 4, when there are multiple maximizing actions, the expression "$\approx \arg\max_u \ldots$" should be interpreted as "approximately equal to one of the maximizing actions."

ALGORITHM 3.5 Approximate policy iteration with Q-functions.

1: initialize policy $\widehat{h}_0$
2: **repeat** at every iteration $\ell = 0, 1, 2, \ldots$
3: find $\widehat{Q}^{\widehat{h}_\ell}$, an approximate Q-function of $\widehat{h}_\ell$ ▷ policy evaluation
4: find $\widehat{h}_{\ell+1}$ so that $\widehat{h}_{\ell+1}(x) \approx \arg\max_u \widehat{Q}^{\widehat{h}_\ell}(x,u), \forall x \in X$ ▷ policy improvement
5: **until** $\widehat{h}_{\ell+1}$ is satisfactory
Output: $\widehat{h}^* = \widehat{h}_{\ell+1}$

Figure 3.8 (repeated from the relevant part of Figure 3.2) illustrates the structure of our upcoming presentation. We first discuss in detail the approximate policy evaluation component, starting in Section 3.5.1 with a class of algorithms that can be derived along the same lines as approximate value iteration. In Section 3.5.2, model-free policy evaluation algorithms with linearly parameterized approximation are introduced, which aim to solve a projected form of the Bellman equation. Section 3.5.3 briefly reviews policy evaluation with nonparametric approximation, and Section 3.5.4 outlines a model-based, direct simulation approach for policy evaluation. In Section 3.5.5, we move on to the policy improvement component and the resulting approximate policy iteration. Theoretical results about approximate policy iteration are reviewed in Section 3.5.6, and a numerical example is provided in Section 3.5.7 (the material of these last two sections is not represented in Figure 3.8).

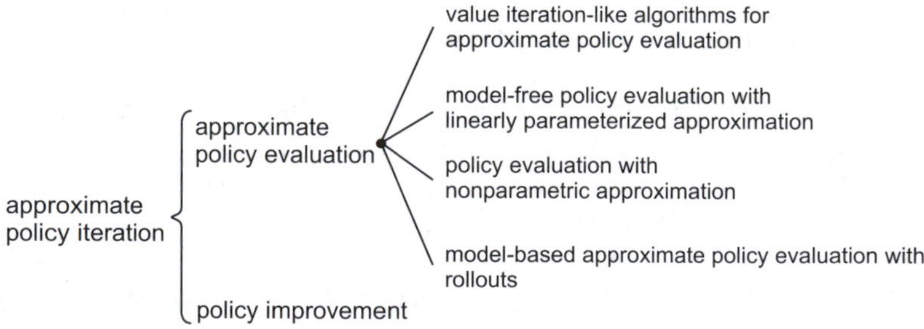

FIGURE 3.8
The organization of the algorithms for approximate policy evaluation and policy improvement presented in the sequel.

3.5.1 Value iteration-like algorithms for approximate policy evaluation

We start our discussion of approximate policy evaluation with a class of algorithms that can be derived along entirely similar lines to approximate value iteration. These algorithms can be model-based or model-free, and can use parametric or nonparametric approximation. We focus here on the parametric case, and discuss two representative algorithms, one model-based and the other model-free. These two algorithms are similar to approximate Q-iteration (Section 3.4.1) and to fitted Q-iteration (Section 3.4.2), respectively. In order to streamline the presentation, we will often refer to these counterparts and to their derivation.

The first algorithm that we develop is based on the model-based, iterative policy evaluation for Q-functions (Section 2.3.1). Denote the policy to be evaluated by h. Recall that policy evaluation for Q-functions starts from an arbitrary Q-function Q_0^h, which is updated at each iteration τ using (2.38), repeated here for easy reference:

$$Q_{\tau+1}^h = T^h(Q_\tau^h)$$

where T^h is the policy evaluation mapping, given by (2.35) in the deterministic case and by (2.36) in the stochastic case. The algorithm asymptotically converges to the Q-function Q^h of the policy h, which is the solution of the Bellman equation (2.39), also repeated here:

$$Q^h = T^h(Q^h) \tag{3.31}$$

Policy evaluation for Q-functions can be extended to the approximate case in a similar way as approximate Q-iteration (see Section 3.4.1). As with approximate Q-iteration, an approximation mapping $F : \mathbb{R}^n \to \mathcal{Q}$ is used to compactly represent Q-functions using parameter vectors $\theta^h \in \mathbb{R}^n$, and a projection mapping $P : \mathcal{Q} \to \mathbb{R}^n$ is used to find parameter vectors that represent the updated Q-functions well.

The iterative, *approximate policy evaluation for Q-functions* starts with an arbitrary (e.g., identically 0) parameter vector θ_0^h, and updates this vector at every iteration τ using the composition of mappings P, T^h, and F:

$$\theta_{\tau+1}^h = (P \circ T^h \circ F)(\theta_\tau^h) \tag{3.32}$$

The algorithm is stopped once a satisfactory parameter vector $\widehat{\theta}^h$ has been found. Under conditions similar to those for value iteration (Section 3.4.4), the composite mapping $P \circ T^h \circ F$ is a contraction, and therefore has a fixed point θ^h to which the update (3.32) asymptotically converges. This is true, e.g., if both F and P are nonexpansions.

As an example, Algorithm 3.6 shows approximate policy evaluation for Q-functions in the case of deterministic MDPs, using the least-squares projection (3.14). In this algorithm, $Q_{\tau+1}^{h,\ddagger}$ denotes the intermediate, updated Q-function:

$$Q_{\tau+1}^{h,\ddagger} = (T^h \circ F)(\theta_\tau^h)$$

Because deterministic MDPs are considered, $Q_{\tau+1}^{h,\ddagger}(x_{l_s}, u_{l_s})$ is computed at line 4 using the policy evaluation mapping (2.35). Note the similarity of Algorithm 3.6 with the approximate *Q-iteration* for MDPs (Algorithm 3.1).

ALGORITHM 3.6 Approximate policy evaluation for Q-functions in deterministic MDPs.

Input: policy h to be evaluated, dynamics f, reward function ρ, discount factor γ, approximation mapping F, samples $\{(x_{l_s}, u_{l_s}) \mid l_s = 1, \dots, n_s\}$

1: initialize parameter vector, e.g., $\theta_0^h \leftarrow 0$

2: **repeat** at every iteration $\tau = 0, 1, 2, \dots$

3:     **for** $l_s = 1, \dots, n_s$ **do**

4:         $Q_{\tau+1}^{h,\ddagger}(x_{l_s}, u_{l_s}) \leftarrow \rho(x_{l_s}, u_{l_s}) + \gamma [F(\theta_\tau^h)](f(x_{l_s}, u_{l_s}), h(f(x_{l_s}, u_{l_s})))$

5:     **end for**

6:     $\theta_{\tau+1}^h \leftarrow \theta^{h,\ddagger}$, where $\theta^{h,\ddagger} \in \arg\min_\theta \sum_{l_s=1}^{n_s} \left(Q_{\tau+1}^{h,\ddagger}(x_{l_s}, u_{l_s}) - [F(\theta)](x_{l_s}, u_{l_s}) \right)^2$

7: **until** $\theta_{\tau+1}^h$ is satisfactory

Output: $\widehat{\theta}^h = \theta_{\tau+1}^h$

The second algorithm that we develop is an analogue of fitted Q-iteration, so it will be called *fitted policy evaluation for Q-functions*. It can also be seen as a model-free variant of the approximate policy evaluation for Q-functions developed above. In this variant, a batch of transition samples is assumed to be available:

$$\{(x_{l_s}, u_{l_s}, x'_{l_s}, r_{l_s}) \mid l_s = 1, \dots, n_s\}$$

where for every l_s, the next state x'_{l_s} and the reward r_{l_s} have been obtained after taking action u_{l_s} in the state x_{l_s}. At every iteration, samples of the updated Q-function $Q_{\tau+1}^{h,\ddagger}$ are computed with:

$$Q_{\tau+1,l_s}^{h,\ddagger} = r_{l_s} + \gamma [F(\theta_\tau)](x'_{l_s}, h(x'_{l_s}))$$

In the deterministic case, the quantity $Q_{\tau+1,l_s}^{h,\ddagger}$ is identical to the updated Q-value $Q_{\tau+1}^{h,\ddagger}(x_{l_s}, u_{l_s})$ (see, e.g., line 4 of Algorithm 3.6). In the stochastic case, $Q_{\tau+1,l_s}^{h,\ddagger}$ is a *sample* of the random variable that has the updated Q-value as its expectation. A complete iteration of the algorithm is obtained by computing an updated parameter vector with a projection mapping, using the samples $((x_{l_s}, u_{l_s}), Q_{\tau+1,l_s}^{h,\ddagger})$.

Algorithm 3.7 presents fitted policy evaluation for Q-functions, using the least-squares projection (3.14). Note that, in the deterministic case, fitted policy evaluation is identical to model-based, approximate policy evaluation (e.g., Algorithm 3.6), if both algorithms use the same approximation and projection mappings, together with the same state-action samples (x_{l_s}, u_{l_s}).

3.5.2 Model-free policy evaluation with linearly parameterized approximation

A different, dedicated framework for approximate policy evaluation can be developed when linearly parameterized approximators are employed. By exploiting the linearity of the approximator in combination with the linearity of the policy evaluation mapping (see Section 2.4.1), it is possible to derive a specific approximate form

ALGORITHM 3.7 Fitted policy evaluation for Q-functions.

Input: policy h to be evaluated, discount factor γ,
approximation mapping F, samples $\{(x_{l_s}, u_{l_s}, x'_{l_s}, r_{l_s}) \mid l_s = 1, \ldots, n_s\}$

initialize parameter vector, e.g., $\theta_0^h \leftarrow 0$
repeat at every iteration $\tau = 0, 1, 2, \ldots$
  for $l_s = 1, \ldots, n_s$ **do**
    $Q^{h,\ddagger}_{\tau+1,l_s} \leftarrow r_{l_s} + \gamma [F(\theta^h_\tau)](x'_{l_s}, h(x'_{l_s}))$
  end for
  $\theta^h_{\tau+1} \leftarrow \theta^{h,\ddagger}$, where $\theta^{h,\ddagger} \in \arg\min_\theta \sum_{l_s=1}^{n_s} \left(Q^{h,\ddagger}_{\tau+1,l_s} - [F(\theta)](x_{l_s}, u_{l_s}) \right)^2$
until $\theta^h_{\tau+1}$ is satisfactory

Output: $\widehat{\theta}^h = \theta^h_{\tau+1}$

of the Bellman equation, called the "projected Bellman equation," which is linear in the parameter vector.[7] Efficient algorithms can be developed to solve this equation. In contrast, in approximate value iteration, the maximum operator leads to nonlinearity even when the approximator is linearly parameterized.

We next introduce the projected Bellman equation, along with several important model-free algorithms that can be used to solve it.

Projected Bellman equation

Assume for now that X and U have a finite number of elements, $X = \{x_1, \ldots, x_{\bar{N}}\}$, $U = \{u_1, \ldots, u_{\bar{M}}\}$. Because the state space is finite, a transition model of the form (2.14) is appropriate, and the policy evaluation mapping T^h can be written as a sum (2.37), repeated here for easy reference:

$$[T^h(Q)](x,u) = \sum_{x'} \bar{f}(x,u,x') \left[\tilde{\rho}(x,u,x') + \gamma Q(x', h(x')) \right] \tag{3.33}$$

In the linearly parameterized case, an approximate Q-function $\widehat{Q}^h$ that has the form (3.3) is sought:

$$\widehat{Q}^h(x,u) = \phi^{\mathrm{T}}(x,u)\theta^h$$

where $\phi(x,u) = [\phi_1(x,u), \ldots, \phi_n(x,u)]^{\mathrm{T}}$ is the vector of BFs and θ^h is the parameter vector. This approximate Q-function satisfies the following approximate version of

[7] Another important class of policy evaluation approaches aims to minimize the *Bellman error* (residual), which is the difference between the two sides of the Bellman equation (Baird, 1995; Antos et al., 2008b; Farahmand et al., 2009b). For instance, in the case of the Bellman equation for Q^h (3.31), the (quadratic) Bellman error is $\int_{X \times U} (\widehat{Q}^h(x,u) - [T^h(\widehat{Q}^h)](x,u))^2 \mathrm{d}(x,u)$. We choose to focus on projected policy evaluation instead, as this class of methods will be required later in the book.

the Bellman equation for Q^h (3.31), called the *projected Bellman equation*:[8]

$$\widehat{Q}^h = (P^w \circ T^h)(\widehat{Q}^h) \tag{3.34}$$

where P^w performs a weighted least-squares projection onto the space of representable (approximate) Q-functions, i.e., the space spanned by the BFs:

$$\{\phi^\mathrm{T}(x,u)\theta \mid \theta \in \mathbb{R}^n\}$$

The projection P^w is defined by:

$$\begin{aligned}&[P^w(Q)](x,u) = \phi^\mathrm{T}(x,u)\theta^\ddagger, \text{ where}\\ &\theta^\ddagger \in \arg\min_{\theta} \sum_{(x,u)\in X\times U} w(x,u)\left(\phi^\mathrm{T}(x,u)\theta - Q(x,u)\right)^2\end{aligned} \tag{3.35}$$

in which the weight function $w : X \times U \to [0,1]$ controls the distribution of the approximation error. The weight function is always interpreted as a probability distribution over the state-action space, so it must satisfy $\sum_{x,u} w(x,u) = 1$. For instance, the distribution given by w will later be used to generate the samples used by some model-free policy evaluation algorithms. Under appropriate conditions, the projected Bellman mapping $P^w \circ T^h$ is a contraction, and so the solution (fixed point) $\widehat{Q}^h$ of the projected Bellman equation exists and is unique (see Bertsekas (2007, Section 6.3) for a discussion of the conditions in the context of V-function approximation).

Figure 3.9 illustrates the projected Bellman equation.

Matrix form of the projected Bellman equation

We will now derive a matrix form of the projected Bellman equation, which is given in terms of the parameter vector. This form will be useful in the sequel, when developing algorithms to solve the projected Bellman equation. To introduce the matrix form, it will be convenient to refer to the state and the actions using explicit indices, e.g., x_i, u_j (recall that the states and actions were temporarily assumed to be discrete).

As a first step, the policy evaluation mapping (3.33) is written in matrix form $\boldsymbol{T}^h : \mathbb{R}^{\bar{N}\bar{M}} \to \mathbb{R}^{\bar{N}\bar{M}}$, as:

$$\boldsymbol{T}^h(\boldsymbol{Q}) = \tilde{\boldsymbol{\rho}} + \gamma \bar{\boldsymbol{f}}\boldsymbol{h}\boldsymbol{Q} \tag{3.36}$$

Denote by $[i,j]$ the scalar index corresponding to i and j, computed with $[i,j] = i + (j-1)\bar{N}$. The vectors and matrices in (3.36) are then defined as follows:[9]

---

[8] A multistep version of this equation can also be given. Instead of the (single-step) policy evaluation mapping T^h, this version uses the following multistep mapping, parameterized by the scalar $\lambda \in [0,1)$:

$$T^h_\lambda(Q) = (1-\lambda)\sum_{k=0}^{\infty} \lambda^k (T^h)^{k+1}(Q)$$

where $(T^h)^k$ denotes the k-times composition of T^h with itself, i.e., $T^h \circ T^h \circ \cdots \circ T^h$. In this chapter, as well as in the remainder of the book, we only consider the single-step case, i.e., the case in which $\lambda = 0$.

[9] Note that boldface notation is used for vector or matrix representations of functions and mappings. Ordinary vectors and matrices are displayed in normal font.

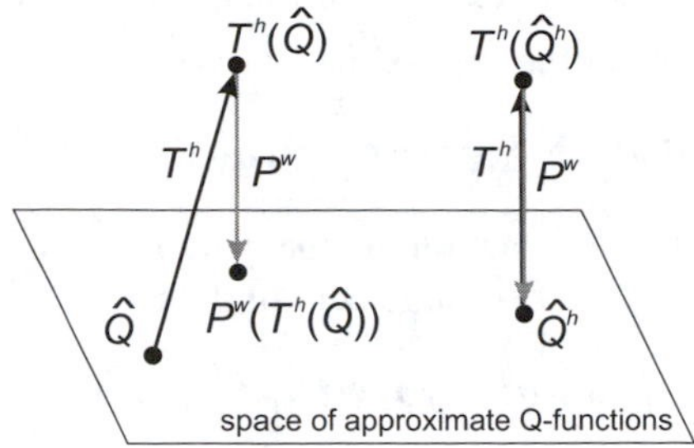

FIGURE 3.9
A conceptual illustration of the projected Bellman equation. Applying T^h and then P^w to an ordinary approximate Q-function $\widehat{Q}$ leads to a different point in the space of approximate Q-functions (left). In contrast, applying T^h and then P^w to the fixed point $\widehat{Q}^h$ of the projected Bellman equation leads back to the same point (right).

- $\boldsymbol{Q} \in \mathbb{R}^{\bar{N}\bar{M}}$ is a vector representation of Q, with $\boldsymbol{Q}_{[i,j]} = Q(x_i, u_j)$.
- $\tilde{\boldsymbol{\rho}} \in \mathbb{R}^{\bar{N}\bar{M}}$ is a vector representation of $\tilde{\rho}$, where the element $\tilde{\boldsymbol{\rho}}_{[i,j]}$ is the expected reward after taking action u_j in state x_i, i.e., $\tilde{\boldsymbol{\rho}}_{[i,j]} = \sum_{i'} \bar{f}(x_i, u_j, x_{i'}) \tilde{\rho}(x_i, u_j, x_{i'})$.
- $\bar{\boldsymbol{f}} \in \mathbb{R}^{\bar{N}\bar{M} \times \bar{N}}$ is a matrix representation of $\bar{f}$, with $\bar{\boldsymbol{f}}_{[i,j],i'} = \bar{f}(x_i, u_j, x_{i'})$. Here, $\bar{\boldsymbol{f}}_{[i,j],i'}$ denotes the element at row $[i, j]$ and column i' of matrix $\bar{\boldsymbol{f}}$.
- $\boldsymbol{h} \in \mathbb{R}^{\bar{N} \times \bar{N}\bar{M}}$ is a matrix representation of h, with $\boldsymbol{h}_{i',[i,j]} = 1$ if $i' = i$ and $h(x_i) = u_j$, and 0 otherwise. Note that stochastic policies can easily be represented, by making $\boldsymbol{h}_{i,[i,j]}$ equal to the probability of taking u_j in x_i, and $\boldsymbol{h}_{i',[i,j]} = 0$ for all $i' \neq i$.

Consider now the setting of approximate policy evaluation. Define the BF matrix $\boldsymbol{\phi} \in \mathbb{R}^{\bar{N}\bar{M} \times n}$ and the diagonal weighting matrix $\boldsymbol{w} \in \mathbb{R}^{\bar{N}\bar{M} \times \bar{N}\bar{M}}$ by:

$$\boldsymbol{\phi}_{[i,j],l} = \phi_l(x_i, u_j)$$
$$\boldsymbol{w}_{[i,j],[i,j]} = w(x_i, u_j)$$

Using $\boldsymbol{\phi}$, the approximate Q-vector corresponding to a parameter θ is:

$$\widehat{\boldsymbol{Q}} = \boldsymbol{\phi}\theta$$

The projected Bellman equation (3.34) can now be written as follows:

$$\boldsymbol{P}^w \boldsymbol{T}^h(\widehat{\boldsymbol{Q}}^h) = \widehat{\boldsymbol{Q}}^h \tag{3.37}$$

where $\boldsymbol{P}^w$ is a matrix representation of the projection operator P^w, which can be written in a closed form (see, e.g., Lagoudakis and Parr, 2003a):

$$\boldsymbol{P}^w = \boldsymbol{\phi}(\boldsymbol{\phi}^{\mathrm{T}} \boldsymbol{w} \boldsymbol{\phi})^{-1} \boldsymbol{\phi}^{\mathrm{T}} \boldsymbol{w}$$

By substituting this closed-form expression for $\boldsymbol{P}^w$, the formula (3.36) for $\boldsymbol{T}^h$, and the expression $\widehat{\boldsymbol{Q}}^h = \boldsymbol{\phi}\,\theta^h$ for the approximate Q-vector into (3.37), we get:

$$\boldsymbol{\phi}(\boldsymbol{\phi}^{\mathrm{T}}\boldsymbol{w}\boldsymbol{\phi})^{-1}\boldsymbol{\phi}^{\mathrm{T}}\boldsymbol{w}(\tilde{\boldsymbol{\rho}} + \gamma\bar{\boldsymbol{f}}\boldsymbol{h}\boldsymbol{\phi}\,\theta^h) = \boldsymbol{\phi}\,\theta^h$$

Notice that this is a linear equation in the parameter vector θ^h. After a left-multiplication with $\boldsymbol{\phi}^{\mathrm{T}}\boldsymbol{w}$ and a rearrangement of the terms, we have:

$$\boldsymbol{\phi}^{\mathrm{T}}\boldsymbol{w}\boldsymbol{\phi}\,\theta^h = \gamma\boldsymbol{\phi}^{\mathrm{T}}\boldsymbol{w}\bar{\boldsymbol{f}}\boldsymbol{h}\boldsymbol{\phi}\,\theta^h + \boldsymbol{\phi}^{\mathrm{T}}\boldsymbol{w}\tilde{\boldsymbol{\rho}}$$

By introducing the matrices $\Gamma, \Lambda \in \mathbb{R}^{n\times n}$ and the vector $z \in \mathbb{R}^n$, given by:

$$\Gamma = \boldsymbol{\phi}^{\mathrm{T}}\boldsymbol{w}\boldsymbol{\phi}, \quad \Lambda = \boldsymbol{\phi}^{\mathrm{T}}\boldsymbol{w}\bar{\boldsymbol{f}}\boldsymbol{h}\boldsymbol{\phi}, \quad z = \boldsymbol{\phi}^{\mathrm{T}}\boldsymbol{w}\tilde{\boldsymbol{\rho}}$$

the projected Bellman equation can be written in the final, matrix form:

$$\Gamma\theta^h = \gamma\Lambda\theta^h + z \tag{3.38}$$

So, instead of the original, high-dimensional Bellman equation (3.31), approximate policy evaluation only needs to solve the low-dimensional system (3.38). A solution θ^h of this system can be employed to find an approximate Q-function using (3.3).

It can also be shown that matrices Γ, Λ and vector z can be written as sums of simpler matrices and vectors (e.g., Lagoudakis and Parr, 2003a):

$$\begin{aligned}
\Gamma &= \sum_{i=1}^{\bar{N}}\sum_{j=1}^{\bar{M}} \Big[\phi(x_i,u_j)w(x_i,u_j)\phi^{\mathrm{T}}(x_i,u_j)\Big] \\
\Lambda &= \sum_{i=1}^{\bar{N}}\sum_{j=1}^{\bar{M}} \left[\phi(x_i,u_j)w(x_i,u_j)\sum_{i'=1}^{\bar{N}}\Big(\bar{f}(x_i,u_j,x_{i'})\phi^{\mathrm{T}}(x_{i'},h(x_{i'}))\Big)\right] \\
z &= \sum_{i=1}^{\bar{N}}\sum_{j=1}^{\bar{M}} \Big[\phi(x_i,u_j)w(x_i,u_j)\sum_{i'=1}^{\bar{N}}\Big(\bar{f}(x_i,u_j,x_{i'})\rho(x_i,u_j,x_{i'})\Big)\Big]
\end{aligned} \tag{3.39}$$

To understand why the summation over i' enters the equation for z, recall that each element $\tilde{\boldsymbol{\rho}}_{[i,j]}$ of the vector $\tilde{\boldsymbol{\rho}}$ is the *expected* reward after taking action u_j in state x_i.

Model-free projected policy evaluation

Some of the most powerful algorithms for approximate policy evaluation solve the matrix form (3.38) of the projected Bellman equation in a model-free fashion, by estimating Γ, Λ, and z from transition samples. Because (3.38) is a linear system of equations, these algorithms are computationally efficient. They are also sample-efficient, i.e., they approach their solution quickly as the number of samples they consider increases, as shown in the context of V-function approximation by Konda (2002, Chapter 6) and by Yu and Bertsekas (2006, 2009).

Consider a set of transition samples:

$$\{(x_{l_s}, u_{l_s}, x'_{l_s} \sim \bar{f}(x_{l_s}, u_{l_s}, \cdot), r_{l_s} = \tilde{\rho}(x_{l_s}, u_{l_s}, x'_{l_s})) \,|\, l_s = 1, \ldots, n_s\}$$

This set is constructed by drawing state-action samples (x,u) from a distribution given by the weight function w: the probability of each pair (x,u) is equal to its weight $w(x,u)$. Using this set of samples, estimates of Γ, Λ, and z can be constructed as follows:

$$\begin{aligned}
&\Gamma_0 = 0, \quad \Lambda_0 = 0, \quad z_0 = 0 \\
&\Gamma_{l_s} = \Gamma_{l_s-1} + \phi(x_{l_s}, u_{l_s})\phi^{\mathrm{T}}(x_{l_s}, u_{l_s}) \\
&\Lambda_{l_s} = \Lambda_{l_s-1} + \phi(x_{l_s}, u_{l_s})\phi^{\mathrm{T}}(x'_{l_s}, h(x'_{l_s})) \\
&z_{l_s} = z_{l_s-1} + \phi(x_{l_s}, u_{l_s}) r_{l_s}
\end{aligned} \tag{3.40}$$

These updates can be derived from (3.39).

The *least-squares temporal difference for Q-functions (LSTD-Q)* (Lagoudakis et al., 2002; Lagoudakis and Parr, 2003a) is a policy evaluation algorithm that processes the samples using (3.40) and then solves the equation:

$$\frac{1}{n_s}\Gamma_{n_s}\widehat{\theta}^h = \gamma\frac{1}{n_s}\Lambda_{n_s}\widehat{\theta}^h + \frac{1}{n_s}z_{n_s} \tag{3.41}$$

to find an approximate parameter vector $\widehat{\theta}^h$. Notice that $\widehat{\theta}^h$ appears on both sides of (3.41), so this equation can be simplified to:

$$\frac{1}{n_s}(\Gamma_{n_s} - \gamma\Lambda_{n_s})\widehat{\theta}^h = \frac{1}{n_s}z_{n_s}$$

Although the division by n_s is not necessary from a formal point of view, it helps to increase the numerical stability of the algorithm (the elements in the Γ_{n_s}, Λ_{n_s}, z_{n_s} can be very large when n_s is large). LSTD-Q is an extension of an earlier, similar algorithm for V-functions, called least-squares temporal difference (Bradtke and Barto, 1996; Boyan, 2002).

Another method, the *least-squares policy evaluation for Q-functions (LSPE-Q)* (e.g., Jung and Polani, 2007a) starts with an arbitrary initial parameter vector θ_0 and updates it incrementally, with:

$$\begin{aligned}
&\theta_{l_s} = \theta_{l_s-1} + \alpha(\theta^{\ddagger}_{l_s} - \theta_{l_s-1}), \text{ where:} \\
&\frac{1}{l_s}\Gamma_{l_s}\theta^{\ddagger}_{l_s} = \gamma\frac{1}{l_s}\Lambda_{l_s}\theta_{l_s-1} + \frac{1}{l_s}z_{l_s}
\end{aligned} \tag{3.42}$$

in which α is a step size parameter. To ensure the invertibility of the matrix Γ at the start of the learning process, when only a few samples have been processed, it can be initialized to a small multiple of the identity matrix. The division by l_s increases the numerical stability of the updates. Like LSTD-Q, LSPE-Q is an extension of an earlier algorithm for V-functions, called least-squares policy evaluation (LSPE) (Bertsekas and Ioffe, 1996).

Algorithms 3.8 and 3.9 present LSTD-Q and LSPE-Q in a procedural form. LSTD-Q is a one-shot algorithm, and the parameter vector it computes does not depend on the order in which the samples are processed. On the other hand, LSPE-Q

ALGORITHM 3.8 Least-squares temporal difference for Q-functions.

Input: policy h to be evaluated, discount factor γ,
BFs $\phi_1, \dots, \phi_n : X \times U \rightarrow \mathbb{R}$, samples $\{(x_{l_s}, u_{l_s}, x'_{l_s}, r_{l_s}) \mid l_s = 1, \dots, n_s\}$

$\Gamma_0 \leftarrow 0$, $\Lambda_0 \leftarrow 0$, $z_0 \leftarrow 0$
for $l_s = 1, \dots, n_s$ **do**
  $\Gamma_{l_s} \leftarrow \Gamma_{l_s-1} + \phi(x_{l_s}, u_{l_s}) \phi^{\mathrm{T}}(x_{l_s}, u_{l_s})$
  $\Lambda_{l_s} \leftarrow \Lambda_{l_s-1} + \phi(x_{l_s}, u_{l_s}) \phi^{\mathrm{T}}(x'_{l_s}, h(x'_{l_s}))$
  $z_{l_s} \leftarrow z_{l_s-1} + \phi(x_{l_s}, u_{l_s}) r_{l_s}$
end for
solve $\frac{1}{n_s} \Gamma_{n_s} \widehat{\theta}^h = \gamma \frac{1}{n_s} \Lambda_{n_s} \widehat{\theta}^h + \frac{1}{n_s} z_{n_s}$ for $\widehat{\theta}^h$

Output: $\widehat{\theta}^h$

ALGORITHM 3.9 Least-squares policy evaluation for Q-functions.

Input: policy h to be evaluated, discount factor γ,
BFs $\phi_1, \dots, \phi_n : X \times U \rightarrow \mathbb{R}$, samples $\{(x_{l_s}, u_{l_s}, x'_{l_s}, r_{l_s}) \mid l_s = 1, \dots, n_s\}$,
step size α, a small constant $\beta_\Gamma > 0$

$\Gamma_0 \leftarrow \beta_\Gamma I$, $\Lambda_0 \leftarrow 0$, $z_0 \leftarrow 0$
for $l_s = 1, \dots, n_s$ **do**
  $\Gamma_{l_s} \leftarrow \Gamma_{l_s-1} + \phi(x_{l_s}, u_{l_s}) \phi^{\mathrm{T}}(x_{l_s}, u_{l_s})$
  $\Lambda_{l_s} \leftarrow \Lambda_{l_s-1} + \phi(x_{l_s}, u_{l_s}) \phi^{\mathrm{T}}(x'_{l_s}, h(x'_{l_s}))$
  $z_{l_s} \leftarrow z_{l_s-1} + \phi(x_{l_s}, u_{l_s}) r_{l_s}$
  $\theta_{l_s} \leftarrow \theta_{l_s-1} + \alpha(\theta^{\ddagger}_{l_s} - \theta_{l_s-1})$, where $\frac{1}{l_s} \Gamma_{l_s} \theta^{\ddagger}_{l_s} = \gamma \frac{1}{l_s} \Lambda_{l_s} \theta_{l_s-1} + \frac{1}{l_s} z_{l_s}$
end for

Output: $\widehat{\theta}^h = \theta_{n_s}$

is an incremental algorithm, so the current parameter vector θ_{l_s} depends on the previous values $\theta_0, \dots, \theta_{l_s-1}$, and therefore the order in which samples are processed is important.

In the context of V-function approximation, such least-squares algorithms have been shown to converge to the fixed point of the projected Bellman equation, namely by Nedić and Bertsekas (2003) for the V-function analogue of LSTD-Q, and by Nedić and Bertsekas (2003); Bertsekas et al. (2004) for the analogue of LSPE-Q. These results also extend to Q-function approximation. To ensure convergence, the weight (probability of being sampled) $w(x,u)$ of each state-action pair (x,u) should be equal to the steady-state probability of this pair along an infinitely-long trajectory generated with the policy h.[10]

Note that collecting samples by using only a *deterministic* policy h is insuffi-

[10]From a practical point of view, note that LSTD-Q is a one-shot algorithm and will produce a solution whenever Γ_{l_s} is invertible. This means the experimenter need not worry excessively about divergence *per se*. Rather, the theoretical results concern the uniqueness and meaning of the solution obtained. LSTD-Q can, in fact, produce meaningful results for many weight functions w, as we illustrate later in Section 3.5.7 and in Chapter 5.

cient for the following reason. If only state-action pairs of the form $(x, h(x))$ were collected, no information about pairs (x, u) with $u \neq h(x)$ would be available (equivalently, the corresponding weights $w(x, u)$ would all be zero). As a result, the approximate Q-values of such pairs would be poorly estimated and could not be relied upon for policy improvement. To alleviate this problem, *exploration* is necessary: sometimes, actions different from $h(x)$ have to be selected, e.g., in a random fashion. Given a stationary (time-invariant) exploration procedure, LSTD-Q and LSPE-Q are simply evaluating the new, exploratory policy, and so they remain convergent.

The following intuitive (albeit informal) line of reasoning is useful to understand the convergence of LSTD-Q and LSPE-Q. Asymptotically, as $n_s \to \infty$, it is true that $\frac{1}{n_s}\Gamma_{n_s} \to \Gamma$, $\frac{1}{n_s}\Lambda_{n_s} \to \Lambda$, and $\frac{1}{n_s}z_{n_s} \to z$, for the following two reasons. First, as the number n_s of state-action samples generated grows, their empirical distribution converges to w. Second, as the number of transition samples involving a given state-action pair (x, u) grows, the empirical distribution of the next states x' converges to the distribution $\bar{f}(x, u, \cdot)$, and the empirical average of the rewards converges to its expected value, given x and u.

Since the estimates of Γ, Λ, and z asymptotically converge to their true values, the equation solved by LSTD-Q asymptotically converges to the projected Bellman equation (3.38). Under the assumptions for convergence, this equation has a unique solution θ^h, so the parameter vector of LSTD-Q asymptotically reaches this solution. For similar reasons, whenever it converges, LSPE-Q asymptotically becomes equivalent to LSTD-Q and the projected Bellman equation. Therefore, if LSPE-Q converges, it must in fact converge to θ^h. In fact, it can additionally be shown that, as n_s grows, the solutions of LSTD-Q and LSPE-Q converge to each other faster than they converge to their limit θ^h. This was proven in the context of V-function approximation by Yu and Bertsekas (2006, 2009).

One possible advantage of LSTD-Q over LSPE-Q may arise when their assumptions are violated, e.g., when the policy to be evaluated changes as samples are being collected. This situation can arise in the important context of optimistic policy iteration, which will be discussed in Section 3.5.5. Violating the assumptions may introduce instability and possibly divergence in the iterative LSPE-Q updates (3.42). In contrast, because it only computes one-shot solutions, LSTD-Q (3.41) may be more resilient to such instabilities. On the other hand, the incremental nature of LSPE-Q offers some advantages over LSTD-Q. For instance, LSPE-Q can benefit from a good initial value of the parameter vector. Additionally, by lowering the step size α, it may be possible to mitigate the destabilizing effects of violating the assumptions. Note that an incremental version of LSTD-Q can also be given, but the benefits of this version are unclear.

While for the derivation above it was assumed that X and U are finite, the updates (3.40), together with LSTD-Q and LSPE-Q, can also be applied without any change in infinite and uncountable (e.g., continuous) state-action spaces.

From a computational point of view, the linear systems in (3.41) and (3.42) can be solved in several ways, e.g., by matrix inversion, by Gaussian elimination, or by incrementally computing the inverse with the Sherman-Morrison formula. The computational cost is $O(n^3)$ for "naive" matrix inversion. More efficient algorithms than matrix inversion can be obtained, e.g., by incrementally computing the inverse, but the cost of solving the linear system will still be larger than $O(n^2)$. In an effort to further reduce the computational costs, variants of the least-squares temporal difference have been proposed in which only a few of the parameters are updated at a given iteration (Geramifard et al., 2006, 2007). Note also that, when the BF vector $\phi(x,u)$ is sparse, the computational efficiency of the updates (3.40) can be improved by exploiting this sparsity.[11]

As already outlined, analogous least-squares algorithms can be given to compute approximate V-functions (Bertsekas and Ioffe, 1996; Bradtke and Barto, 1996; Boyan, 2002; Bertsekas, 2007, Chapter 6). However, as explained in Section 2.2, policy improvement is more difficult to perform using V-functions. Namely, a model of the MDP is required, and in the stochastic case, expectations over the transitions must be estimated.

Gradient-based policy evaluation

Gradient-based algorithms for policy evaluation historically precede the least-squares methods discussed above (Sutton, 1988). However, under appropriate conditions, they find, in fact, a solution of the projected Bellman equation (3.34). These algorithms are called temporal-difference learning in the literature, and are more popular in the context of V-function approximation (Sutton, 1988; Jaakkola et al., 1994; Tsitsiklis and Van Roy, 1997). Nevertheless, given the focus of this chapter, we will present gradient-based policy evaluation for the case of Q-function approximation.

We use SARSA as a starting point in developing such an algorithm. Recall that SARSA (Algorithm 2.7) uses tuples $(x_k, u_k, r_{k+1}, x_{k+1}, u_{k+1})$ to update a Q-function online (2.40):

$$Q_{k+1}(x_k,u_k) = Q_k(x_k,u_k) + \alpha_k[r_{k+1} + \gamma Q_k(x_{k+1},u_{k+1}) - Q_k(x_k,u_k)] \tag{3.43}$$

where α_k is the learning rate. When u_k is chosen according to a fixed policy h, SARSA actually performs policy evaluation (see also Section 2.4.2). We exploit this property and combine (3.43) with gradient-based updates to obtain the desired policy evaluation algorithm. As before, linearly parameterized approximation is considered. By a derivation similar to that given for gradient-based Q-learning in Section 3.4.2, the following update rule is obtained:

$$\theta_{k+1} = \theta_k + \alpha_k \left[r_{k+1} + \gamma \phi^{\mathrm{T}}(x_{k+1},u_{k+1})\theta_k - \phi^{\mathrm{T}}(x_k,u_k)\theta_k \right] \phi(x_k,u_k) \tag{3.44}$$

where the quantity in square brackets is an approximation of the temporal difference. The resulting algorithm for policy evaluation is called *temporal difference for Q-functions (TD-Q)* . Note that TD-Q can be seen as an extension of a corresponding algorithm for V-functions, which is called temporal difference (TD) (Sutton, 1988).

Like the least-squares algorithms presented earlier, TD-Q requires exploration to

---

[11] The BF vector is sparse, e.g., for the discrete-action approximator described in Example 3.1. This is because the BF vector contains zeros for all the discrete actions that are different from the current discrete action.

obtain samples (x,u) with $u \neq h(x)$. Algorithm 3.10 presents TD-Q with ε-greedy exploration. In this algorithm, because the update at step k involves the action u_{k+1} at the next step, this action is chosen prior to updating the parameter vector.

ALGORITHM 3.10 Temporal difference for Q-functions, with ε-greedy exploration.

Input: discount factor γ, policy h to be evaluated,
BFs $\phi_1,\dots,\phi_n : X \times U \to \mathbb{R}$,
exploration schedule $\{\varepsilon_k\}_{k=0}^{\infty}$, learning rate schedule $\{\alpha_k\}_{k=0}^{\infty}$

initialize parameter vector, e.g., $\theta_0 \leftarrow 0$
measure initial state x_0

$$u_0 \leftarrow \begin{cases} h(x_0) & \text{with probability } 1-\varepsilon_0 \\ \text{a uniform random action in } U & \text{with probability } \varepsilon_0 \text{ (explore)} \end{cases}$$

for every time step $k = 0,1,2,\dots$ **do**
    apply u_k, measure next state x_{k+1} and reward r_{k+1}

$$u_{k+1} \leftarrow \begin{cases} h(x_{k+1}) & \text{with probability } 1-\varepsilon_{k+1} \\ \text{a uniform random action in } U & \text{with probability } \varepsilon_{k+1} \end{cases}$$

$$\theta_{k+1} \leftarrow \theta_k + \alpha_k \left[r_{k+1} + \gamma \phi^{\mathrm{T}}(x_{k+1},u_{k+1})\theta_k - \phi^{\mathrm{T}}(x_k,u_k)\theta_k \right] \phi(x_k,u_k)$$

end for

A comprehensive convergence analysis of gradient-based policy evaluation was provided by Tsitsiklis and Van Roy (1997) in the context of V-function approximation. This analysis extends to Q-function approximation under appropriate conditions. An important condition is that the stochastic policy $\tilde{h}$ resulting from the combination of h with exploration should be time-invariant, which can be achieved by simply making the exploration time-invariant, e.g., in the case of ε-greedy exploration, by making ε_k the same for all steps k. The main result is that TD-Q asymptotically converges to the solution of the projected Bellman equation for the exploratory policy $\tilde{h}$, for a weight function given by the steady-state distribution of the state-action pairs under $\tilde{h}$.

Gradient-based algorithms such as TD-Q are less computationally demanding than least-squares algorithms such as LSTD-Q and LSPE-Q. The time and memory complexity of TD-Q are both $\mathrm{O}(n)$, since they store and update vectors of length n. The memory complexity of LSTD-Q and LSPE-Q is at least $\mathrm{O}(n^2)$ (since they store matrices of size n) and their time complexity is $\mathrm{O}(n^3)$ (when "naive" matrix inversion is used to solve the linear system). On the other hand, gradient-based algorithms typically require more samples than least-squares algorithms to achieve a similar accuracy (Konda, 2002; Yu and Bertsekas, 2006, 2009), and are more sensitive to the learning rate (step size) schedule. LSTD-Q has no step size at all, and LSPE-Q works for a wide range of constant step sizes, as shown in the context of V-functions by Bertsekas et al. (2004) (this range includes $\alpha = 1$, leading to a nonincremental variant of LSPE-Q).

Efforts have been made to extend gradient-based policy evaluation algorithms to off-policy learning, i.e., evaluating one policy while using another policy to generate the samples (Sutton et al., 2009b,a). These extensions perform gradient descent on error measures that are different from the measure used in the basic temporal-difference algorithms such as TD-Q (i.e., different from the squared value function error for the current sample).

3.5.3 Policy evaluation with nonparametric approximation

Nonparametric approximators have been combined with a number of algorithms for approximate policy evaluation. For instance, kernel-based approximators were combined with LSTD by Xu et al. (2005), with LSTD-Q by Xu et al. (2007); Jung and Polani (2007b); Farahmand et al. (2009b), and with LSPE-Q by Jung and Polani (2007a,b). Rasmussen and Kuss (2004) and Engel et al. (2003, 2005) used the related framework of Gaussian processes to approximate V-functions in policy evaluation. Taylor and Parr (2009) showed that, in fact, the algorithms in (Rasmussen and Kuss, 2004; Engel et al., 2005; Xu et al., 2005) produce the same solution when they use the same samples and the same kernel function. Fitted policy evaluation (Algorithm 3.7) can be extended to the nonparametric case along the same lines as fitted Q-iteration in Section 3.4.3. Such an algorithm was proposed by Jodogne et al. (2006), who employed ensembles of extremely randomized trees to approximate the Q-function.

As explained in Section 3.3.2, a kernel-based approximator can be seen as linearly parameterized if all the samples are known in advance. In certain cases, this property can be exploited to extend the theoretical guarantees about approximate policy evaluation from the parametric case to the nonparametric case (Xu et al., 2007). Farahmand et al. (2009b) provided performance guarantees for their kernel-based LSTD-Q variant for the case when only a finite number of samples is available.

An important concern in the nonparametric case is controlling the complexity of the approximator. Originally, the computational demands of many nonparametric approximators, including kernel-based methods and Gaussian processes, grow with the number of samples considered. Many of the approaches mentioned above employ kernel sparsification techniques to limit the number of samples that contribute to the solution (Xu et al., 2007; Engel et al., 2003, 2005; Jung and Polani, 2007a,b).

3.5.4 Model-based approximate policy evaluation with rollouts

All the policy evaluation algorithms discussed above obtain a value function by solving the Bellman equation (3.31) approximately. While this is a powerful approach, it also has its drawbacks. A core problem is that a good value function approximator is required, which is often difficult to find. Nonparametric approximation alleviates this problem to some extent. Another problem is that the convergence requirements of the algorithms, such as the linearity of the approximate Q-function in the parameters, can sometimes be too restrictive.

Another class of policy evaluation approaches sidesteps these difficulties by avoiding an explicit representation of the value function. Instead, the value function is evaluated on demand, by Monte Carlo simulations. A model is required to perform the simulations, so these approaches are model-based. For instance, to estimate the Q-value $\widehat{Q}^h(x,u)$ of a given state-action pair (x,u), a number N_{MC} of trajectories are simulated, where each trajectory is generated using the policy h, has length K, and starts from the pair (x,u). The estimated Q-value is then the average of the sample returns obtained along these trajectories:

$$\widehat{Q}^h(x,u) = \frac{1}{N_{\mathrm{MC}}} \sum_{i_0=1}^{N_{\mathrm{MC}}} \left[\tilde{\rho}(x,u,x_{i_0,1}) + \sum_{k=1}^{K} \gamma^k \tilde{\rho}(x_{i_0,k}, h(x_{i_0,k}), x_{i_0,k+1}) \right] \tag{3.45}$$

where N_{MC} is the number of trajectories to simulate. For each trajectory i_0, the first state-action pair is fixed to (x,u) and leads to a next state $x_{i_0,1} \sim \tilde{f}(x,u,\cdot)$. Thereafter, actions are chosen using the policy h, which means that for $k \geq 1$:

$$x_{i_0,k+1} \sim f(x_{i_0,k}, h(x_{i_0,k}), \cdot)$$

Such a simulation-based estimation procedure is called a *rollout* (Lagoudakis and Parr, 2003b; Bertsekas, 2005b; Dimitrakakis and Lagoudakis, 2008). The length K of the trajectories can be chosen using (2.41) to ensure $\varepsilon_{\mathrm{MC}}$-accurate returns, where $\varepsilon_{\mathrm{MC}} > 0$. Note that if the MDP is deterministic, a single trajectory suffices. In the stochastic case, an appropriate value for the number N_{MC} of trajectories will depend on the problem.

Rollouts can be computationally expensive, especially in the stochastic case. Their computational cost is proportional to the number of points at which the value function must be evaluated. Therefore, rollouts are most beneficial when this number is small. If the value function must be evaluated at many (or all) points of the state(-action) space, then methods that solve the Bellman equation approximately (Sections 3.5.1 – 3.5.3) may be computationally less costly than rollouts.

3.5.5 Policy improvement and approximate policy iteration

Up to this point, approximate policy evaluation has been considered. To obtain a complete algorithm for approximate policy iteration, a method to perform policy improvement is also required.

Exact and approximate policy improvement

Consider first policy improvement in the case where the policy is not represented explicitly. Instead, greedy actions are computed on demand from the value function, for every state where a control action is required. For instance, when Q-functions are employed, an improved action for the state x can be found with:

$$h_{\ell+1}(x) = u, \text{ where } u \in \arg\max_{\bar{u}} \widehat{Q}^{h_\ell}(x,\bar{u}) \tag{3.46}$$

The policy is thus implicitly defined by the value function. In (3.46), it was assumed that a greedy action can be computed exactly. This is true, e.g., when the action space only contains a small, discrete set of actions, and the maximization in the policy improvement step is solved by enumeration. In this situation, policy improvement is exact, but if greedy actions cannot be computed exactly, then the result of the maximization is approximate, and the (implicitly defined) policy thus becomes an approximation.

Alternatively, the policy can also be represented explicitly, in which case it generally must be approximated. The policy can be approximated, e.g., by a linear parametrization (3.12):

$$\widehat{h}(x) = \sum_{i=1}^{\mathcal{N}} \varphi_i(x)\vartheta_i = \varphi^{\mathrm{T}}(x)\vartheta$$

where $\varphi_i(x)$, $i = 1, \ldots, \mathcal{N}$ are the state-dependent BFs and ϑ is the policy parameter vector (see Section 3.3.4 for a discussion of the notation used for policy approximation). A scalar action was assumed, but the parametrization can easily be extended to multiple action variables. For this parametrization, approximate policy improvement can be performed by solving the linear least-squares problem:

$$\vartheta_{\ell+1} = \vartheta^{\ddagger}, \text{ where } \vartheta^{\ddagger} \in \arg\min_{\vartheta} \sum_{i_s=1}^{\mathcal{N}_s} \left(\varphi^{\mathrm{T}}(x_{i_s})\vartheta - u_{i_s}\right)^2 \tag{3.47}$$

to find a parameter vector $\vartheta_{\ell+1}$, where $\{x_1, \ldots, x_{\mathcal{N}_s}\}$ is a set of state samples to be used for policy improvement, and $u_1, \ldots, u_{\mathcal{N}_s}$ are corresponding greedy actions:

$$u_{i_s} \in \arg\max_{u} \widehat{Q}^{\widehat{h}_\ell}(x_{i_s}, u) \tag{3.48}$$

Note that the previous policy $\widehat{h}_\ell$ is now also an approximation. In (3.48), it was implicitly assumed that greedy actions can be computed exactly; if this is not the case, then u_{i_s} will only be approximations of the true greedy actions.

Such a policy improvement is therefore a two-step procedure: first, greedy actions u_{i_s} are chosen using (3.48), and then these actions are used to solve the least-squares problem (3.47). The solution depends on the greedy actions chosen, but remains meaningful for any combination of choices, since for any such combination, it approximates one of the possible greedy policies in the Q-function.

Alternatively, policy improvement could be performed with:

$$\vartheta_{\ell+1} = \vartheta^{\ddagger}, \text{ where } \vartheta^{\ddagger} \in \arg\max_{\vartheta} \sum_{i_s=1}^{\mathcal{N}_s} \widehat{Q}^{\widehat{h}_\ell}(x_{i_s}, \varphi^{\mathrm{T}}(x_{i_s})\vartheta) \tag{3.49}$$

which maximizes the approximate Q-values of the actions chosen by the policy in the state samples. However, (3.49) is generally a difficult nonlinear optimization problem, whereas (3.47) is (once greedy actions have been chosen) a convex optimization problem, which is easier to solve.

More generally, for any policy representation (e.g., for a nonlinear parametrization), a regression problem generalizing either (3.47) or (3.49) must be solved to perform policy improvement.

Offline approximate policy iteration

Approximate policy iteration algorithms can be obtained by combining a policy evaluation algorithm (e.g., one of those described in Sections 3.5.1 – 3.5.3) with a policy improvement technique (e.g., one of those described above); see again Algorithm 3.5 for a generic template of approximate policy iteration. In the offline case, the approximate policy evaluation is run until (near) convergence, to ensure the accuracy of the value function and therefore an accurate policy improvement.

For example, the algorithm resulting from combining LSTD-Q (Algorithm 3.8) with exact policy improvement is called *least-squares policy iteration (LSPI)*. LSPI was proposed by Lagoudakis et al. (2002) and by Lagoudakis and Parr (2003a), and has been studied often since then (e.g., Mahadevan and Maggioni, 2007; Xu et al., 2007; Farahmand et al., 2009b). Algorithm 3.11 shows LSPI, in a simple variant that uses the same set of transition samples at every policy evaluation. In general, different sets of samples can be used at different iterations. The explicit policy improvement at line 4 is included for clarity. In practice, the policy $h_{\ell+1}$ does not have to be computed and stored for every state. Instead, it is computed on demand from the current Q-function, only for those states where an improved action is necessary. In particular, LSTD-Q only evaluates the policy at the state samples x'_{l_s}.

ALGORITHM 3.11 Least-squares policy iteration.

Input: discount factor γ,
BFs $\phi_1, \dots, \phi_n : X \times U \to \mathbb{R}$, samples $\{(x_{l_s}, u_{l_s}, x'_{l_s}, r_{l_s}) \,|\, l_s = 1, \dots, n_s\}$
1: initialize policy h_0
2: **repeat** at every iteration $\ell = 0, 1, 2, \dots$
3: evaluate h_ℓ using LSTD-Q (Algorithm 3.8), yielding θ_ℓ ▷ policy evaluation
4: $h_{\ell+1}(x) \leftarrow u$, $u \in \arg\max_{\bar{u}} \phi^{\mathrm{T}}(x, \bar{u})\theta_\ell$ for each $x \in X$ ▷ policy improvement
5: **until** $h_{\ell+1}$ is satisfactory
Output: $\widehat{h}^* = h_{\ell+1}$

Policy iteration with rollout policy evaluation (Section 3.5.4) was studied, e.g., by Lagoudakis and Parr (2003b) and by Dimitrakakis and Lagoudakis (2008), who employed nonparametric approximation to represent the policy. Note that rollout policy evaluation (which represents value functions implicitly) should not be combined with implicit policy improvement. Such an algorithm would be impractical, because neither the value function nor the policy would be represented explicitly.

Online, optimistic approximate policy iteration

In online learning, the performance should improve once every few transition samples. This is in contrast to the offline case, in which only the performance at the end of the learning process is important. One way in which policy iteration can take this requirement into account is by performing policy improvements once every few transition samples, before an accurate evaluation of the current policy can be completed. Such a variant is sometimes called *optimistic* policy iteration (Bertsekas and Tsitsiklis, 1996, Section 6.4; Sutton 1988; Tsitsiklis 2002). In the extreme case, the policy is improved after every transition, and then applied to obtain a new transition sample that is fed into the policy evaluation algorithm. Then, another policy improvement takes place, and the cycle repeats. This variant is called fully optimistic. In general, the policy is improved once every several (but not too many) transitions; this variant is partially optimistic. As in any online RL algorithm, exploration is also necessary in optimistic policy iteration.

Optimistic policy iteration was already outlined in Section 2.4.2, where it was also explained that SARSA (Algorithm 2.7) belongs to this class. So, an approximate version of SARSA will naturally be optimistic, as well. A gradient-based version of SARSA can be easily obtained from TD-Q (Algorithm 3.10), by choosing actions with a policy that is greedy in the current Q-function, instead of with a fixed policy as in TD-Q. Of course, exploration is required in addition to greedy action selection. Algorithm 3.12 presents approximate SARSA with an ε-greedy exploration procedure. Approximate SARSA has been studied, e.g., by Sutton (1996); Santamaria et al. (1998); Gordon (2001); Melo et al. (2008).

ALGORITHM 3.12 SARSA with a linear parametrization and ε-greedy exploration.

Input: discount factor γ,
BFs $\phi_1, \dots, \phi_n : X \times U \to \mathbb{R}$,
exploration schedule $\{\varepsilon_k\}_{k=0}^{\infty}$, learning rate schedule $\{\alpha_k\}_{k=0}^{\infty}$

initialize parameter vector, e.g., $\theta_0 \leftarrow 0$

measure initial state x_0

$$u_0 \leftarrow \begin{cases} u \in \arg\max_{\bar{u}} \left(\phi^{\mathrm{T}}(x_0, \bar{u})\theta_0\right) & \text{with probability } 1-\varepsilon_0 \text{ (exploit)} \\ \text{a uniform random action in } U & \text{with probability } \varepsilon_0 \text{ (explore)} \end{cases}$$

for every time step $k = 0, 1, 2, \dots$ **do**

apply u_k, measure next state x_{k+1} and reward r_{k+1}

$$u_{k+1} \leftarrow \begin{cases} u \in \arg\max_{\bar{u}} \left(\phi^{\mathrm{T}}(x_{k+1}, \bar{u})\theta_k\right) & \text{with probability } 1-\varepsilon_{k+1} \\ \text{a uniform random action in } U & \text{with probability } \varepsilon_{k+1} \end{cases}$$

$$\theta_{k+1} \leftarrow \theta_k + \alpha_k \left[r_{k+1} + \gamma \phi^{\mathrm{T}}(x_{k+1}, u_{k+1})\theta_k - \phi^{\mathrm{T}}(x_k, u_k)\theta_k\right] \phi(x_k, u_k)$$

end for

Other policy evaluation algorithms can also be used in optimistic policy iteration. For instance, optimistic policy iteration with LSPE-Q was applied by Jung and Polani (2007a,b), while a V-function based algorithm similar to approximate SARSA was proposed by Jung and Uthmann (2004). In Chapter 5 of this book, an online, optimistic variant of LSPI will be introduced in detail and evaluated experimentally.

3.5.6 Theoretical guarantees

Under appropriate assumptions, *offline* policy iteration eventually produces policies with a bounded suboptimality. However, in general it cannot be guaranteed to converge to a fixed policy. The theoretical understanding of *optimistic* policy iteration is currently limited, and guarantees can only be provided in some special cases. We first discuss the properties of policy iteration in the offline setting, and then continue to the online, optimistic setting.

Theoretical guarantees for offline approximate policy iteration

As long as the policy evaluation and improvement errors are bounded, offline approximate policy iteration eventually produces policies with a bounded suboptimality. This result applies to any type of value function or policy approximator, and can be formalized as follows.

Consider the general case where both the value functions and the policies are approximated. Consider also the case where Q-functions are used, and assume that the error at every policy evaluation step is bounded by ς_Q:

$$\|\widehat{Q}^{\widehat{h}_\ell} - Q^{\widehat{h}_\ell}\|_\infty \leq \varsigma_Q, \quad \text{for any } \ell \geq 0$$

and that the error at every policy improvement step is bounded by ς_h, in the following sense:

$$\|T^{\widehat{h}_{\ell+1}}(\widehat{Q}^{\widehat{h}_\ell}) - T(\widehat{Q}^{\widehat{h}_\ell})\|_\infty \leq \varsigma_h, \quad \text{for any } \ell \geq 0$$

where $T^{\widehat{h}_{\ell+1}}$ is the policy evaluation mapping for the improved (approximate) policy, and T is the Q-iteration mapping (2.22). Then, approximate policy iteration eventually produces policies with performances that lie within a bounded distance from the optimal performance (e.g., Lagoudakis and Parr, 2003a):

$$\limsup_{\ell\to\infty} \left\|\widehat{Q}^{\widehat{h}_\ell} - Q^*\right\|_\infty \leq \frac{\varsigma_h + 2\gamma\varsigma_Q}{(1-\gamma)^2} \tag{3.50}$$

For an algorithm that performs exact policy improvements, such as LSPI, $\varsigma_h = 0$ and the bound is tightened to:

$$\limsup_{\ell\to\infty} \|\widehat{Q}^{h_\ell} - Q^*\|_\infty \leq \frac{2\gamma\varsigma_Q}{(1-\gamma)^2} \tag{3.51}$$

where $\|\widehat{Q}^{h_\ell} - Q^{h_\ell}\|_\infty \leq \varsigma_Q$, for any $\ell \geq 0$. Note that finding ς_Q and (when approximate policies are used) ς_h may be difficult in practice, and the existence of these bounds may require additional assumptions.

These guarantees do not necessarily imply the convergence to a fixed policy. For instance, both the value function and policy parameters might converge to limit cycles, so that every point on the cycle yields a policy that satisfies the bound. Convergence to limit cycles can indeed happen, as will be seen in the upcoming example of Section 3.5.7. Similarly, when exact policy improvements are used, the value function parameter may oscillate, implicitly leading to an oscillating policy. This is a disadvantage with respect to offline approximate value iteration, which under appropriate assumptions converges monotonically to a unique fixed point (Section 3.4.4).

Similar results hold when V-functions are used instead of Q-functions (Bertsekas and Tsitsiklis, 1996, Section 6.2).

Theoretical guarantees for online, optimistic policy iteration

The performance guarantees given above for offline policy iteration rely on bounded policy evaluation errors. Because optimistic policy iteration improves the policy before an accurate value function is available, the policy evaluation error can be very large, and the performance guarantees for offline policy iteration are not useful in the online case.

The behavior of optimistic policy iteration has not been properly understood yet, and can be very complicated. Optimistic policy iteration can, e.g., exhibit a phenomenon called chattering, whereby the value function converges to a stationary function, while the policy sequence oscillates, because the limit of the value function parameter corresponds to multiple policies (Bertsekas and Tsitsiklis, 1996, Section 6.4).

Theoretical guarantees can, however, be provided in certain special cases. Gordon (2001) showed that the parameter vector of approximate SARSA cannot diverge when the MDP has terminal states and the policy is only improved in-between trials (see Section 2.2.1 for the meaning of terminal states and trials). Melo et al. (2008) improved on this result, by showing that approximate SARSA converges with probability 1 to a fixed point, if the dependence of the policy on the parameter vector satisfies a certain Lipschitz continuity condition. This condition prohibits using fully greedy policies, because those generally depend on the parameters in a discontinuous fashion.

These theoretical results concern the gradient-based SARSA algorithm. However, in practice, least-squares algorithms may be preferable due to their improved sample efficiency. While no theoretical guarantees are available when using least-squares algorithms in the optimistic setting, some promising empirical results have been reported (Jung and Polani, 2007a,b); see also Chapter 5 for an empirical evaluation of optimistic LSPI.

3.5.7 Example: Least-squares policy iteration for a DC motor

In this example, approximate policy iteration is applied to the DC motor problem introduced in Section 3.4.5. In a first experiment, the original LSPI (Algorithm 3.11) is applied. This algorithm represents policies implicitly and performs exact policy improvements. The results of this experiment are compared with the results of approximate Q-iteration from Section 3.4.5. In a second experiment, LSPI is modified to use approximate policies and sample-based, approximate policy improvements. The resulting solution is compared with the solution found with exact policy improvements.

In both experiments, the policies are evaluated using their Q-functions, which are approximated with a discrete-action parametrization of the type described in Example 3.1. Recall that such an approximator replicates state-dependent BFs for every discrete action, and in order to obtain the state-action BFs, it sets to 0 all the BFs that do not correspond to the current discrete action. Like in Section 3.4.5, the action space is discretized into the set $U_{\mathrm{d}} = \{-10, 0, 10\}$, so the number of discrete actions is $M = 3$. The state-dependent BFs are axis-aligned, normalized Gaussian RBFs (see Example 3.1). The centers of the RBFs are arranged on a 9×9 equidistant grid over the state space, so there are $N = 81$ RBFs in total. All the RBFs are identical in shape, and their width b_d along each dimension d is equal to $b_d'^2/2$, where b_d' is the distance between adjacent RBFs along that dimension (the grid step). These RBFs yield a smooth interpolation of the Q-function over the state space. Recalling that the domains of the state variables are $[-\pi, \pi]$ for the angle and $[-16\pi, 16\pi]$ for the angular velocity, we obtain $b_1' = \frac{2\pi}{9-1} \approx 0.79$ and $b_2' = \frac{32\pi}{9-1} \approx 12.57$, which lead to $b_1 \approx 0.31$ and $b_2 \approx 78.96$. The parameter vector θ contains $n = NM = 243$ parameters.

Least-squares policy iteration with exact policy improvement

In the first part of the example, the original LSPI algorithm is applied to the DC motor problem. Recall that LSPI combines LSTD-Q policy evaluation with exact policy improvement.

The same set of $n_s = 7500$ samples is used at every LSTD-Q policy evaluation. The samples are random, uniformly distributed over the state-discrete action space $X \times U_d$. The initial policy h_0 is identically equal to -10 throughout the state space. To illustrate the results of LSTD-Q, Figure 3.10 presents the first improved policy found by the algorithm, h_1, and its approximate Q-function, computed with LSTD-Q. Note that this Q-function is the *second* found by LSPI; the first Q-function evaluates the initial policy h_0.

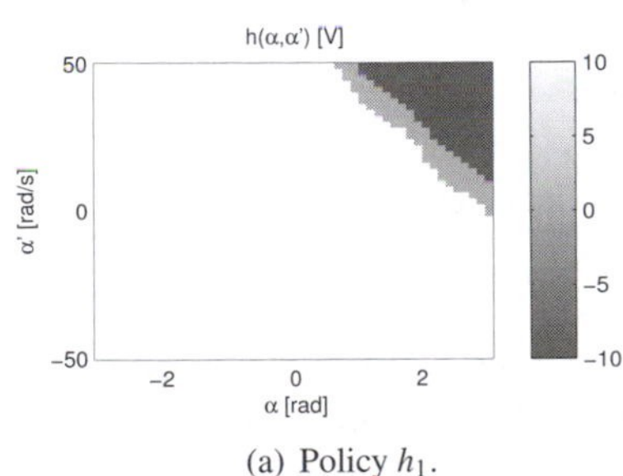

(a) Policy h_1.

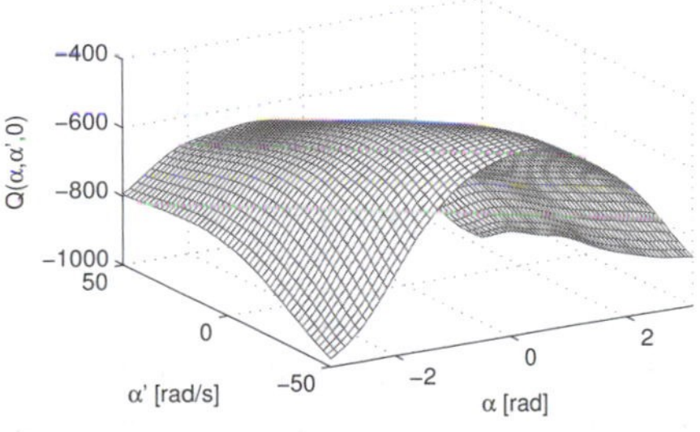

(b) Slice through the Q-function for $u = 0$. Note the difference in vertical scale from the other Q-functions shown in this chapter.

FIGURE 3.10
An early policy and its approximate Q-function, for LSPI with exact policy improvements.

In this problem, LSPI fully converged in 11 iterations. Figure 3.11 shows the resulting policy and Q-function, together with a representative controlled trajectory. The policy and the Q-function in Figure 3.11 are good approximations of the near-optimal solution in Figure 3.5.

Compared to the results of grid Q-iteration in Figure 3.6, LSPI needs fewer BFs (81 rather than 400 or 160000) while still being able to find a similarly accurate approximation of the policy. This is mainly because the Q-function is largely smooth (see Figure 3.5(a)), and thus can be represented more easily by the wide RBFs of

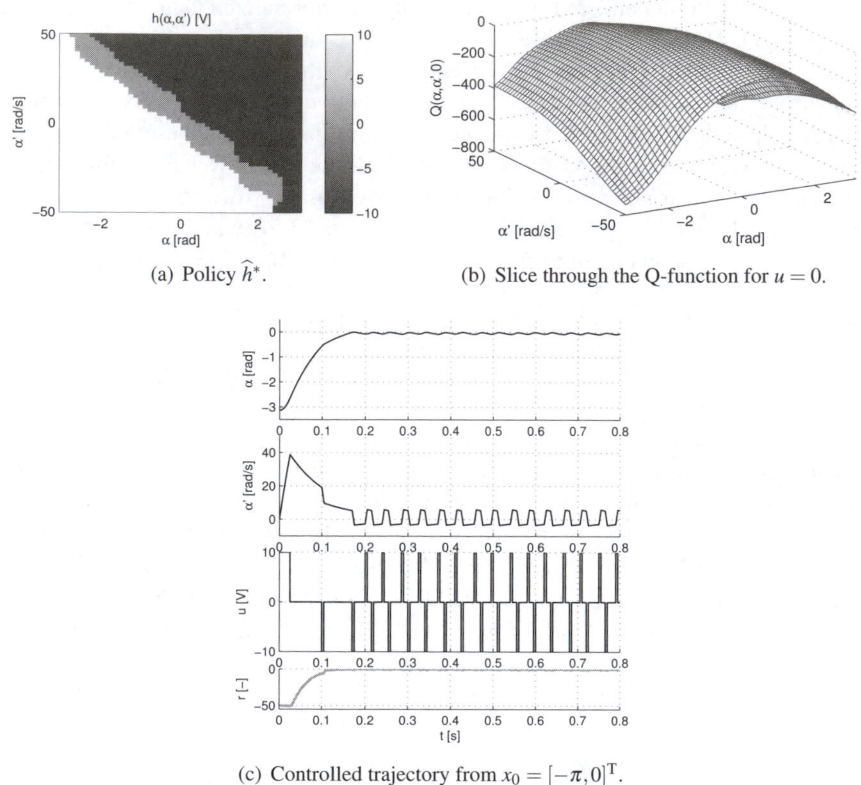

(a) Policy $\widehat{h}^*$.

(b) Slice through the Q-function for $u = 0$.

(c) Controlled trajectory from $x_0 = [-\pi, 0]^{\mathrm{T}}$.

FIGURE 3.11 Results of LSPI with exact policy improvements for the DC motor.

the approximator employed in LSPI. In contrast, the grid BFs give a discontinuous approximate Q-function, which is less appropriate for this problem. Although certain types of continuous BFs can be used with Q-iteration, using wide RBFs such as these in combination with the least-squares projection (3.14) is unfortunately not possible, because they do not satisfy the assumptions for convergence, and indeed lead to divergence when they are too wide. The controlled trajectory in Figure 3.11(c) is comparable in quality with the trajectory controlled by the fine-grid policy, shown in Figure 3.6(f); however, it does produce more chattering.

Another observation is that LSPI converged in significantly fewer iterations than grid Q-iteration did in Section 3.4.5 (12 iterations for LSPI, instead of 160 for grid Q-iteration with the coarse grid, and 123 with the fine grid). Such a convergence rate advantage of policy iteration over value iteration is often observed in practice. However, while LSPI did converge faster, it was actually more computationally intensive than grid Q-iteration: it required approximately 23 s to run, whereas grid Q-iteration required only 0.06 s for the coarse grid and 7.80 s for the fine grid. Some insight into this difference can be obtained by examining the asymptotic complexity of the two algorithms. The complexity of policy evaluation with LSTD-Q is larger than $\mathrm{O}(n^2)$ due to solving a linear system of size n. For grid Q-iteration, when binary search is used to locate the position of a state on the grid, the cost is $\mathrm{O}(n\log(N))$, where $n = NM$, N is the number of elements on the grid, and M the number of discrete actions. On the other hand, while the convergence of grid Q-iteration to a fixed point was guaranteed by the theory, this is not the case for LSPI (although for this problem LSPI did, in fact, fully converge).

Compared to the results of fitted Q-iteration in Figure 3.7, the LSPI solution is of a similar quality. LSPI introduces some curved artifacts in the policy, due to the limitations of the wide RBFs employed. On the other hand, the execution time of 2151 s for fitted Q-iteration is much larger than the 23 s for LSPI.

Least-squares policy iteration with policy approximation

The aim of the second part of the example is to illustrate the effects of approximating policies. To this end, LSPI is modified to work with approximate policies and sample-based, approximate policy improvement.

The policy approximator is linearly parameterized (3.12) and uses the same RBFs as the Q-function approximator. Such an approximate policy produces *continuous* actions, which must be quantized (into discrete actions belonging to U_{d}) before performing policy evaluation, because the Q-function approximator only works for discrete actions. Policy improvement is performed with the linear least-squares procedure (3.47), using a number $\mathcal{N}_{\mathrm{s}} = 2500$ of random, uniformly distributed state samples. The same samples are used at every iteration. As before, policy evaluation employs $N_{\mathrm{s}} = 7500$ samples.

In this experiment, both the Q-functions and the policies *oscillate* in the steady state of the algorithm, with a period of 2 iterations. The execution time until the oscillation was detected was 58 s. The differences between the two distinct policies and Q-functions on the limit cycle are too small to be noticed in a figure. Instead, Figure 3.12 shows the evolution of the policy parameter that changes the most in steady state, for which the oscillation is clearly visible. The appearance of oscillations may be related to the fact that the weaker suboptimality bound (3.50) applies when approximate policies are used, rather than the stronger bound (3.51), which applies for exact policy improvements.

Figure 3.13 presents one of the two policies from the limit cycle, one of the Q-functions, and a representative controlled trajectory. The policy and Q-function have a similar accuracy to those computed with exact, discrete-action policy improvements. One advantage of the approximate policy is that it produces continuous actions. The beneficial effects of continuous actions on the control performance are apparent in the trajectory shown in Figure 3.13(c), which is very close to the near-optimal trajectory of Figure 3.5(c).

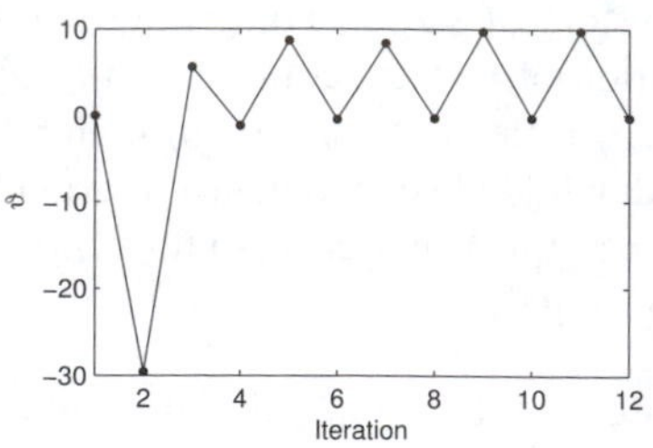

FIGURE 3.12
The variation of one of the policy parameters for LSPI with policy approximation on the DC motor.

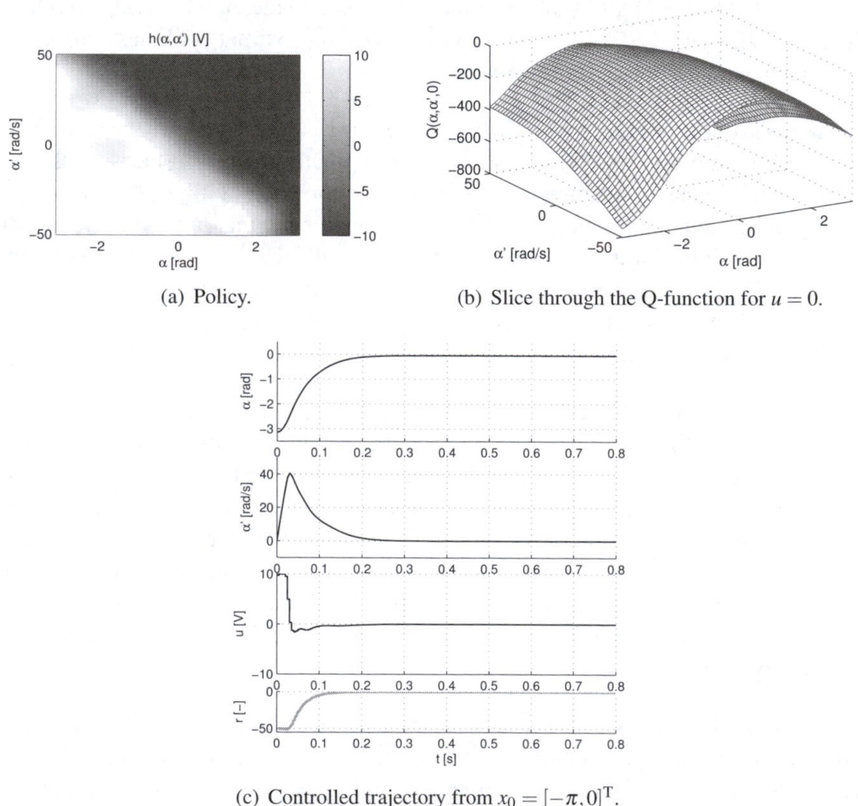

(a) Policy.

(b) Slice through the Q-function for $u = 0$.

(c) Controlled trajectory from $x_0 = [-\pi, 0]^T$.

FIGURE 3.13 Results of LSPI with policy approximation for the DC motor.

3.6 Finding value function approximators automatically

Parametric approximators of the value function play an important role in approximate value iteration and approximate policy iteration, as seen in Sections 3.4 and 3.5. Given the functional form of such an approximator, the DP/RL algorithm computes its parameters. However, there still remains the problem of finding a good functional form, well suited to the problem at hand. For concreteness, we will consider linearly parameterized approximators (3.3), in which case a good set of BFs has to be found. This focus is motivated by the fact that many methods to find good approximators work in such a linear setting.

The most straightforward solution is to design the BFs in advance, in which case two approaches are possible. The first is to design the BFs so that a uniform resolution is obtained over the entire state space (for V-functions) or over the entire state-action space (for Q-functions). Unfortunately, such an approach suffers from the curse of dimensionality: the complexity of a uniform approximator grows exponentially with the number of state variables, and in the case of Q-functions, also with the number of action variables. The second approach is to focus the resolution on certain parts of the state (or state-action) space, where the value function has a more complex shape, or where it is more important to approximate it accurately. Prior knowledge about the shape of the value function or about the importance of certain regions of the state (or state-action) space is necessary in this case. Unfortunately, such prior knowledge is often nonintuitive and very difficult to obtain without actually computing the value function.

A more general alternative is to devise a method to automatically find BFs suited to the problem at hand, rather than designing them manually. Two major categories of methods to find BFs automatically are BF optimization and BF construction. *BF optimization* methods search for the best placement and shape of a (usually fixed) number of BFs. *BF construction* methods are not constrained by a fixed number of BFs, but add new or remove old BFs to improve the approximation accuracy. The newly added BFs may have different shapes, or they may all have the same shape. Several subcategories of BF construction can be distinguished, some of the most important of which are defined next.

- BF refinement methods work in a top-down fashion. They start with a few BFs (a coarse resolution) and refine them as needed.
- BF selection methods work oppositely, in a bottom-up fashion. Starting from a large number of BFs (a fine resolution), they select a small subset of BFs that still ensure a good accuracy.
- Bellman error methods for BF construction define new BFs using the Bellman error of the value function represented with the current BFs. The Bellman error (or Bellman residual) is the difference between the two sides of the Bellman equation, where the current value function has been filled in (see also the upcoming Section 3.6.1 and, e.g., (3.52)).

Figure 3.14 summarizes this taxonomy.

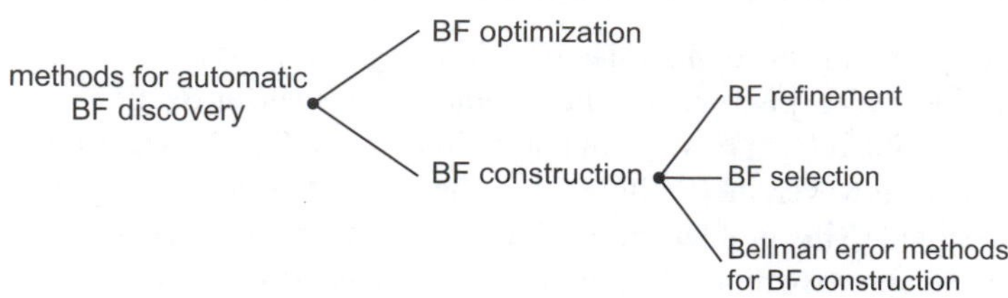

FIGURE 3.14 A taxonomy of methods for the automatic discovery of BFs.

In the remainder of this section, we first describe BF optimization, in Section 3.6.1, followed by BF construction in Section 3.6.2, and by some additional remarks in Section 3.6.3.

3.6.1 Basis function optimization

BF optimization methods search for the best placement and shape of a (typically fixed) number of BFs. Consider, e.g., the linear parametrization (3.3) of the Q-function. To optimize the n BFs, they are parameterized by a vector of BF parameters ξ that encodes their locations and shapes. The approximate Q-function is:

$$\widehat{Q}(x,u) = \phi^{\mathrm{T}}(x,u;\xi)\theta$$

where the parameterized BFs have been denoted by:

$$\phi^{\mathrm{T}}(x,u;\xi) : X \times U \to \mathbb{R}, \quad l = 1,\ldots,n$$

to highlight their dependence on ξ. For instance, an RBF is characterized by its center and width, so for an RBF approximator, the vector ξ contains the centers and widths of all the RBFs.

The BF optimization algorithm searches for an optimal parameter vector ξ^* that optimizes a criterion related to the accuracy of the value function approximator. Many optimization algorithms can be applied to this problem. For instance, gradient-based optimization has been used for policy evaluation with temporal difference (Singh et al., 1995), with LSTD (Menache et al., 2005; Bertsekas and Yu, 2009), and with LSPE (Bertsekas and Yu, 2009). Among these works, Bertsekas and Yu (2009) gave a general framework for gradient-based BF optimization in approximate policy evaluation, and provided an efficient recursive procedure to estimate the gradient. The cross-entropy method has been applied to LSTD (Menache et al., 2005). In Chapter 4 of this book, we will employ the cross-entropy method to optimize approximators for Q-iteration.

The most widely used optimization criterion (score function) is the *Bellman error*, also called Bellman residual (Singh et al., 1995; Menache et al., 2005; Bertsekas and Yu, 2009). This error measures how much the estimated value function violates the Bellman equation, which would be precisely satisfied by the exact value function. For instance, in the context of policy evaluation for a policy h, the Bellman error for an estimate $\widehat{Q}^h$ of the Q-function Q^h can be derived from the Bellman equation (3.31) as:

$$[T^h(\widehat{Q}^h)](x,u) - \widehat{Q}^h(x,u) \tag{3.52}$$

at the state-action pair (x,u), where T^h is the policy evaluation mapping. This error was derived from the Bellman equation (3.31). A quadratic Bellman error over the entire state-action space can therefore be defined as:

$$\int_{X \times U} \left([T^h(\widehat{Q}^h)](x,u) - \widehat{Q}^h(x,u) \right)^2 \mathrm{d}(x,u) \tag{3.53}$$

In the context of value iteration, the quadratic Bellman error for an estimate $\widehat{Q}$ of the optimal Q-function Q^* can be defined similarly:

$$\int_{X \times U} \left([T(\widehat{Q})](x,u) - \widehat{Q}(x,u) \right)^2 \mathrm{d}(x,u) \tag{3.54}$$

where T is the Q-iteration mapping. In practice, approximations of the Bellman errors are computed using a finite set of samples. A weight function can additionally be used to adjust the contribution of the errors according to the importance of each region of the state-action space.

In the context of policy evaluation, the distance between an approximate Q-function $\widehat{Q}^h$ and Q^h is related to the infinity norm of the Bellman error as follows (Williams and Baird, 1994):

$$\|\widehat{Q}^h - Q^h\|_\infty \leq \frac{1}{1-\gamma} \|T^h(\widehat{Q}^h) - \widehat{Q}^h\|_\infty$$

A similar result holds in the context of value iteration, where the suboptimality of an approximate Q-function $\widehat{Q}$ satisfies (Williams and Baird, 1994; Bertsekas and Tsitsiklis, 1996, Section 6.10):

$$\|\widehat{Q} - Q^*\|_\infty \leq \frac{1}{1-\gamma} \|T(\widehat{Q}) - \widehat{Q}\|_\infty$$

Furthermore, the suboptimality of $\widehat{Q}$ is related to the suboptimality of the resulting policy by (3.25), hence, in principle, minimizing the Bellman error is useful. However, in practice, *quadratic* Bellman errors (3.53), (3.54) are often employed. Because minimizing such quadratic errors may still lead to large *infinity-norm* Bellman errors, it is unfortunately unclear whether this procedure leads to accurate Q-functions.

Other optimization criteria can, of course, be used. For instance, in approximate value iteration, the return of the policy obtained by the DP/RL algorithm can be directly maximized:

$$\sum_{x_0 \in X_0} w(x_0) R^h(x_0) \tag{3.55}$$

where h is obtained by running approximate value iteration to (near-)convergence using the current approximator, X_0 is a finite set of representative initial states, and $w : X_0 \rightarrow (0,\infty)$ is a weight function. The set X_0 and the weight function w determine the performance of the resulting policy, and an appropriate choice of X_0 and w depends on the problem at hand. The returns $R^h(x_0)$ can be estimated by simulation, as in approximate policy search, see Section 3.7.2.

In approximate policy evaluation, if accurate Q-values $Q^h(x_{l_s}, u_{l_s})$ can be obtained for a set of n_s samples (x_{l_s}, u_{l_s}), then the following error measure can be minimized instead of the Bellman error (Menache et al., 2005; Bertsekas and Yu, 2009):

$$\sum_{l_s=1}^{n_s} \left(Q^h(x_{l_s}, u_{l_s}) - \widehat{Q}^h(x_{l_s}, u_{l_s}) \right)^2$$

The Q-values $Q^h(x_{l_s}, u_{l_s})$ can be obtained by simulation, as explained in Section 3.5.4.

3.6.2 Basis function construction

From the class of BF construction methods, we discuss in turn BF refinement, BF selection, and Bellman error methods for BF construction (see again Figure 3.14). Additionally, we explain how some nonparametric approximators can be seen as techniques to construct BFs automatically.

Basis function refinement

BF refinement is a widely used subclass of BF construction methods. Refinement methods work in a top-down fashion, by starting with a few BFs (a coarse resolution) and refining them as needed. They can be further classified into two categories:

- Local refinement (splitting) methods evaluate whether the value function is represented with a sufficient accuracy in a particular region of the state space (corresponding to one or several neighboring BFs), and add new BFs when the accuracy is deemed insufficient. Such methods have been proposed, e.g., for Q-learning (Reynolds, 2000; Ratitch and Precup, 2004; Waldock and Carse, 2008), V-iteration (Munos and Moore, 2002), and Q-iteration (Munos, 1997; Uther and Veloso, 1998).

- Global refinement methods evaluate the global accuracy of the representation and, if the accuracy is deemed insufficient, they refine the BFs using various techniques. All the BFs may be refined uniformly (Chow and Tsitsiklis, 1991), or the algorithm may decide that certain regions of the state space require more resolution (Munos and Moore, 2002; Grüne, 2004). For instance, Chow and Tsitsiklis (1991); Munos and Moore (2002); and Grüne (2004) applied global refinement to V-iteration, while Szepesvári and Smart (2004) used it for Q-learning.

A variety of criteria are used to decide when the BFs should be refined. An overview of typical criteria, and a comparison between them in the context of

V-iteration, was given by Munos and Moore (2002). For instance, local refinement in a certain region can be performed:

- when the value function is not (approximately) constant in that region (Munos and Moore, 2002; Waldock and Carse, 2008);
- when the value function is not (approximately) linear in that region (Munos and Moore, 2002; Munos, 1997);
- when the Bellman error (see Section 3.6.1) is large in that region (Grüne, 2004);
- using various other heuristics (Uther and Veloso, 1998; Ratitch and Precup, 2004).

Global refinement can be performed, e.g., until a desired level of solution accuracy is met (Chow and Tsitsiklis, 1991). The approach of Munos and Moore (2002) works for discrete-action problems, and globally identifies the regions of the state space that must be more accurately approximated to find a better policy. To this end, it refines regions that satisfy two conditions: (i) the V-function is poorly approximated in these regions, and (ii) this poor approximation affects, in a certain sense, (other) regions where the actions that are dictated by the policy change.

BF refinement methods increase the memory and computational demands of the DP/RL algorithm when they increase the resolution. Thus, care must be taken to prevent the memory and computation costs from becoming prohibitive, especially in the online case. This is an important concern in both approximate DP and approximate RL. Equally important in approximate RL are the restrictions imposed on BF refinement by the limited amount of data available. Increasing the power of the approximator means that more data will be required to compute an accurate solution, so the resolution cannot be refined to arbitrary levels for a given amount of data.

Basis function selection

BF selection methods work in a bottom-up fashion, by starting from a large number of BFs (a fine resolution), and then selecting a smaller subset of BFs that still provide a good accuracy. When using this type of methods, care should be taken to ensure that selecting the BFs and running the DP/RL algorithm with the selected BFs is less expensive than running the DP/RL algorithm with the original BFs. The cost may be expressed in terms of computational complexity or in terms of the number of samples required.

Kolter and Ng (2009) employed regularization to select BFs for policy evaluation with LSTD. Regularization is a technique that penalizes functional complexity in the approximate value function. In practice, the effect of regularization in the linear case is to drive some of the value function parameters (close) to 0, which means that the corresponding BFs can be ignored. By incrementally selecting the BFs, Kolter and Ng (2009) obtained a computational complexity that is linear in the total number of BFs, in contrast to the original complexity of LSTD which is at least quadratic (see Section 3.5.2).

Bellman error basis functions

Another class of BF construction approaches define new BFs by employing the Bellman error of the value function represented with the currently available BFs (3.53), (3.54). For instance, Bertsekas and Castañon (1989) proposed a method to interleave automatic state aggregation steps with iterations of a model-based policy evaluation algorithm. The aggregation steps group together states with similar Bellman errors. In this work, convergence speed was the main concern, rather than limited representation power, so the value function and the Bellman error function were assumed to be exactly representable.

More recently, Keller et al. (2006) proposed a method that follows similar lines, but that explicitly addresses the approximate case, by combining LSTD with Bellman-error based BF construction. At every BF construction step, this method computes a linear projection of the state space onto a space in which points with similar Bellman errors are close to each other. Several new BFs are defined in this projected space. Then, the augmented set of BFs is used to generate a new LSTD solution, and the cycle repeats. Parr et al. (2008) showed that in policy evaluation with linear parametrization, the Bellman error can be decomposed into two components: a transition error component and a reward error component, and proposed adding new BFs defined in terms of these error components.

Nonparametric approximators as methods for basis function construction

As previously explained in Section 3.3, some nonparametric approximators can be seen as methods to automatically generate BFs from the data. A typical example is kernel-based approximation, which, in its original form, generates a BF for every sample considered. An interesting effect of nonparametric approximators is that they adapt the complexity of the approximator to the amount of available data, which is beneficial in situations where obtaining data is costly.

When techniques to control the complexity of the nonparametric approximator are applied, they can sometimes be viewed as BF selection. For instance, regularization techniques were used in LSTD-Q by Farahmand et al. (2009a) and in fitted Q-iteration by Farahmand et al. (2009b). (In both of these cases, however, the advantage of regularization is a reduced functional complexity of the solution, while the computational complexity is not reduced.) Kernel sparsification techniques also fall in this category (Xu et al., 2007; Engel et al., 2003, 2005), as well as sample selection methods for regression tree approximators (Ernst, 2005).

3.6.3 Remarks

Some of the methods for automatic BF discovery work offline (e.g., Menache et al., 2005; Mahadevan and Maggioni, 2007), while others adapt the BFs while the DP/RL algorithm is running (e.g., Munos and Moore, 2002; Ratitch and Precup, 2004). Since convergence guarantees for approximate value iteration and approximate policy evaluation typically rely on a fixed set of BFs, adapting the BFs online invalidates these guarantees. Convergence guarantees can be recovered by ensuring that BF adaptation is stopped after a finite number of updates; fixed-BF proofs can then be applied to guarantee asymptotic convergence (Ernst et al., 2005).

The presentation above has not been exhaustive, and BFs can also be found using various other methods. For instance, in (Mahadevan, 2005; Mahadevan and Maggioni, 2007), a spectral analysis of the MDP transition dynamics is performed to find BFs for use with LSPI. Because the BFs represent the underlying topology of the state transitions, they provide a good accuracy in representing the value function. Moreover, while we have focused above on the popular approach of finding linearly parameterized approximators, nonlinearly parameterized approximators can also be found automatically. For example, Whiteson and Stone (2006) introduced an approach to optimize the parameters *and* the structure of neural network approximators for a tailored variant of Q-learning. This approach works in episodic tasks, and optimizes the total reward accumulated along episodes.

Finally, note that a fully worked-out example of finding an approximator automatically is beyond the scope of this chapter. Instead, we direct the interested reader to Section 4.4, where an approach to optimize the approximator for a value iteration algorithm is developed in detail, and to Section 4.5.4, where this approach is empirically evaluated.

3.7 Approximate policy search

Algorithms for approximate policy search represent the policy approximately, most often using a parametric approximator. An optimal parameter vector is then sought using optimization techniques. In some special cases, the policy parametrization may represent an optimal policy exactly. For instance, when the transition dynamics are linear in the state and action variables and the reward function is quadratic, the optimal policy is linear in the state variables. So, a linear parametrization in the state variables can exactly represent this optimal policy. However, in general, optimal policies can only be represented approximately.

Figure 3.15 (repeated from the relevant part of Figure 3.2) shows in a graphical form how our upcoming presentation of approximate policy search is organized. In Section 3.7.1, gradient-based methods for policy search are described, including the important category of actor-critic techniques. Then, in Section 3.7.2, gradient-free policy optimization methods are discussed.

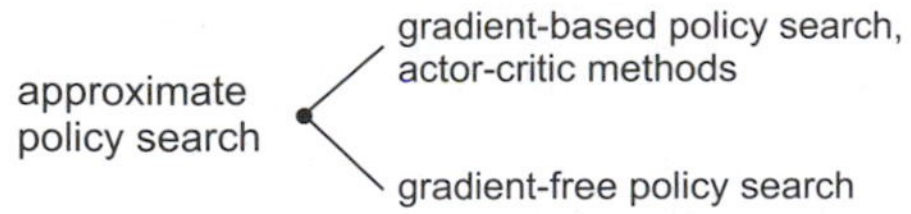

FIGURE 3.15
The organization of the algorithms for approximate policy search presented next.

Having completed our review, we then provide a numerical example involving policy search for a DC motor in Section 3.7.3.

3.7.1 Policy gradient and actor-critic algorithms

An important class of methods for approximate policy search relies on gradient-based optimization. In such *policy gradient* methods, the policy is represented using a differentiable parametrization, and gradient updates are performed to find parameters that lead to (locally) maximal returns. Some policy gradient methods estimate the gradient without using a value function (Marbach and Tsitsiklis, 2003; Munos, 2006; Riedmiller et al., 2007). Other methods compute an approximate value function of the current policy and use it to form the gradient estimate. These are called *actor-critic* methods, where the actor is the approximate policy and the critic is the approximate value function. By extension, policy gradient methods that do not use value functions are sometimes called *actor-only* methods (Bertsekas, 2007, Section 6.7).

Actor-critic algorithms were introduced by Barto et al. (1983) and have been investigated often since then (Berenji and Khedkar, 1992; Sutton et al., 2000; Konda and Tsitsiklis, 2003; Berenji and Vengerov, 2003; Borkar, 2005; Nakamura et al., 2007). Many actor-critic algorithms approximate the policy and the value function using neural networks (Prokhorov and Wunsch, 1997; Pérez-Uribe, 2001; Liu et al., 2008). Actor-critic methods are similar to policy iteration, which also improves the policy on the basis of its value function. The main difference is that in policy iteration, the improved policy is greedy in the value function, i.e., it *fully maximizes* this value function over the action variables (3.46). In contrast, actor-critic methods employ gradient rules to update the policy in a *direction* that increases the received returns. The gradient estimate is constructed using the value function.

Some important results for policy gradient methods have been developed under the expected average return criterion for optimality. We therefore discuss this setting first, in a temporary departure from the main focus of the book, which is the discounted return. We then return to the discounted setting, and present an online actor-critic algorithm for this setting.

Policy gradient and actor-critic methods for average returns

Policy gradient and actor-critic methods have often been given in the average return setting (see also Section 2.2.1). We therefore introduce these methods in the average-return case, mainly following the derivation of Bertsekas (2007, Section 6.7). We assume that the MDP has a finite state-action space, but under appropriate conditions these methods can also be extended to continuous state-action spaces (see, e.g., Konda and Tsitsiklis, 2003).

Consider a stochastic MDP with a finite state space $X = \{x_1, \dots, x_{\bar{N}}\}$, a finite action space $U = \{u_1, \dots, u_{\bar{M}}\}$, a transition function $\bar{f}$ of the form (2.14), and a reward function $\tilde{\rho}$. A stochastic policy of the form $\tilde{h} : X \times U \to [0,1]$ is employed, parameterized by the vector $\vartheta \in \mathbb{R}^{\mathcal{N}}$. This policy takes an action u in state x with the probability:

$$\mathrm{P}(u \mid x) = \tilde{h}(x, u; \vartheta)$$

The functional dependence of the policy on the parameter vector must be designed in advance, and must be differentiable.

The *expected average return* of state x_0 under the policy parameterized by ϑ is:

$$R^{\vartheta}(x_0) = \lim_{K \to \infty} \frac{1}{K} \mathrm{E}_{\substack{u_k \sim \tilde{h}(x_k, \cdot; \vartheta) \\ x_{k+1} \sim \bar{f}(x_k, u_k, \cdot)}} \left\{ \sum_{k=0}^{K} \tilde{\rho}(x_k, u_k, x_{k+1}) \right\}$$

Note that we have directly highlighted the dependence of the return on the parameter vector ϑ, rather than on the policy $\tilde{h}$. A similar notation will be used for other policy-dependent quantities in this section.

Under certain conditions (see, e.g., Bertsekas, 2007, Chapter 4), the average return is the same for every initial state, i.e., $R^{\vartheta}(x_0) = \mathscr{R}^{\vartheta}$ for all $x_0 \in X$, and together with the so-called *differential* V-function, $V^{\vartheta} : X \to \mathbb{R}$, satisfies the Bellman equation:

$$\mathscr{R}^{\vartheta} + V^{\vartheta}(x_i) = \tilde{\rho}^{\vartheta}(x_i) + \sum_{i'=1}^{\bar{N}} \bar{f}^{\vartheta}(x_i, x_{i'}) V^{\vartheta}(x_{i'}) \tag{3.56}$$

The differential value of a state x can be interpreted as the expected excess return, on top of the average return, obtained from x (Konda and Tsitsiklis, 2003). The other quantities appearing in (3.56) are defined as follows:

- $\bar{f}^{\vartheta} : X \times X \to [0, 1]$ gives the state transition probabilities under the policy considered, from which the influence of the actions has been integrated out.[12] These probabilities can be computed with:

$$\bar{f}^{\vartheta}(x_i, x_{i'}) = \sum_{j=1}^{\bar{M}} \left[\tilde{h}(x_i, u_j; \vartheta) \bar{f}(x_i, u_j, x_{i'}) \right]$$

- $\tilde{\rho}^{\vartheta} : X \to \mathbb{R}$ gives the expected rewards obtained from every state by the policy considered, and can be computed with:

$$\tilde{\rho}^{\vartheta}(x_i) = \sum_{j=1}^{\bar{M}} \left[\tilde{h}(x_i, u_j; \vartheta) \sum_{i'=1}^{\bar{N}} \left(\bar{f}(x_i, u_j, x_{i'}) \tilde{\rho}(x_i, u_j, x_{i'}) \right) \right]$$

Policy gradient methods aim to find a (locally) optimal policy within the class of parameterized policies considered. An optimal policy maximizes the average return, which is the same for every initial state. So, a parameter vector that (locally)

[12]For simplicity, a slight abuse of notation is made by using $\bar{f}$ to denote both the original transition function and the transition probabilities from which the actions have been factored out. Similarly, the expected rewards are denoted by $\tilde{\rho}$, like the original reward function.

maximizes the average return must be found. To this end, policy gradient methods perform gradient ascent on the average return:

$$\vartheta \leftarrow \vartheta + \alpha \frac{\partial \mathscr{R}^{\vartheta}}{\partial \vartheta} \tag{3.57}$$

where α is the step size. When a local optimum has been reached, the gradient is zero, i.e., $\frac{\partial \mathscr{R}^{\vartheta}}{\partial \vartheta} = 0$.

The core problem is to estimate the gradient $\frac{\partial \mathscr{R}^{\vartheta}}{\partial \vartheta}$. By differentiating the Bellman equation (3.56) with respect to ϑ and after some calculations (see Bertsekas, 2007, Section 6.7), the following formula for the gradient is obtained:

$$\frac{\partial \mathscr{R}^{\vartheta}}{\partial \vartheta} = \sum_{i=1}^{\bar{N}} \zeta^{\vartheta}(x_i) \left[\frac{\partial \tilde{\rho}^{\vartheta}(x_i)}{\partial \vartheta} + \sum_{i'=1}^{\bar{N}} \left(\frac{\partial \bar{f}^{\vartheta}(x_i, x_{i'})}{\partial \vartheta} V^{\vartheta}(x_{i'}) \right) \right] \tag{3.58}$$

where $\zeta^{\vartheta}(x_i)$ is the steady-state probability of encountering the state x_i when using the policy given by ϑ. Note that all the gradients in (3.58) are $\mathcal{N}$-dimensional vectors.

The right-hand side of (3.58) can be estimated using simulation, as proposed, e.g., by Marbach and Tsitsiklis (2003), and the convergence of the resulting policy gradient algorithms to a locally optimal parameter vector can be ensured under mild conditions. An important concern is controlling the variance of the gradient estimate, and Marbach and Tsitsiklis (2003) focused on this problem. Munos (2006) considered policy gradient methods in the continuous-time setting. Because the usual methods to estimate the gradient lead to a variance that grows very large as the sampling time decreases, other methods are necessary to keep the variance small in the continuous-time case (Munos, 2006).

Actor-critic methods explicitly approximate the V-function in (3.58). This approximate V-function can be found, e.g., by using variants of the TD, LSTD, and LSPE techniques adapted to the average return setting (Bertsekas, 2007, Section 6.6).

The gradient can also be expressed in terms of a Q-function, which can be defined in the average return setting by using the differential V-function, as follows:

$$Q^{\vartheta}(x_i, u_j) = \sum_{i'=1}^{\bar{N}} \left[\bar{f}(x_i, u_j, x_{i'}) \left(\tilde{\rho}(x_i, u_j, x_{i'}) - \mathscr{R}^{\vartheta} + V^{\vartheta}(x_{i'}) \right) \right]$$

Using the Q-function, the gradient of the average return can be written as (Sutton et al., 2000; Konda and Tsitsiklis, 2000, 2003):

$$\frac{\partial \mathscr{R}^{\vartheta}}{\partial \vartheta} = \sum_{i=1}^{\bar{N}} \sum_{j=1}^{\bar{M}} \left[w^{\vartheta}(x_i, u_j) Q^{\vartheta}(x_i, u_j) \phi^{\vartheta}(x_i, u_j) \right] \tag{3.59}$$

where $w^{\vartheta}(x_i, u_j) = \zeta^{\vartheta}(x_i) \tilde{h}(x_i, u_j; \vartheta)$ is the steady-state probability of encountering the state-action pair (x_i, u_j) when using the policy considered, and:

$$\phi^{\vartheta} : X \times U \to \mathbb{R}^{\mathcal{N}}, \quad \phi^{\vartheta}(x_i, u_j) = \frac{1}{\tilde{h}(x_i, u_j; \vartheta)} \frac{\partial \tilde{h}(x_i, u_j; \vartheta)}{\partial \vartheta} \tag{3.60}$$

The function ϕ^{ϑ} is regarded as a *vector of state-action BFs*, for reasons that will become clear shortly. It can be shown that (3.59) is equal to (Sutton et al., 2000; Konda and Tsitsiklis, 2003):

$$\frac{\partial \mathscr{R}^{\vartheta}}{\partial \vartheta} = \sum_{i=1}^{\bar{N}} \sum_{j=1}^{\bar{M}} \left[w(x_i, u_j)[P^{w^{\vartheta}}(Q^{\vartheta})](x_i, u_j)\phi^{\vartheta}(x_i, u_j) \right]$$

where the exact Q-function has been substituted by its weighted least-squares projection (3.35) onto the space spanned by the BFs ϕ^{ϑ}. So, in order to find the *exact* gradient, it is sufficient to compute an *approximate* Q-function – provided that the BFs ϕ^{ϑ}, computed with (3.60) from the policy parametrization, are used. In the literature, such BFs are sometimes called "compatible" with the policy parametrization (Sutton et al., 2000) or "essential features" (Bertsekas, 2007, Section 6.7). Note that other BFs can be used in addition to these.

Using this property, actor-critic algorithms that linearly approximate the Q-function using the BFs (3.60) can be given. These algorithms converge to a locally optimal policy, as shown by Sutton et al. (2000); Konda and Tsitsiklis (2000, 2003). Konda and Tsitsiklis (2003) additionally extended their analysis to the case of continuous state-action spaces. This theoretical framework was used by Berenji and Vengerov (2003) to prove the convergence of an actor-critic algorithm relying on fuzzy approximation.

Kakade (2001) proposed an improvement to the gradient update formula (3.57), by scaling it with the inverse of the (expected) Fisher information matrix of the stochastic policy (Schervish, 1995, Section 2.3.1), and thereby obtaining the so-called natural policy gradient. Peters and Schaal (2008) and Bhatnagar et al. (2009) employed this idea to develop some natural actor-critic algorithms. Riedmiller et al. (2007) provided an experimental comparison of several policy gradient methods, including the natural policy gradient.

An online actor-critic algorithm for discounted returns

We now come back to the discounted return criterion for optimality, and describe an actor-critic algorithm for this discounted setting (rather than in the average-return setting, as above). This algorithm works online, in problems with continuous states and actions. Denote by $\widehat{h}(x;\vartheta)$ the (deterministic) approximate policy, parameterized by $\vartheta \in \mathbb{R}^{\mathscr{N}}$, and by $\widehat{V}(x;\theta)$ the approximate V-function, parameterized by $\theta \in \mathbb{R}^{N}$. The algorithm does not distinguish between the value functions of different policies, so the value function notation is not superscripted by the policy. Although a deterministic approximate policy is considered, a stochastic policy could also be used.

At each time step, an action u_k is chosen by adding a random, exploratory term to the action recommended by the policy $\widehat{h}(x;\vartheta)$. This term could be drawn, e.g., from a zero-mean Gaussian distribution. After the transition from x_k to x_{k+1}, an approximate temporal difference is computed with:

$$\delta_{\mathrm{TD},k} = r_{k+1} + \gamma \widehat{V}(x_{k+1};\theta_k) - \widehat{V}(x_k;\theta_k)$$

This temporal difference can be obtained from the Bellman equation for the policy

V-function (2.20). It is analogous to the temporal difference for Q-functions, used, e.g., in approximate SARSA (Algorithm 3.12). Once the temporal difference is computed, the policy and V-function parameters are updated with the following gradient formulas:

$$\vartheta_{k+1} = \vartheta_k + \alpha_{\mathrm{A},k} \frac{\partial \widehat{h}(x_k; \vartheta_k)}{\partial \vartheta} [u_k - \widehat{h}(x_k; \vartheta_k)] \, \delta_{\mathrm{TD},k} \tag{3.61}$$

$$\theta_{k+1} = \theta_k + \alpha_{\mathrm{C},k} \frac{\partial \widehat{V}(x_k; \theta_k)}{\partial \theta} \delta_{\mathrm{TD},k} \tag{3.62}$$

where $\alpha_{\mathrm{A},k}$ and $\alpha_{\mathrm{C},k}$ are the (possibly time-varying) step sizes for the actor and the critic, respectively. Note that the action signal is assumed to be scalar, but the method can be extended to multiple action variables.

In the actor update (3.61), due to exploration, the actual action u_k applied at step k can be different from the action recommended by the policy. When the exploratory action u_k leads to a positive temporal difference, the policy is adjusted towards this action. Conversely, when $\delta_{\mathrm{TD},k}$ is negative, the policy is adjusted away from u_k. This is because the temporal difference is interpreted as a correction of the predicted performance, so that, e.g., if the temporal difference is positive, the obtained performance is considered to be better than the predicted one. In the critic update (3.62), the temporal difference takes the place of the prediction error $V(x_k) - \widehat{V}(x_k; \theta_k)$, where $V(x_k)$ is the exact value of x_k, given the current policy. Since this exact value is not available, it is replaced by the estimate $r_{k+1} + \gamma \widehat{V}(x_{k+1}; \theta_k)$ suggested by the Bellman equation (2.20), thus leading to the temporal difference.

This actor-critic method is summarized in Algorithm 3.13, which generates exploratory actions using a Gaussian density with a standard deviation that can vary over time.

ALGORITHM 3.13 Actor-critic with Gaussian exploration.

Input: discount factor γ,
policy parametrization $\widehat{h}$, V-function parametrization $\widehat{V}$,
exploration schedule $\{\sigma_k\}_{k=0}^{\infty}$, step size schedules $\{\alpha_{\mathrm{A},k}\}_{k=0}^{\infty}$, $\{\alpha_{\mathrm{C},k}\}_{k=0}^{\infty}$

initialize parameter vectors, e.g., $\vartheta_0 \leftarrow 0$, $\theta_0 \leftarrow 0$
measure initial state x_0
for every time step $k = 0, 1, 2, \ldots$ **do**
  $u_k \leftarrow \widehat{h}(x_k; \vartheta_k) + \bar{u}$, where $\bar{u} \sim \mathcal{N}(0, \sigma_k)$
  apply u_k, measure next state x_{k+1} and reward r_{k+1}
  $\delta_{\mathrm{TD},k} = r_{k+1} + \gamma \widehat{V}(x_{k+1}; \theta_k) - \widehat{V}(x_k; \theta_k)$
  $\vartheta_{k+1} = \vartheta_k + \alpha_{\mathrm{A},k} \frac{\partial \widehat{h}(x_k; \vartheta_k)}{\partial \vartheta} [u_k - \widehat{h}(x_k; \vartheta_k)] \, \delta_{\mathrm{TD},k}$
  $\theta_{k+1} = \theta_k + \alpha_{\mathrm{C},k} \frac{\partial \widehat{V}(x_k; \theta_k)}{\partial \theta} \delta_{\mathrm{TD},k}$
end for

3.7.2 Gradient-free policy search

Gradient-based policy optimization is based on the assumption that the locally optimal parameters found by the gradient method are good enough. This may be true when the policy parametrization is simple and well suited to the problem at hand. However, in order to design such a parametrization, prior knowledge about a (near-)optimal policy is required.

When prior knowledge about the policy is not available, a richer policy parametrization must be used. In this case, the optimization criterion is likely to have many local optima, and may also be nondifferentiable. This means that gradient-based algorithms are unsuitable, and global, gradient-free optimization algorithms are required. Even when a simple policy parametrization can be designed, global optimization can help by avoiding local optima.

Consider the DP/RL problem under the expected discounted return criterion. Denote by $\widehat{h}(x;\vartheta)$ the approximate policy, parameterized by $\vartheta \in \mathbb{R}^{\mathcal{N}}$. Policy search algorithms look for an optimal parameter vector that maximizes the return $R^{\widehat{h}(\cdot;\vartheta)}(x)$ for all $x \in X$. When X is large or continuous, computing the return for every initial state is not possible. A practical procedure to circumvent this difficulty requires choosing a finite set X_0 of representative initial states. Returns are estimated only for the states in X_0, and the score function (optimization criterion) is the weighted average return over these states:

$$s(\vartheta) = \sum_{x_0 \in X_0} w(x_0) R^{\widehat{h}(\cdot;\vartheta)}(x_0) \tag{3.63}$$

where $w : X_0 \to (0,1]$ is the weight function.[13] The return from each representative state is estimated by simulation. A number of $N_{\text{MC}} \geq 1$ independent trajectories are simulated from every representative state, and an estimate of the expected return is obtained by averaging the returns obtained along these sample trajectories:

$$R^{\widehat{h}(\cdot;\vartheta)}(x_0) = \frac{1}{N_{\text{MC}}} \sum_{i_0=1}^{N_{\text{MC}}} \sum_{k=0}^{K} \gamma^k \tilde{\rho}(x_{i_0,k}, h(x_{i_0,k};\vartheta), x_{i_0,k+1}) \tag{3.64}$$

For each trajectory i_0, the initial state $x_{i_0,0}$ is equal to x_0, and actions are chosen with the policy h, which means that for $k \geq 0$:

$$x_{i_0,k+1} \sim f(x_{i_0,k}, h(x_{i_0,k};\vartheta), \cdot)$$

If the system is deterministic, a single trajectory suffices, i.e., $N_{\text{MC}} = 1$. In the stochastic case, a good value for N_{MC} will depend on the problem at hand. Note that this Monte Carlo estimation procedure is similar to a rollout (3.45).

The infinite-horizon return is approximated by truncating each simulated trajectory after K steps. A value of K that guarantees that this truncation introduces an

[13] More generally, a density $\tilde{w}$ over the initial states can be considered, and the score function is then $\mathrm{E}_{x_0 \sim \tilde{w}(\cdot)}\left\{R^{h(\cdot;\xi,\vartheta)}(x_0)\right\}$, i.e., the expected value of the return when $x_0 \sim \tilde{w}(\cdot)$.

error of at most $\varepsilon_{MC} > 0$ can be chosen using (2.41), repeated here:

$$K = \left\lceil \log_\gamma \frac{\varepsilon_{MC}(1-\gamma)}{\|\tilde{\rho}\|_\infty} \right\rceil \tag{3.65}$$

In the stochastic context, Ng and Jordan (2000) assumed the availability of a simulation model that offers access to the random variables driving the stochastic transitions. They proposed to pregenerate sequences of values for these random variables, and to use the same sequences when evaluating every policy. This leads to a deterministic optimization problem.

Representative set of initial states and weight function. The set X_0 of representative states, together with the weight function w, determines the performance of the resulting policy. Of course, this performance is in general only approximately optimal, since maximizing the returns from states in X_0 cannot guarantee that returns from other states in X are maximal. A good choice of X_0 and w will depend on the problem at hand. For instance, if the process only needs to be controlled starting from a known set X_{init} of initial states, then X_0 should be equal to X_{init}, or included in it when X_{init} is too large. Initial states that are deemed more important can be assigned larger weights. When all initial states are equally important, the elements of X_0 should be uniformly spread over the state space and identical weights equal to $\frac{1}{|X_0|}$ should be assigned to every element of X_0.

A wide range of gradient-free, global optimization techniques can be employed in policy search, including evolutionary optimization (e.g., genetic algorithms, see Goldberg, 1989), tabu search (Glover and Laguna, 1997), pattern search (Torczon, 1997; Lewis and Torczon, 2000), the cross-entropy method (Rubinstein and Kroese, 2004), etc. For instance, evolutionary computation was applied to policy search by Barash (1999); Chin and Jafari (1998); Gomez et al. (2006); Chang et al. (2007, Chapter 3), and cross-entropy optimization was applied by Mannor et al. (2003). Chang et al. (2007, Chapter 4) described an approach to find a policy by using the so-called "model-reference adaptive search," which is closely related to the cross-entropy method. In Chapter 6 of this book, we will employ the cross-entropy method to develop a policy search algorithm. A dedicated algorithm that optimizes the parameters and structure of neural network policy approximators was given by Whiteson and Stone (2006). General policy modification heuristics were proposed by Schmidhuber (2000).

In another class of model-based policy search approaches, near-optimal actions are sought online, by executing at every time step a search over open-loop sequences of actions (Hren and Munos, 2008). The controller selects a sequence leading to a maximal estimated return and applies the first action in this sequence. Then, the entire cycle repeats.[14] The total number of open-loop action sequences grows exponentially with the time horizon considered, but by limiting the search to promising sequences only, such an approach can avoid incurring excessive computational costs.

[14]This is very similar to how model-predictive control works (Maciejowski, 2002; Camacho and Bordons, 2004).

Hren and Munos (2008) studied this method of limiting the computational cost in a deterministic setting. In a stochastic setting, open-loop sequences are suboptimal. However, some approaches exist to extend this open-loop philosophy to the stochastic case. These approaches model the sequences of random transitions by scenario trees (Birge and Louveaux, 1997; Dupacová et al., 2000) and optimize the actions attached to the tree nodes (Defourny et al., 2008, 2009).

3.7.3 Example: Gradient-free policy search for a DC motor

In this example, approximate, gradient-free policy search is applied to the DC motor problem introduced in Section 3.4.5. In a first experiment, a general policy parametrization is used that does not rely on prior knowledge, whereas in a second experiment, a tailored policy parametrization is derived from prior knowledge. The results obtained with these two parametrizations are compared.

To compute the score function (3.63), a set X_0 of representative states and a weight function w have to be selected. We aim to obtain a uniform performance across the state space, so a regular grid of representative states is chosen:

$$X_0 = \{-\pi, -2\pi/3, -\pi/3, \dots, \pi\} \times \{-16\pi, -12\pi, -8\pi, \dots, 16\pi\}$$

and these initial states are weighted uniformly by $w(x_0) = \frac{1}{|X_0|}$, where the number of states is $|X_0| = 63$. A maximum error $\varepsilon_{\text{MC}} = 0.01$ is imposed in the estimation of the return. A bound on the reward function (3.28) for the DC motor problem can be computed with:

$$\begin{aligned}
\|\rho\|_\infty &= \sup_{x,u} \left| -x_k^{\mathrm{T}} Q_{\text{rew}} x_k - R_{\text{rew}} u_k^2 \right| \\
&= \left| -[\pi \;\, 16\pi] \begin{bmatrix} 5 & 0 \\ 0 & 0.01 \end{bmatrix} \begin{bmatrix} \pi \\ 16\pi \end{bmatrix} - 0.01 \cdot 10^2 \right| \\
&\approx 75.61
\end{aligned}$$

To find the trajectory length K required to achieve the precision ε_{MC}, the values of ε_{MC}, $\|\rho\|_\infty$, and $\gamma = 0.95$ are substituted into (3.65); this yields $K = 233$. Because the problem is deterministic, simulating multiple trajectories from every initial state is not necessary; instead, a single trajectory from every initial state will suffice.

We use the global, gradient-free pattern search algorithm to optimize the policy (Torczon, 1997; Lewis and Torczon, 2000). The algorithm is considered convergent when the score variation decreases below the threshold $\varepsilon_{\text{PS}} = 0.01$ (equal to ε_{MC}).[15]

Policy search with a general parametrization

Consider first the case in which no prior knowledge about the optimal policy is available, which means that a general policy parametrization must be used. The linear

[15]We use the pattern search algorithm from the *Genetic Algorithm and Direct Search Toolbox* of MATLAB 7.4.0. The algorithm is configured to use the threshold ε_{PS} and to cache the score values for the parameter vectors it already evaluated, in order to avoid recomputing them. Besides these changes, the default settings of the algorithm are employed.

policy parametrization (3.12) is chosen:

$$\widehat{h}(x) = \sum_{i=1}^{\mathscr{N}} \varphi_i(x)\vartheta_i = \varphi^{\mathrm{T}}(x)\vartheta$$

Axis-aligned, normalized RBFs (see Example 3.1) are defined, with their centers arranged on an equidistant 7×7 grid in the state space. All the RBFs are identical in shape, and their width b_d along each dimension d is equal to $b_d'^2/2$, where b_d' is the distance between adjacent RBFs along that dimension (the grid step). Namely, $b_1' = \frac{2\pi}{7-1} \approx 1.05$ and $b_2' = \frac{32\pi}{7-1} \approx 16.76$, which lead to $b_1 \approx 0.55$ and $b_2 \approx 140.37$. In total, 49 parameters (for 7×7 RBFs) must be optimized.

Pattern search optimization is applied to find an optimal parameter vector ϑ^*, starting from an identically zero parameter vector. Figure 3.16 shows the policy obtained and a representative trajectory that is controlled by this policy. The policy is largely linear in the state variables (within the saturation limits), and leads to a good convergence to the zero state.

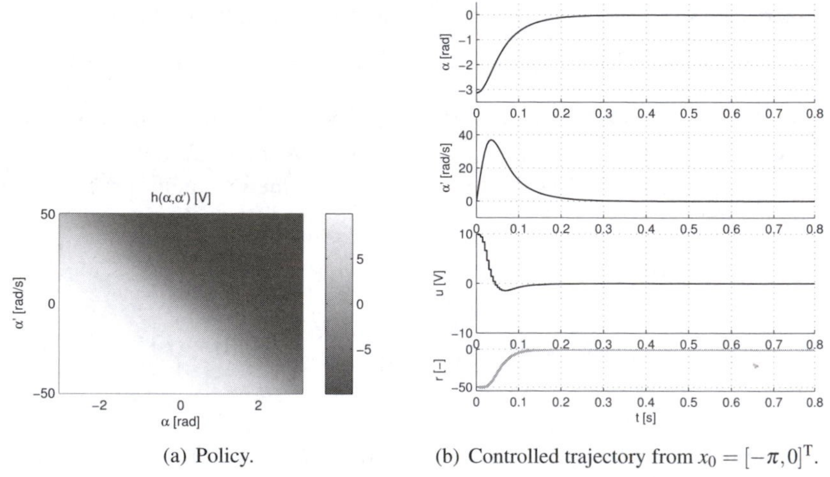

(a) Policy. (b) Controlled trajectory from $x_0 = [-\pi, 0]^{\mathrm{T}}$.

FIGURE 3.16
Results of policy search with the general policy parametrization for the DC motor.

In this experiment, the pattern search algorithm required 18173 s to converge. This execution time is larger than for all other algorithms applied earlier to the DC motor (grid Q-iteration and fitted Q-iteration in Section 3.4.5, and LSPI in Section 3.5.7), illustrating the large computational demands of policy search with general parametrizations.

Policy search spends the majority of its execution time estimating the score function (3.63), which is a computationally expensive operation. For this experiment, the score of 11440 different parameter vectors had to be computed until convergence. The computational cost of evaluating each parameter vector can be decreased by taking a smaller X_0 or larger $\varepsilon_{\mathrm{MC}}$ and $\varepsilon_{\mathrm{PS}}$, at the expense of a possible decrease in control performance.

Policy search with a tailored parametrization

In this second part of the example, we employ a simple policy parametrization that is well suited to the DC motor problem. This parametrization is derived by using prior knowledge. Because the system is linear and the reward function is quadratic, the optimal policy would be a linear state feedback if the constraints on the state and action variables were disregarded (Bertsekas, 2007, Section 3.2).[16] Now taking into account the constraints on the action, we assume that a good approximation of an optimal policy is linear in the state variables, within the constraints on the action:

$$\widehat{h}(x;\vartheta) = \mathrm{sat}\{\vartheta_1 x_1 + \vartheta_2 x_2, -10, 10\} \tag{3.66}$$

where "sat" denotes saturation. In fact, an examination of the near-optimal policy in Figure 3.5(b) on page 67 reveals that this assumption is largely correct: the only nonlinearities appear in the top-left and bottom-right corners of the figure; they are probably due to the constraints on the state variables, which were not taken into account when deriving the parametrization (3.66). We employ this tailored parametrization to perform policy search. Note that only 2 parameters must be optimized, significantly fewer than the 49 parameters required by the general parametrization used earlier.

Figure 3.17 shows the policy obtained by pattern search optimization, together with a representative controlled trajectory. As expected, the policy closely resembles the near-optimal policy of Figure 3.5(b), with the exception of the nonlinearities in the corners of the state space. The trajectory obtained is also close to the near-optimal one in Figure 3.5(c). Compared to the general-parametrization solution of Figure 3.16, the policy varies more quickly in the linear portion, which results in a more aggressive control signal. This is because the tailored parametrization can lead to a large slope of the policy, whereas the wide RBFs used in Figure 3.16 lead to a smoother interpolation. The score obtained by the policy of Figure 3.17 is -229.25, slightly better than the score of -230.69 obtained by the RBF policy of Figure 3.16.

The execution time of pattern search with the tailored parametrization was approximately 75 s. As expected, the computational cost is much smaller than for the general parametrization, because only 2 parameters must be optimized, instead of 49. This illustrates the benefits of using a compact policy parametrization that is appropriate for the problem at hand. Unfortunately, deriving an appropriate parametrization requires prior knowledge, which is not always available. The execution time is larger than that of grid Q-iteration in Section 3.4.5, which was 7.80 s for the fine grid and 0.06 s for the coarse grid. It has the same order of magnitude as the execution time of LSPI in Section 3.5.7, which was 23 s when using exact policy improvements, and 58 s with approximate policy improvements; but it is smaller than the execution

[16] This optimal linear state feedback is given by:

$$h(x) = Kx = -\gamma(\gamma B^{\mathrm{T}} Y B + R_{\mathrm{rew}})^{-1} B^{\mathrm{T}} Y A x$$

where Y is the stabilizing solution of the Riccati equation:

$$Y = A^{\mathrm{T}}[\gamma Y - \gamma^2 Y B(\gamma B^{\mathrm{T}} Y B + R_{\mathrm{rew}})^{-1} B^{\mathrm{T}}]A + Q_{\mathrm{rew}}$$

Substituting A, B, Q_{rew}, R_{rew}, and γ in these equations leads to a state feedback gain of $K \approx [-11.16, -0.67]^{\mathrm{T}}$ for the DC motor.

time 2151 s of fitted Q-iteration. To enable an easy comparison of all these execution times, they are collected in Table 3.1.[17]

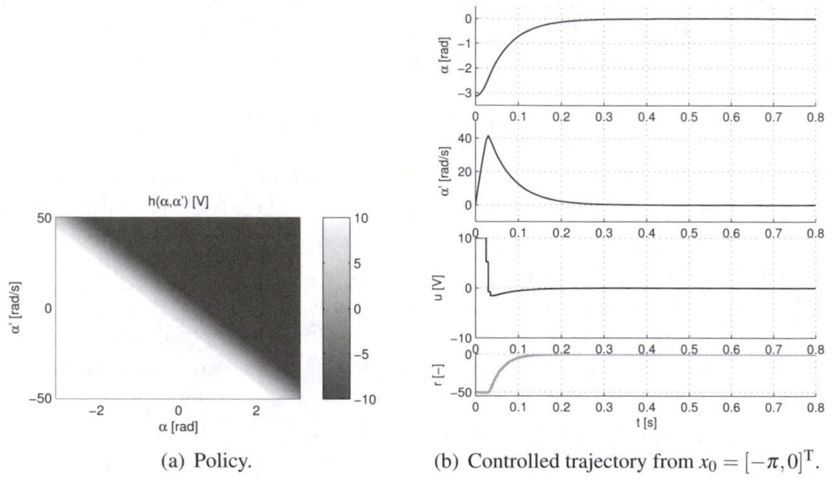

(a) Policy. (b) Controlled trajectory from $x_0 = [-\pi, 0]^T$.

FIGURE 3.17
Results of policy search with the tailored policy parametrization (3.66) on the DC motor. The policy parameter is $\widehat{\vartheta}^* = [-16.69, -1]^T$.

TABLE 3.1
Execution time of approximate DP and RL algorithms for the DC motor problem.

Algorithm	**Execution time** [s]
grid Q-iteration with a coarse grid	0.06
grid Q-iteration with a fine grid	7.80
fitted Q-iteration	2151
LSPI with exact policy improvement	23
LSPI with exact policy approximation	58
policy search with a general parametrization	18173
policy search with a tailored parametrization	75

[17]Recall that all these execution times were recorded on a PC with an Intel Core 2 Duo T9550 2.66 GHz CPU and with 3 GB RAM.

3.8 Comparison of approximate value iteration, policy iteration, and policy search

This section provides a general, qualitative comparison of approximate value iteration, approximate policy iteration, and approximate policy search. A more specific comparison would of course depend on the particular algorithms considered and on the problem at hand.

Approximate value iteration versus approximate policy iteration

Offline approximate policy iteration often converges in a small number of iterations, possibly smaller than the number of iterations taken by offline approximate value iteration. This was illustrated for the DC motor example, in which LSPI (Section 3.5.7) converged faster than grid Q-iteration (Section 3.4.5). However, this does not mean that approximate policy iteration is computationally less demanding than approximate value iteration, since approximate policy evaluation is a difficult problem by itself, which must be solved at every single policy iteration. One advantage of approximate value iteration is that it usually guarantees convergence to a unique solution, whereas approximate policy iteration is generally only guaranteed to converge to a sequence of policies that all provide a guaranteed level of performance. This was illustrated in Section 3.5.7, where LSPI with policy approximation converged to a limit cycle.

Consider now the approximate policy evaluation step of policy iteration, in comparison to approximate value iteration. Some approximate policy evaluation algorithms closely parallel approximate value iteration and converge under similar conditions (Section 3.5.1). However, approximate policy evaluation can additionally benefit from the linearity of the Bellman equation for a policy's value function, e.g., (2.7), whereas the Bellman optimality equation, which characterizes the optimal value function, e.g., (2.8), is highly nonlinear due to the maximization in the right-hand side. A class of algorithms for approximate policy evaluation exploit this linearity property by solving a projected form of the Bellman equation (Section 3.5.2). One advantage of such algorithms is that they only require the approximator to be linearly parameterized, whereas in approximate value iteration the approximator must lead to contracting updates (Section 3.4.4). Moreover, some of these algorithms, such as LSTD-Q and LSPE-Q, are highly sample-efficient. However, a disadvantage of these algorithms is that their convergence guarantees typically require a sample distribution identical with the steady-state distribution under the policy being evaluated.

Approximate policy search versus approximate value iteration and policy iteration

For some problems, deriving a good policy parametrization using prior knowledge may be easier and more natural than deriving a good value function parametrization. If a good policy parametrization is available and this parametrization is differentiable, policy gradient algorithms can be used (Section 3.7.1). Such algorithms are backed by useful convergence guarantees and have moderate computational demands. Policy gradient algorithms have the disadvantage that they can only find local optima in the class of parameterized policies considered, and may also suffer from slow convergence.

Note that the difficulty of designing a good value function parametrization can be alleviated either by automatically finding the parametric approximator (Section 3.6) or by using nonparametric approximators. Both of these options require less tuning than a predefined parametric approximator, but may increase the computational demands of the algorithm.

Even when prior knowledge is not available and a good policy parametrization cannot be obtained, approximate policy search can still be useful in its gradient-free forms, which do not employ value functions (Section 3.7.2). One situation in which value functions are undesirable is when value-function based algorithms fail to obtain a good solution, or require too restrictive assumptions. In such situations, a general policy parametrization can be defined, and a global, gradient-free optimization technique can be used to search for optimal parameters. These techniques are usually free from numerical problems – such as divergence to infinity – even when used with general nonlinear parametrizations, which is not the case for value and policy iteration. However, because of its generality, this approach typically incurs large computational costs.

3.9 Summary and discussion

In this chapter, we have introduced approximate dynamic programming (DP) and approximate reinforcement learning (RL) for large or continuous-space problems. After explaining the need for approximation in such problems, parametric and nonparametric approximation architectures have been presented. Then, approximate versions for the three main categories of algorithms have been described: value iteration, policy iteration, and policy search. Theoretical results have been provided and the behavior of representative algorithms has been illustrated using numerical examples. Additionally, techniques to automatically determine value function approximators have been reviewed, and the three categories of algorithms have been compared. Extensive accounts of approximate DP and RL, presented from different perspectives, can also be found in the books of Bertsekas and Tsitsiklis (1996); Powell (2007); Chang et al. (2007); Cao (2007).

Approximate DP/RL is a young, but active and rapidly expanding, field of research. Important challenges still remain to be overcome in this field, some of which are pointed out next.

When the problem considered is high-dimensional and prior knowledge is not available, it is very difficult to design a good parametrization that does not lead to excessive computational costs. An additional, related difficulty arises in the model-free (RL) setting, when only a limited amount of data is available. In this case, if the approximator is too complex, the data may be insufficient to compute its parameters. One alternative to designing the approximator in advance is to find a good parametrization automatically, while another option is to exploit the powerful framework of nonparametric approximators, which can also be viewed as deriving a parametrization from the data. Adaptive and nonparametric approximators are often studied in the context of value iteration and policy iteration (Sections 3.4.3, 3.5.3, and 3.6). In policy search, finding good approximators automatically is a comparatively underexplored but promising idea.

Actions that take continuous values are important in many problems of practical interest. For instance, in the context of automatic control, stabilizing a system around an unstable equilibrium requires continuous actions to avoid chattering, which would otherwise damage the system in the long run. However, in DP and RL, continuous-action problems are more rarely studied than discrete-action problems. A major difficulty of value iteration and policy iteration in the continuous-action case is that they rely on solving many potentially difficult, nonconcave maximization problems over the action variables (Section 3.2). Continuous actions are easier to handle in actor-critic and policy search algorithms, in the sense that explicit maximization over the action variables is not necessary.

Theoretical results about approximate value iteration traditionally rely on the requirement of nonexpansive approximation. To satisfy this requirement, the approximators are often confined to restricted subclasses of linear parameterizations. Analyzing approximate value iteration without assuming nonexpansiveness can be very beneficial, e.g., by allowing powerful nonlinearly parameterized approximators, which may alleviate the difficulties of designing a good parametrization in advance. The work on finite-sample performance guarantees, outlined in Section 3.4.4, provides encouraging results in this direction.

In the context of approximate policy iteration, least-squares techniques for policy evaluation are very promising, owing to their sample efficiency and ease of tuning. However, currently available performance guarantees for these algorithms require that they process relatively many samples generated using a fixed policy. From a learning perspective, it would be very useful to analyze how these techniques behave in online, optimistic policy iteration, in which the policy is not kept fixed for a long time, but is improved once every few samples. Promising empirical results have been reported using such algorithms, but their theoretical understanding is still limited (see Section 3.5.6).

The material in this chapter provides a broad understanding of approximate value iteration, policy iteration, and policy search. In order to deepen and strengthen this understanding, in each of the upcoming three chapters we treat in detail a particular algorithm from one of these three classes. Namely, in Chapter 4, a model-based value iteration algorithm with fuzzy approximation is introduced, theoretically analyzed, and experimentally evaluated. The theoretical analysis illustrates how convergence and consistency guarantees can be developed for approximate DP. In Chapter 5, least-squares policy iteration is revisited, and several extensions to this algorithm are introduced and empirically studied. In particular, an online variant is developed, and some important issues that appear in online RL are emphasized along the way. In Chapter 6, a policy search approach relying on the gradient-free cross-entropy method for optimization is described and experimentally evaluated. This approach highlights one possibility for developing techniques that scale better to high-dimensional state spaces, by focusing the computation only on important initial states.

4

Approximate value iteration with a fuzzy representation

This chapter introduces fuzzy Q-iteration, an algorithm for approximate value iteration that relies on a fuzzy representation of the Q-function. This representation combines a fuzzy partition defined over the state space with a discretization of the action space. The convergence and consistency of fuzzy Q-iteration are analyzed. As an alternative to designing the membership functions for the fuzzy partition in advance, a technique to optimize the membership functions using the cross-entropy method is described. The performance of fuzzy Q-iteration is evaluated in an extensive experimental study.

4.1 Introduction

Value iteration algorithms (introduced in Section 2.3) search for the optimal value function, and then employ a policy that is greedy in this value function to control the system. In large or continuous spaces, the value function must be approximated, leading to approximate value iteration, which was introduced in Section 3.4.

In this chapter, we design and study in detail an algorithm for approximate value iteration, building on – and at the same time adding depth to – the knowledge gained in the previous chapters. We exploit the fuzzy approximation paradigm (Fantuzzi and Rovatti, 1996) to develop *fuzzy Q-iteration*: an algorithm for approximate Q-iteration that represents Q-functions using a fuzzy partition of the state space and a discretization of the action space. Fuzzy Q-iteration requires a model and works for problems with deterministic dynamics. The fuzzy sets in the partition are described by membership functions (MFs), and the discrete actions are selected beforehand from the (possibly large or continuous) original action space. The Q-value of a given state-discrete action pair is computed as a weighted sum of parameters, where the weights are given by the MFs. The fuzzy representation can therefore also be seen as a linearly parameterized approximator, and in this context the MFs are state-dependent basis functions.

In addition to the fuzzy approximator, an important new development in this chapter is a variant of fuzzy Q-iteration that works asynchronously, by employing the most recently updated values of the parameters at each step of the computation.

This variant is called *asynchronous fuzzy Q-iteration*. The original algorithm, which keeps the parameters unchanged while performing the computations of the current iteration, is called *synchronous fuzzy Q-iteration*, in order to differentiate it from the asynchronous variant. For the sake of conciseness, the name "fuzzy Q-iteration" is used to refer collectively to both of these variants; e.g., from the statement "fuzzy Q-iteration converges," it should be understood that both the asynchronous and synchronous variants are convergent. Whenever the distinction between the two variants is important, we use the "synchronous" and "asynchronous" qualifiers.

Two desirable properties of algorithms for approximate value iteration are convergence to a near-optimal value function and consistency. Consistency means the asymptotical convergence to the optimal value function as the approximation accuracy increases. By using the theoretical framework of nonexpansive approximators developed in Section 3.4.4, and by extending this framework to handle the asynchronous case, we show that fuzzy Q-iteration asymptotically converges to a fixed point. This fixed point corresponds to an approximate Q-function that lies within a bounded distance from the optimal Q-function; moreover, the suboptimality of the Q-function obtained after a finite number of iterations is also bounded. Both of these Q-functions lead to greedy policies with a bounded suboptimality. Additionally, in a certain sense, the asynchronous algorithm converges at least as fast as the synchronous one. In a second part of our analysis, we also show that fuzzy Q-iteration is consistent: under appropriate continuity assumptions on the process dynamics and on the reward function, the approximate Q-function converges to the optimal one as the approximation accuracy increases.

The accuracy of the solution found by fuzzy Q-iteration crucially depends on the MFs. In its original form, fuzzy Q-iteration requires the MFs to be designed beforehand. Either prior knowledge about the optimal Q-function is required to design good MFs, or many MFs must be defined to provide a good coverage and resolution over the entire state space. Neither of these approaches always works well. As an alternative to designing the MFs in advance, we consider a method to optimize the location and shape of a fixed number of MFs. This method belongs to the class of approximator optimization techniques introduced in Section 3.6.1. To evaluate each configuration of the MFs, a policy is computed with fuzzy Q-iteration using these MFs, and the performance of this policy is estimated by simulation. Using the cross-entropy method for optimization, we design an algorithm to optimize triangular MFs.

The theoretical analysis of fuzzy Q-iteration provides confidence in its results. We complement this analysis with an extensive numerical and experimental study, which is organized in four parts, each focusing on different aspects relevant to the practical application of the algorithm. The first example illustrates the convergence and consistency of fuzzy Q-iteration, using a DC motor problem. The second example employs a two-link manipulator to demonstrate the effects of interpolating the actions, and also to compare fuzzy Q-iteration with fitted Q-iteration (Algorithm 3.4). In the third example, the real-life control performance of fuzzy Q-iteration is illustrated using an inverted pendulum swing-up problem. For these three examples, the MFs are designed in advance. In the fourth and final example, the effects of optimizing the MFs are studied in the classical car-on-the-hill benchmark (Moore and Atkeson, 1995; Munos and Moore, 2002; Ernst et al., 2005).

Next, Section 4.2 describes fuzzy Q-iteration. In Section 4.3, the convergence, consistency, and computational demands of fuzzy Q-iteration are analyzed, while Section 4.4 presents our approach to optimize the MFs. Section 4.5 describes the experimental evaluation outlined above, and Section 4.6 closes the chapter with a summary and discussion.

4.2 Fuzzy Q-iteration

Fuzzy Q-iteration belongs to the class of value iteration algorithms with parametric approximation (Section 3.4.1). Similarly to other algorithms in this class, it works by combining the Q-iteration mapping (2.22) with an approximation mapping and a projection mapping. After introducing the fuzzy approximation mapping and the projection mapping used by fuzzy Q-iteration, the two versions of the algorithm are described, namely synchronous and asynchronous fuzzy Q-iteration.

4.2.1 Approximation and projection mappings of fuzzy Q-iteration

Consider a deterministic Markov decision process (MDP) (see Section 2.2.1). The state space X and the action space U of the MDP may be either continuous or discrete, but they are assumed to be subsets of Euclidean spaces, such that the Euclidean norm of the states and actions is well-defined.

Fuzzy approximation

The proposed approximator relies on a fuzzy partition of the state space and on a discretization of the action space. The fuzzy partition of X contains N fuzzy sets χ_i, each described by a membership function (MF):

$$\mu_i : X \rightarrow [0, 1]$$

where $i = 1, \ldots, N$. A state x then belongs to each set i with a degree of membership $\mu_i(x)$. The following requirement is imposed on the MFs:

Requirement 4.1 *Each MF has its maximum at a single point, i.e., for every i there exists a unique x_i for which $\mu_i(x_i) > \mu_i(x) \;\; \forall x \neq x_i$. Additionally, the other MFs take zero values at x_i, i.e., $\mu_{i'}(x_i) = 0 \;\; \forall i' \neq i$.*

This requirement will be useful later on for obtaining a projection mapping that helps with the convergence of fuzzy Q-iteration, as well as for proving consistency. Because the other MFs take zero values in x_i, it can be assumed without loss of generality that $\mu_i(x_i) = 1$, and hence μ_i is normal. The state x_i is then called the *core* of the ith MF.

Example 4.1 Triangular fuzzy partitions. A simple type of fuzzy partition that satisfies Requirement 4.1 can be obtained as follows. For each state variable x_d, where $d \in \{1, \dots, D\}$ and $D = \dim(X)$, a number N_d of triangular MFs are defined as follows:

$$\phi_{d,1}(x_d) = \max\left(0, \frac{c_{d,2} - x_d}{c_{d,2} - c_{d,1}}\right)$$

$$\phi_{d,i}(x_d) = \max\left[0, \min\left(\frac{x_d - c_{d,i-1}}{c_{d,i} - c_{d,i-1}}, \frac{c_{d,i+1} - x_d}{c_{d,i+1} - c_{d,i}}\right)\right], \quad \text{for } i = 2, \dots, N_d - 1$$

$$\phi_{d,N_d}(x_d) = \max\left(0, \frac{x_d - c_{d,N_d-1}}{c_{d,N_d} - c_{d,N_d-1}}\right)$$

where $c_{d,1}, \dots, c_{d,N_d}$ are the cores along dimension d and must satisfy $c_{d,1} < \dots < c_{d,N_d}$. These cores fully determine the shape of the MFs. The state space should be contained in the support of the MFs, i.e., $x_d \in [c_{d,1}, c_{d,N_d}]$ for $d = 1, \dots, D$. Adjacent single-dimensional MFs always intersect at a level of 0.5. The product of each combination of single-dimensional MFs thus gives a pyramid-shaped D-dimensional MF in the fuzzy partition of X. Examples of single-dimensional and two-dimensional triangular partitions are given in Figure 4.1. □

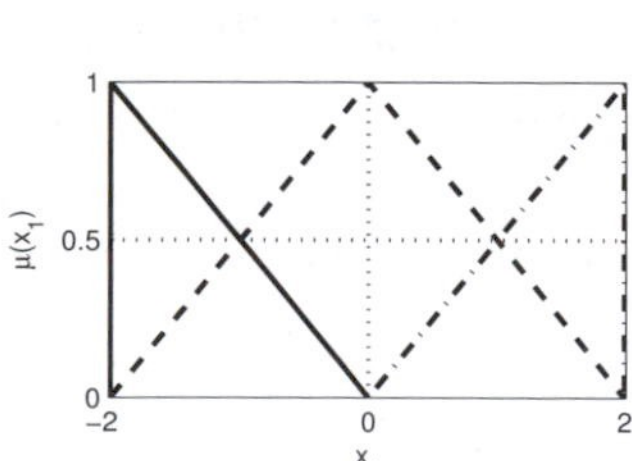

(a) A set of single-dimensional triangular MFs, each shown in a different line style.

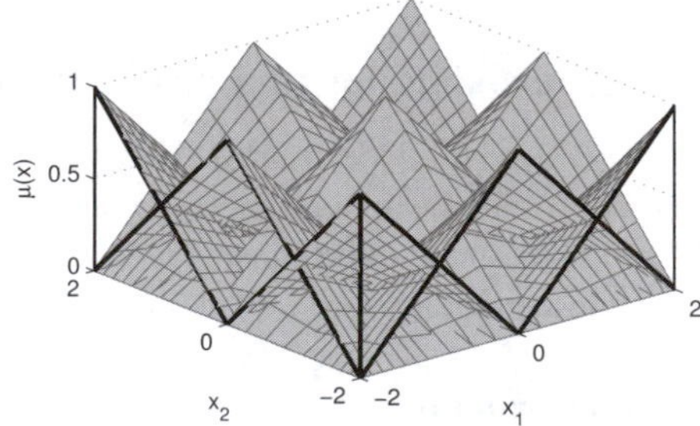

(b) Two-dimensional MFs, obtained by combining two sets of single-dimensional MFs, each identical to the set in Figure 4.1(a).

FIGURE 4.1 Examples of triangular MFs.

Other types of MFs that satisfy Requirement 4.1 can be obtained, e.g., by using higher-order B-splines (Brown and Harris, 1994, Ch. 8) (triangular MFs are second-order B-splines), or Kuhn triangulations combined with barycentric interpolation (Munos and Moore, 2002; Abonyi et al., 2001). Kuhn triangulations can lead to a smaller number of MFs than triangular or B-spline partitions; in the latter types of partitions, the number of MFs grows exponentially with the dimensionality of the state space. Although fuzzy Q-iteration is not limited to triangular MFs, these will nevertheless be used in the examples, because they are the simplest MFs that satisfy the requirements for the convergence and consistency of the algorithm.

Requirement 4.1 can be relaxed so that other MFs can take nonzero values at the core x_i of a given MF i. If these values are sufficiently small, fuzzy Q-iteration can still be proven to converge to a near-optimal solution by extending the results of Tsitsiklis and Van Roy (1996). This relaxation allows other types of localized MFs, such as Gaussian MFs. Note that, in practice, fuzzy Q-iteration can indeed diverge when the other MFs have too large values at x_i.

Until now, the approximation over the state space was discussed. To approximate over the (continuous or discrete) action space U, a discrete subset of actions U_{d} is chosen:

$$U_{\mathrm{d}} = \{u_j | u_j \in U, j = 1, \ldots, M\} \tag{4.1}$$

The fuzzy approximator stores a parameter vector θ with $n = NM$ elements. Each parameter $\theta_{[i,j]}$ corresponds to the MF-discrete action pair (μ_i, u_j), where $[i, j] = i + (j-1)N$ denotes the scalar index corresponding to i and j. To compute the Q-value of the state-action pair (x, u), first the action u is discretized by selecting a discrete action $u_j \in U_{\mathrm{d}}$ that is closest to u:

$$j \in \arg\min_{j'} \|u - u_{j'}\|_2$$

where $\|\cdot\|_2$ denotes the Euclidean norm of the argument. Then, the approximate Q-value is computed as a weighted sum of the parameters $\theta_{[1,j]}, \ldots, \theta_{[N,j]}$:

$$\widehat{Q}(x, u) = \sum_{i=1}^{N} \phi_i(x) \theta_{[i,j]}$$

where the weights $\phi_i(x)$ are the normalized MFs (degrees of fulfillment):[1]

$$\phi_i(x) = \frac{\mu_i(x)}{\sum_{i'=1}^{N} \mu_{i'}(x)} \tag{4.2}$$

This entire procedure can be written concisely as the following *approximation mapping*:

$$\widehat{Q}(x, u) = [F(\theta)](x, u) = \sum_{i=1}^{N} \phi_i(x) \theta_{[i,j]} \quad \text{where } j \in \arg\min_{j'} \|u - u_{j'}\|_2 \tag{4.3}$$

To ensure that $F(\theta)$ is a well-defined function, any ties in the minimization from (4.3) have to be broken consistently. In the sequel we assume that they are broken in favor of the smallest index that satisfies the condition. For a fixed x, such an approximator is constant over each subset of actions U_j, $j = 1, \ldots, M$, defined by:

$$u \in U_j \text{ if } \begin{cases} \|u - u_j\|_2 \leq \|u - u_{j'}\|_2 \text{ for all } j' \neq j, \text{ and:} \\ j < j' \text{ for any } j' \neq j \text{ such that } \|u - u_j\|_2 = \|u - u_{j'}\|_2 \end{cases} \tag{4.4}$$

[1]The MFs are already normal (see the discussion after Requirement 4.1), but they may not yet be *normalized*, because their sum may be different from 1 for some values of x. The sum of normalized MFs must be 1 for any value of x.

where the second condition is due to the manner in which ties are broken. The sets U_j form a partition of U.

Note that (4.3) describes a linearly parameterized approximator (3.3). More specifically, the fuzzy approximator is closely related to the discrete-action, linearly parameterized approximators introduced in Example 3.1. It extends these approximators by introducing an explicit action discretization procedure. In this context, the normalized MFs can be seen as state-dependent basis functions or features (Bertsekas and Tsitsiklis, 1996).

Interpretation of the approximator as a fuzzy rule base

We next provide an interpretation of the Q-function approximator as the output of a fuzzy rule base (Kruse et al., 1994; Klir and Yuan, 1995; Yen and Langari, 1999). Consider a so-called Takagi-Sugeno fuzzy rule base (Takagi and Sugeno, 1985; Kruse et al., 1994, Section 4.2.2), which describes the relationship between inputs and outputs using if-then rules of the form:

$$R_i : \textbf{ if } x \textbf{ is } \chi_i \textbf{ then } y = g_i(x) \tag{4.5}$$

where $i \in 1,\dots,N$ is the index of the rule, $x \in X$ is the input variable (which for now does not need to be the state of an MDP), $\chi_1,\dots,\chi_N$ are the input fuzzy sets, $y \in Y$ is the output variable, and $g_1,\dots,g_N : X \to Y$ are the (algebraic) output functions. The input and output variables can be scalars or vectors. Each fuzzy set χ_i is defined by an MF $\mu_i : X \to [0,1]$ and can be seen as describing a fuzzy region in the input space, in which the corresponding consequent expression holds. A particular input x belongs to each fuzzy set (region) χ_i with membership degree $\mu_i(x)$. The output of the rule base (4.5) is a weighted sum of the output functions g_i, where the weights are the normalized MFs ϕ_i (see again (4.2)):

$$y = \sum_{i=1}^{N} \phi_i(x) g_i(x) \tag{4.6}$$

In this expression, if y is a vector (which means that $g_i(x)$ is also a vector), algebraic operations are understood to be performed element-wise. Note that, more generally than in (4.5), the consequents can also be propositions of the form y **is** $\mathscr{Y}_i$, where $\mathscr{Y}_1,\dots,\mathscr{Y}_N$ are fuzzy sets defined over the output space. The fuzzy rule base resulting from this change is called a Mamdani rule base (Mamdani, 1977; Kruse et al., 1994, Section 4.2.1).

With this framework in place, we can now interpret the Q-function approximator as a Takagi-Sugeno fuzzy rule base that takes the state x as input and produces as outputs the Q-values $q_1,\dots,q_M$ of the M discrete actions:

$$R_i : \textbf{ if } x \textbf{ is } \chi_i \textbf{ then } q_1 = \theta_{[i,1]}; q_2 = \theta_{[i,2]}; \dots; q_M = \theta_{[i,M]} \tag{4.7}$$

where the M outputs have been shown separately to enhance readability. The output functions are in this case constant and consist of the parameters $\theta_{[i,j]}$. To obtain the approximate Q-value (4.3), the action is discretized and the output q_j corresponding to this discretized action is selected. This output is computed with (4.6), thereby leading to the approximation mapping (4.3).

In classical fuzzy theory, the fuzzy sets are associated with linguistic terms describing the corresponding antecedent regions. For instance, if x represents a temperature, the fuzzy sets could be associated with linguistic values such as "cold," "warm," and "hot." In fuzzy Q-iteration, the rule base is simply used as an approximator, and the fuzzy sets do not necessarily have to be associated with meaningful linguistic terms. Nevertheless, if prior knowledge is available on the shape of the optimal Q-function, then fuzzy sets with meaningful linguistic terms can be defined. However, such knowledge is typically difficult to obtain without actually computing the optimal Q-function.

Projection mapping

The *projection mapping* of fuzzy Q-iteration is a special case of the least-squares projection mapping (3.14), repeated here for easy reference:

$$P(Q) = \theta^{\ddagger}, \text{ where } \theta^{\ddagger} \in \arg\min_{\theta} \sum_{l_s=1}^{n_s} \Big(Q(x_{l_s}, u_{l_s}) - [F(\theta)](x_{l_s}, u_{l_s}) \Big)^2 \tag{4.8}$$

in which a set of state action samples $\{(x_{l_s}, u_{l_s}) \mid l_s = 1, \ldots, n_s\}$ is used. In fuzzy Q-iteration, NM samples are used, obtained as the cross-product between the set of MF cores $\{x_1, \ldots, x_N\}$ and the set of discrete actions U_d. Due to Requirement 4.1, with these samples the least-squares projection (4.8) reduces to an assignment of the form (3.26), specifically:

$$\theta_{[i,j]} = [P(Q)]_{[i,j]} = Q(x_i, u_j) \tag{4.9}$$

The parameter vector θ given by (4.9) reduces the least-squares error for the samples considered to zero:

$$\sum_{i=1}^{N} \sum_{j=1}^{M} \Big(Q(x_i, u_j) - [F(\theta)](x_i, u_j) \Big)^2 = 0$$

4.2.2 Synchronous and asynchronous fuzzy Q-iteration

The *synchronous fuzzy Q-iteration* algorithm is obtained by using the approximation mapping (4.3) and the projection mapping (4.9) in the approximate Q-iteration updates given by (3.15), and also repeated here:

$$\theta_{\ell+1} = (P \circ T \circ F)(\theta_\ell) \tag{4.10}$$

The algorithm starts with an arbitrary initial parameter vector $\theta_0 \in \mathbb{R}^n$ and stops when the difference between two consecutive parameter vectors decreases below a threshold ε_{QI}, i.e., when $\|\theta_{\ell+1} - \theta_\ell\|_\infty \leq \varepsilon_{\text{QI}}$. A near-optimal parameter vector $\widehat{\theta}^* = \theta_{\ell+1}$ is obtained.

Because all the Q-functions considered by fuzzy Q-iteration are of the form $F(\theta)$, they are constant in every region U_j given by (4.4). Therefore, when computing maximal Q-values, it suffices to consider only the discrete actions in U_d:[2]

$$\max_u [F(\theta)](x,u) = \max_j [F(\theta)](x,u_j)$$

By exploiting this property, the following discrete-action version of the Q-iteration mapping can be used in practical implementations of fuzzy Q-iteration:

$$[T_\mathrm{d}(Q)](x,u) = \rho(x,u) + \gamma \max_j Q(f(x,u),u_j) \tag{4.11}$$

Each iteration (4.10) can be implemented as:

$$\theta_{\ell+1} = (P \circ T_\mathrm{d} \circ F)(\theta_\ell) \tag{4.12}$$

without changing the sequence of parameter vectors obtained. The maximization over U in the original updates has been replaced with an easier maximization over the discrete set U_d, which can be solved by enumeration. Furthermore, the norms in (4.3) no longer have to be computed to implement (4.12).

Synchronous fuzzy Q-iteration using the update (4.12) can be written in a procedural form as Algorithm 4.1. To establish the equivalence between Algorithm 4.1 and the form (4.12), notice that the right-hand side of line 4 in Algorithm 4.1 corresponds to $[T_\mathrm{d}(F(\theta_\ell))](x_i,u_j)$. Hence, line 4 can be written as $\theta_{\ell+1,[i,j]} \leftarrow [(P \circ T_\mathrm{d} \circ F)(\theta_\ell)]_{[i,j]}$ and the entire **for** loop described by lines 3–5 is equivalent to (4.12).

ALGORITHM 4.1 Synchronous fuzzy Q-iteration.

Input: dynamics f, reward function ρ, discount factor γ,
MFs ϕ_i, $i = 1,\dots,N$, set of discrete actions U_d, threshold ε_QI
1: initialize parameter vector, e.g., $\theta_0 \leftarrow 0$
2: **repeat** at every iteration $\ell = 0,1,2,\dots$
3: **for** $i = 1,\dots,N$, $j = 1,\dots,M$ **do**
4: $\theta_{\ell+1,[i,j]} \leftarrow \rho(x_i,u_j) + \gamma \max_{j'} \sum_{i'=1}^{N} \phi_{i'}(f(x_i,u_j))\theta_{\ell,[i',j']}$
5: **end for**
6: **until** $\|\theta_{\ell+1} - \theta_\ell\|_\infty \le \varepsilon_\mathrm{QI}$
Output: $\widehat{\theta}^* = \theta_{\ell+1}$

Algorithm 4.1 computes the new parameters $\theta_{\ell+1}$ using the parameters θ_ℓ found at the previous iteration, which remain unchanged throughout the current iteration. Algorithm 4.2 is a different version of fuzzy Q-iteration that makes more efficient use of the updates: at each step of the computation, the latest updated values of

[2]This property would also hold if, instead of pure discretization, a triangular fuzzy partition would be defined over the action space, since the maximal Q-values would always be attained in the cores of the triangular MFs (recall that triangular MFs lead to multilinear interpolation). Such a partition may be helpful in a model-free context, to extract information from action samples that do not fall precisely on the MF cores. In this chapter, however, we remain within a model-based context, where action samples can be generated at will.

ALGORITHM 4.2 Asynchronous fuzzy Q-iteration.

Input: dynamics f, reward function ρ, discount factor γ,
MFs ϕ_i, $i = 1, \ldots, N$, set of discrete actions U_d, threshold ε_QI
initialize parameter vector, e.g., $\theta_0 \leftarrow 0$
repeat at every iteration $\ell = 0, 1, 2, \ldots$
$\quad \theta \leftarrow \theta_\ell$
$\quad$ **for** $i = 1, \ldots, N, j = 1, \ldots, M$ **do**
$\quad\quad \theta_{[i,j]} \leftarrow \rho(x_i, u_j) + \gamma \max_{j'} \sum_{i'=1}^{N} \phi_{i'}(f(x_i, u_j)) \theta_{[i',j']}$
$\quad$ **end for**
$\quad \theta_{\ell+1} \leftarrow \theta$
until $\|\theta_{\ell+1} - \theta_\ell\|_\infty \leq \varepsilon_\mathrm{QI}$
Output: $\widehat{\theta}^* = \theta_{\ell+1}$

the parameters are employed. Since the parameters are updated in an asynchronous fashion, this version is called *asynchronous fuzzy Q-iteration*. In Algorithm 4.2 the parameters are shown being updated in sequence, but our analysis still holds even if they are updated in any order.

Either of the two variants of fuzzy Q-iteration produces an approximately optimal Q-function $F(\widehat{\theta}^*)$. A greedy policy in this Q-function can then be employed to control the system, i.e., a policy that satisfies (3.16):

$$\hat{\hat{h}}^*(x) \in \arg\max_u [F(\widehat{\theta}^*)](x, u)$$

Like before, because the Q-function $F(\widehat{\theta}^*)$ is constant in every region U_j, the computation of the greedy policy can be simplified by only considering the discrete actions in U_d:

$$\hat{\hat{h}}^*(x) = u_{j^*}, \quad \text{where } j^* \in \arg\max_j [F(\widehat{\theta}^*)](x, u_j) \tag{4.13}$$

The notation $\hat{\hat{h}}^*$ is used to differentiate this policy from a policy $\widehat{h}^*$ that is greedy in $F(\theta^*)$, where θ^* is the parameter vector obtained asymptotically, as $\ell \to \infty$ (see Section 4.3.1):

$$\widehat{h}^*(x) = u_{j^*}, \quad \text{where } j^* \in \arg\max_j [F(\theta^*)](x, u_j) \tag{4.14}$$

It is also possible to obtain a continuous-action policy using the following heuristic. For any state, an action is computed by interpolating between the best local actions for every MF core, using the MFs as weights:

$$h(x) = \sum_{i=1}^{N} \phi_i(x) u_{j_i^*}, \quad \text{where } j_i^* \in \arg\max_j [F(\widehat{\theta}^*)](x_i, u_j) \tag{4.15}$$

The index j_i^* corresponds to a locally optimal action for the core x_i. For instance, when used with triangular MFs (cf. Example 4.1), the interpolation procedure (4.15) is well suited to problems where (near-)optimal policies are locally affine with respect to the state. Interpolated policies may, however, be a poor choice for other problems, as illustrated in the example below. Theoretical guarantees about policies of the form (4.15) are therefore difficult to provide, and the analysis of Section 4.3 only considers discrete-action policies of the form (4.13).

Example 4.2 Interpolated policies may perform poorly. Consider the problem schematically represented in Figure 4.2(a), in which a robot must avoid an obstacle.

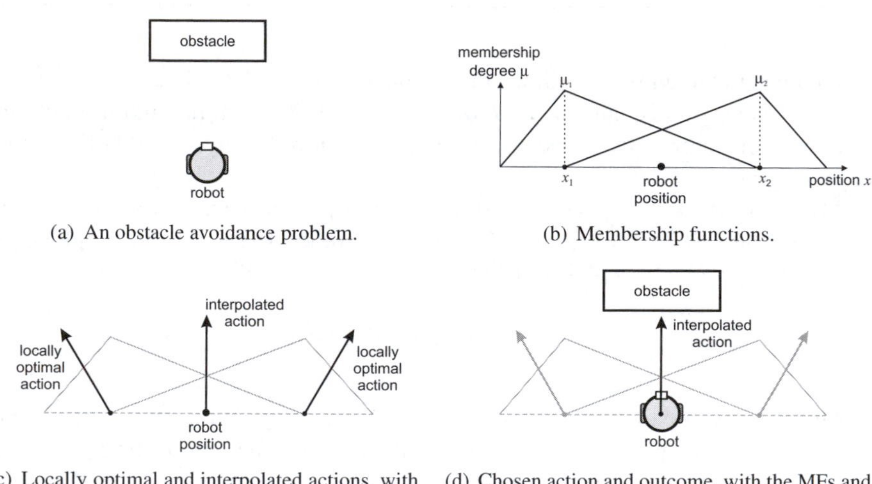

(a) An obstacle avoidance problem.

(b) Membership functions.

(c) Locally optimal and interpolated actions, with the MFs repeated in gray color.

(d) Chosen action and outcome, with the MFs and the locally optimal actions repeated in gray color.

FIGURE 4.2

An obstacle avoidance problem, where interpolation between two locally optimal actions leads to undesirable behavior.

Two MFs are defined for the position variable, and the robot is located at the midpoint of the distance between the cores x_1 and x_2 of these two MFs, see Figure 4.2(b). As shown in Figure 4.2(c), the action of steering left is locally optimal for the MF core to the left of the robot, while the action of steering right is locally optimal for the MF core to the right. These actions are locally optimal because they would take the robot around the obstacle, if the robot were located in the respective cores. However, interpolating between these two actions at the robot's current position makes it (incorrectly) move forward and collide with the obstacle, as seen in Figure 4.2(d). In this problem, rather than interpolating, a good policy would apply either of the two locally optimal actions, e.g., by randomly picking one of them. □

4.3 Analysis of fuzzy Q-iteration

Next, we analyze the convergence, consistency, and computational complexity of (synchronous and asynchronous) fuzzy Q-iteration. Specifically, in Section 4.3.1, we show that fuzzy Q-iteration is convergent and we characterize the suboptimality of its solution, making use of the theoretical framework developed earlier, in Section 3.4.4. In Section 4.3.2, we prove that fuzzy Q-iteration is consistent, i.e., that its solution asymptotically converges to Q^* as the approximation accuracy increases. These results show that fuzzy Q-iteration is a theoretically sound algorithm. In Section 4.3.3, the computational complexity of fuzzy Q-iteration is briefly examined.

4.3.1 Convergence

In this section, the following theoretical results about fuzzy Q-iteration are established:

- Synchronous and asynchronous fuzzy Q-iteration asymptotically converge to a fixed point (parameter vector) θ^* as the number of iterations grows.
- Asynchronous fuzzy Q-iteration converges faster than synchronous fuzzy Q-iteration, in a well-defined sense that will be described later.
- For any strictly positive convergence threshold ε_{QI}, synchronous and asynchronous fuzzy Q-iteration terminate in a finite number of iterations.
- The asymptotically obtained parameter vector θ^* yields an approximate Q-function that is within a bound of the optimal Q-function, and the corresponding greedy policy has a bounded suboptimality. Similar bounds hold for the parameter vector $\widehat{\theta}^*$ obtained in a finite number of iterations.

Theorem 4.1 (Convergence of synchronous fuzzy Q-iteration) *Synchronous fuzzy Q-iteration (Algorithm 4.1) converges to a unique fixed point.*

Proof: To prove convergence, we use the framework of nonexpansive approximators developed in Section 3.4.4. In this framework, the convergence of approximate Q-iteration is guaranteed by ensuring that the composite mapping $P \circ T \circ F$ is a contraction in the infinity norm. For synchronous fuzzy Q-iteration, this will be done by showing that F and P are nonexpansions.

Since the approximation mapping F (4.3) is a weighted linear combination of normalized MFs, it is a nonexpansion. Formally:

$$\begin{aligned}\left|[F(\theta)](x,u) - [F(\theta')](x,u)\right| &= \left|\sum_{i=1}^{N} \phi_i(x)\theta_{[i,j]} - \sum_{i=1}^{N} \phi_i(x)\theta'_{[i,j]}\right| \\ &\qquad (\text{where } j \in \arg\min_{j'} \|u - u_{j'}\|_2) \\ &\leq \sum_{i=1}^{N} \phi_i(x) \left|\theta_{[i,j]} - \theta'_{[i,j]}\right|\end{aligned}$$

$$\leq \sum_{i=1}^{N} \phi_i(x) \|\theta - \theta'\|_\infty$$
$$\leq \|\theta - \theta'\|_\infty$$

where the last step is true because the sum of the normalized MFs $\phi_i(x)$ is 1. Since P consists of a set of assignments (4.9), it is also a nonexpansion.

Additionally, T is a contraction with factor γ (see Section 2.3), and therefore $P \circ T \circ F$ is also a contraction with factor γ, i.e., for any θ, θ':

$$\|(P \circ T \circ F)(\theta) - (P \circ T \circ F)(\theta')\|_\infty \leq \gamma \|\theta - \theta'\|_\infty$$

So, $P \circ T \circ F$ has a unique fixed point θ^*, and synchronous fuzzy Q-iteration converges to this fixed point as $\ell \to \infty$. □

In the sequel, a concise notation of asynchronous fuzzy Q-iteration is needed. Recall that $n = NM$, and that $[i, j] = i + (j-1)N$, with $[i, j] \in \{1, \ldots, n\}$ for $i \in \{1, \ldots, N\}$ and $j \in \{1, \ldots, M\}$. Define for all $l = 0, \ldots, n$, recursively, the mappings $S_l : \mathbb{R}^n \to \mathbb{R}^n$ as:

$$S_0(\theta) = \theta$$
$$[S_l(\theta)]_{l'} = \begin{cases} [(P \circ T \circ F)(S_{l-1}(\theta))]_{l'} & \text{if } l' = l \\ [S_{l-1}(\theta)]_{l'} & \text{if } l' \in \{1, \ldots, n\} \setminus \{l\} \end{cases}$$

So, S_l for $l > 0$ corresponds to updating the first l parameters using approximate asynchronous Q-iteration, and S_n is a complete iteration of the algorithm.

Theorem 4.2 (Convergence of asynchronous fuzzy Q-iteration) *Asynchronous fuzzy Q-iteration (Algorithm 4.2) converges to the same fixed point as synchronous fuzzy Q-iteration.*

Proof: We first show that S_n is a contraction with factor $\gamma < 1$, i.e., that for any θ, θ':

$$\|S_n(\theta) - S_n(\theta')\|_\infty \leq \gamma \|\theta - \theta'\|_\infty$$

This is done by induction, element by element. By the definition of S_l, the first element is only updated by S_1:

$$\begin{aligned} \left|[S_n(\theta)]_1 - [S_n(\theta')]_1\right| &= \left|[S_1(\theta)]_1 - [S_1(\theta')]_1\right| \\ &= \left|[(P \circ T \circ F)(\theta)]_1 - [(P \circ T \circ F)(\theta')]_1\right| \\ &\leq \gamma \|\theta - \theta'\|_\infty \end{aligned}$$

The last step in this equation is true because $P \circ T \circ F$ is a contraction. Furthermore, S_1 is a nonexpansion:

$$\begin{aligned} \|S_1(\theta) - S_1(\theta')\|_\infty &= \max\left\{ \left|[(P \circ T \circ F)(\theta)]_1 - [(P \circ T \circ F)(\theta')]_1\right|, \right. \\ &\qquad \left. \left|\theta_2 - \theta_2'\right|, \ldots, \left|\theta_n - \theta_n'\right| \right\} \\ &\leq \max\left\{\gamma \|\theta - \theta'\|_\infty, \|\theta - \theta'\|_\infty, \ldots, \|\theta - \theta'\|_\infty\right\} \\ &\leq \|\theta - \theta'\|_\infty \end{aligned}$$

We now prove the lth step, thus completing the induction. Assume that for $l' = 1,\dots,l-1$, the following relationships hold:

$$\left|[S_n(\theta)]_{l'} - [S_n(\theta')]_{l'}\right| = \left|[S_{l'}(\theta)]_{l'} - [S_{l'}(\theta')]_{l'}\right| \leq \gamma \|\theta - \theta'\|_\infty$$
$$\|S_{l'}(\theta) - S_{l'}(\theta')\|_\infty \leq \|\theta - \theta'\|_\infty$$

Then, the contraction property for the lth element of S_n can be proven as follows:

$$\begin{aligned}
\left|[S_n(\theta)]_l - [S_n(\theta')]_l\right| &= \left|[S_l(\theta)]_l - [S_l(\theta')]_l\right| \\
&= \left|[(P\circ T\circ F)(S_{l-1}(\theta))]_l - [(P\circ T\circ F)(S_{l-1}(\theta'))]_l\right| \\
&\leq \gamma \|S_{l-1}(\theta) - S_{l-1}(\theta')\|_\infty \\
&\leq \gamma \|\theta - \theta'\|_\infty
\end{aligned}$$

Furthermore, the lth intermediate mapping S_l is a nonexpansion:

$$\begin{aligned}
\|S_l(\theta) - S_l(\theta')\|_\infty = \max\Big\{ &\left|[S_1(\theta)]_1 - [S_1(\theta')]_1\right|, \dots, \left|[S_{l-1}(\theta)]_{l-1} - [S_{l-1}(\theta')]_{l-1}\right|, \\
&\left|[(P\circ T\circ F)(S_{l-1}(\theta))]_l - [(P\circ T\circ F)(S_{l-1}(\theta'))]_l\right|, \\
&\left|\theta_{l+1} - \theta'_{l+1}\right|, \dots, \left|\theta_n - \theta'_n\right| \Big\} \\
\leq \max\big\{ &\gamma\|\theta - \theta'\|_\infty, \dots, \gamma\|\theta - \theta'\|_\infty, \\
&\gamma\|\theta - \theta'\|_\infty, \\
&\|\theta - \theta'\|_\infty, \dots, \|\theta - \theta'\|_\infty \big\} \\
\leq \|&\theta - \theta'\|_\infty
\end{aligned}$$

So, for any l, $|[S_n(\theta)]_l - [S_n(\theta')]_l| \leq \gamma\|\theta - \theta'\|_\infty$, which means that S_n is a contraction with factor $\gamma < 1$. Therefore, asynchronous fuzzy Q-iteration has a unique fixed point.

The fixed point θ^* of $P\circ T\circ F$ can be shown to be a fixed point of S_n using a very similar, element-by-element procedure, which is not given here. Since S_n has a *unique* fixed point, this has to be θ^*. Therefore, asynchronous fuzzy Q-iteration asymptotically converges to θ^*, and the proof is complete. □

This proof is actually more general, showing that approximate asynchronous Q-iteration converges for any approximation mapping F and projection mapping P for which $P\circ T\circ F$ is a contraction. Note that a similar result holds for *exact* asynchronous V-iteration (Bertsekas, 2007, Sec. 1.3.2).

We show next that asynchronous fuzzy Q-iteration converges at least as fast as the synchronous version, i.e., that in a given number of iterations, the asynchronous algorithm takes the parameter vector at least as close to the fixed point as the synchronous one. For that, we first need the following monotonicity lemma. In the sequel, vector and vector function inequalities are understood to be satisfied element-wise.

Lemma 4.1 (Monotonicity) *If* $\theta \leq \theta'$, *then* $(P\circ T\circ F)(\theta) \leq (P\circ T\circ F)(\theta')$ *and* $S_n(\theta) \leq S_n(\theta')$.

Proof: To show that $P\circ T\circ F$ is monotonic, we will show in turn that (i) P, (ii) T, and (iii) F are monotonic.

(i) Given $Q \leq Q'$, it follows that for all i, j:

$$Q(x_i, u_j) \leq Q'(x_i, u_j)$$

This is equivalent to $[P(Q)]_{[i,j]} \leq [P(Q')]_{[i,j]}$, so P is monotonic.

(ii) Given $Q \leq Q'$, it follows that, for any state-action pair (x,u):

$$\max_{u'} Q(f(x,u),u') \leq \max_{\bar{u}} Q'(f(x,u),\bar{u})$$

By multiplying both sides of the equation by γ and then adding $\rho(x,u)$, we obtain:

$$\rho(x,u) + \gamma \max_{u'} Q(f(x,u),u') \leq \rho(x,u) + \gamma \max_{\bar{u}} Q'(f(x,u),\bar{u})$$

which is equivalent to $[T(Q)](x,u) \leq [T(Q')](x,u)$, so T is monotonic.

(iii) Given $\theta \leq \theta'$, it follows that:

$$\sum_{i=1}^{N} \phi_i(x)\theta_{[i,j]} \leq \sum_{i=1}^{N} \phi_i(x)\theta'_{[i,j]}, \quad \text{where } j \in \arg\min_{j'} \|u - u_{j'}\|_2$$

This is equivalent to $[F(\theta)](x,u) \leq [F(\theta')](x,u)$, so F is monotonic. Because P, T, and F are all monotonic, so is $P \circ T \circ F$.

Next, the asynchronous Q-iteration mapping S_n is shown to be monotonic by induction, using a derivation similar to that in the proof of Theorem 4.2. For the first element of S_n:

$$\begin{aligned}[S_n(\theta)]_1 = [S_1(\theta)]_1 &= [(P \circ T \circ F)(\theta)]_1 \\ &\leq [(P \circ T \circ F)(\theta')]_1 = [S_1(\theta')]_1 = [S_n(\theta')]_1\end{aligned}$$

where the monotonicity property of $P \circ T \circ F$ was used. The intermediate mapping S_1 is monotonic:

$$\begin{aligned}S_1(\theta) &= [[(P \circ T \circ F)(\theta)]_1, \theta_2, \ldots, \theta_n]^{\mathrm{T}} \\ &\leq [[(P \circ T \circ F)(\theta')]_1, \theta'_2, \ldots, \theta'_n]^{\mathrm{T}} = S_1(\theta')\end{aligned}$$

We now prove the lth step, thus completing the induction. Assume that for $l' = 1, \ldots, l-1$, the mappings $S_{l'}$ are monotonic:

$$S_{l'}(\theta) \leq S_{l'}(\theta')$$

Then, the monotonicity property for the lth element of S_n can be proven as follows:

$$\begin{aligned}[S_n(\theta)]_l = [S_l(\theta)]_l &= [(P \circ T \circ F)(S_{l-1}(\theta))]_l \\ &\leq [(P \circ T \circ F)(S_{l-1}(\theta'))]_l = [S_l(\theta')]_l = [S_n(\theta')]_l\end{aligned}$$

where the monotonicity of $P \circ T \circ F$ and S_{l-1} was used. Furthermore, the lth intermediate mapping S_l is also monotonic:

$$\begin{aligned}S_l(\theta) &= [[S_1(\theta)]_1 \ldots, [S_{l-1}(\theta)]_{l-1}, [(P \circ T \circ F)(S_{l-1}(\theta))]_l, \theta_{l+1}, \ldots, \theta_n]^{\mathrm{T}} \\ &\leq [[S_1(\theta')]_1 \ldots, [S_{l-1}(\theta')]_{l-1}, [(P \circ T \circ F)(S_{l-1}(\theta'))]_l, \theta'_{l+1}, \ldots, \theta'_n]^{\mathrm{T}} = S_l(\theta')\end{aligned}$$

Therefore, for any l, $[S_n(\theta)]_l \leq [S_n(\theta')]_l$, i.e., S_n is monotonic, which concludes the proof. □

Asynchronous fuzzy Q-iteration converges at least as fast as the synchronous algorithm, in the sense that ℓ iterations of the asynchronous algorithm take the parameter vector at least as close to θ^* as ℓ iterations of the synchronous algorithm. This is stated formally as follows.

Theorem 4.3 (Convergence rate) *If a parameter vector θ satisfies $\theta \leq (P \circ T \circ F)(\theta) \leq \theta^*$, then:*

$$(P \circ T \circ F)^{\ell}(\theta) \leq S_n^{\ell}(\theta) \leq \theta^* \quad \forall \ell \geq 1$$

where $S_n^{\ell}(\theta)$ denotes the composition of $S_n(\theta)$ ℓ-times with itself, i.e., $S_n^{\ell}(\theta) = (S_n \circ S_n \circ \cdots \circ S_n)(\theta)$, and similarly for $(P \circ T \circ F)^{\ell}(\theta)$.

Proof: The theorem will be proven by induction on ℓ. First, take $\ell = 1$, for which the following must be proven:

$$(P \circ T \circ F)(\theta) \leq S_n(\theta) \leq \theta^*$$

By assumption, $(P \circ T \circ F)(\theta) \leq \theta^*$. It remains to be shown that:

$$S_n(\theta) \leq \theta^* \tag{4.16}$$

$$(P \circ T \circ F)(\theta) \leq S_n(\theta) \tag{4.17}$$

Equation (4.16) follows by applying S_n to each side of the inequality $\theta \leq \theta^*$, which is true by assumption. Because S_n is monotonic, we obtain:

$$S_n(\theta) \leq S_n(\theta^*) = \theta^*$$

where the last step is true because θ^* is the fixed point of S_n. The inequality (4.17) can be shown element-wise, in a similar way to the proof that S_n is monotonic (Lemma 4.1). So, this part of the proof is omitted.

We now prove the ℓth step, thus completing the induction. Assuming that:

$$(P \circ T \circ F)^{\ell-1}(\theta) \leq S_n^{\ell-1}(\theta) \leq \theta^*$$

we intend to prove:

$$(P \circ T \circ F)^{\ell}(\theta) \leq S_n^{\ell}(\theta) \leq \theta^*$$

which can be split into three inequalities:

$$(P \circ T \circ F)^{\ell}(\theta) \leq \theta^* \tag{4.18}$$

$$S_n^{\ell}(\theta) \leq \theta^* \tag{4.19}$$

$$(P \circ T \circ F)^{\ell}(\theta) \leq S_n^{\ell}(\theta) \tag{4.20}$$

To obtain (4.18), we apply $P \circ T \circ F$ to both sides of the equation $(P \circ T \circ F)^{\ell-1}(\theta) \leq \theta^*$, and use the fact that $P \circ T \circ F$ is monotonic and has the fixed point θ^*. The inequality (4.19) can be obtained in a similar way, by exploiting the properties of S_n. To derive (4.20), we start from:

$$(P \circ T \circ F)^{\ell-1}(\theta) \leq S_n^{\ell-1}(\theta)$$

and apply S_n to both sides of this equation, using the fact that it is monotonic:

$$S_n\left((P \circ T \circ F)^{\ell-1}(\theta)\right) \leq S_n\left(S_n^{\ell-1}(\theta)\right), \quad \text{i.e.,}$$
$$S_n\left((P \circ T \circ F)^{\ell-1}(\theta)\right) \leq S_n^{\ell}(\theta) \tag{4.21}$$

Notice that $(P \circ T \circ F)^{\ell-1}(\theta)$ satisfies a relationship similar to that assumed for θ in the body of the theorem, namely:

$$(P \circ T \circ F)^{\ell-1}(\theta) \leq (P \circ T \circ F)\left((P \circ T \circ F)^{\ell-1}(\theta)\right) \leq \theta^*$$

So, by replacing θ with $(P \circ T \circ F)^{\ell-1}(\theta)$ in (4.17), the following (valid) relationship is obtained:

$$(P \circ T \circ F)\left((P \circ T \circ F)^{\ell-1}(\theta)\right) \leq S_n\left((P \circ T \circ F)^{\ell-1}(\theta)\right), \quad \text{i.e.,}$$
$$(P \circ T \circ F)^{\ell}(\theta) \leq S_n\left((P \circ T \circ F)^{\ell-1}(\theta)\right)$$

Combining this inequality with (4.21) leads to the desired result (4.20), and the proof is complete. □

Note that a similar result holds in the context of exact (synchronous versus asynchronous) V-iteration (Bertsekas, 2007, Sec. 1.3.2). The result of Theorem 4.3 will not be needed for the analysis in the sequel.

In the remainder of this section, in addition to examining the asymptotical properties of fuzzy Q-iteration, we also consider an implementation that stops when $\|\theta_{\ell+1} - \theta_\ell\|_\infty \leq \varepsilon_{\text{QI}}$, with a convergence threshold $\varepsilon_{\text{QI}} > 0$ (see Algorithms 4.1 and 4.2). This implementation returns the solution $\widehat{\theta}^* = \theta_{\ell+1}$. Such an implementation was given in Algorithm 4.1 for synchronous fuzzy Q-iteration, and in Algorithm 4.2 for asynchronous fuzzy Q-iteration.

Theorem 4.4 (Finite termination) *For any choice of threshold $\varepsilon_{\text{QI}} > 0$ and any initial parameter vector $\theta_0 \in \mathbb{R}^n$, synchronous and asynchronous fuzzy Q-iteration stop in a finite number of iterations.*

Proof: Consider synchronous fuzzy Q-iteration. Because the mapping $P \circ T \circ F$ is a contraction with factor $\gamma < 1$ and fixed point θ^*, we have:

$$\begin{aligned}\|\theta_{\ell+1} - \theta^*\|_\infty &= \|(P \circ T \circ F)(\theta_\ell) - (P \circ T \circ F)(\theta^*)\|_\infty \\ &\leq \gamma \|\theta_\ell - \theta^*\|_\infty\end{aligned}$$

By induction, $\|\theta_\ell - \theta^*\|_\infty \leq \gamma^\ell \|\theta_0 - \theta^*\|_\infty$ for any $\ell > 0$. By the Banach fixed point theorem (see, e.g., Istratescu, 2002, Ch. 3), θ^* is bounded. Because the initial parameter vector θ_0 is also bounded, $\|\theta_0 - \theta^*\|_\infty$ is bounded. Using the notation

$B_0 = \|\theta_0 - \theta^*\|_\infty$, it follows that B_0 is bounded and that $\|\theta_\ell - \theta^*\|_\infty \le \gamma^\ell B_0$ for any $\ell > 0$. Therefore, we have:

$$\begin{aligned}\|\theta_{\ell+1} - \theta_\ell\|_\infty &\le \|\theta_{\ell+1} - \theta^*\|_\infty + \|\theta_\ell - \theta^*\|_\infty \\ &\le \gamma^\ell(\gamma+1)B_0\end{aligned}$$

Using this inequality, for any $\varepsilon_{\mathrm{QI}} > 0$, a number of iterations, L, that guarantees $\|\theta_{L+1} - \theta_L\|_\infty \le \varepsilon_{\mathrm{QI}}$ can be chosen as:

$$L = \left\lceil \log_\gamma \frac{\varepsilon_{\mathrm{QI}}}{(\gamma+1)B_0} \right\rceil$$

Therefore, the algorithm stops in at most L iterations. Because B_0 is bounded, L is finite.

The proof for asynchronous fuzzy Q-iteration proceeds in the same way, because the asynchronous Q-iteration mapping S_n is also a contraction with factor $\gamma < 1$ and fixed point θ^*. □

The following bounds on the suboptimality of the resulting approximate Q-function and policy hold.

Theorem 4.5 (Near-optimality) *Denote by $\mathscr{F}_{F\circ P} \subset \mathscr{Q}$ the set of fixed points of the mapping $F \circ P$, and define the minimum distance[3] between Q^* and any fixed point of $F \circ P$: $\varsigma^*_{\mathrm{QI}} = \min_{Q' \in \mathscr{F}_{F\circ P}} \|Q^* - Q'\|_\infty$. The convergence point θ^* of asynchronous and synchronous fuzzy Q-iteration satisfies:*

$$\|Q^* - F(\theta^*)\|_\infty \le \frac{2\varsigma^*_{\mathrm{QI}}}{1-\gamma} \tag{4.22}$$

Additionally, the parameter vector $\widehat{\theta}^*$ obtained by asynchronous or synchronous fuzzy Q-iteration in a finite number of iterations, with a threshold $\varepsilon_{\mathrm{QI}}$, satisfies:

$$\|Q^* - F(\widehat{\theta}^*)\|_\infty \le \frac{2\varsigma^*_{\mathrm{QI}} + \gamma\varepsilon_{\mathrm{QI}}}{1-\gamma} \tag{4.23}$$

Furthermore:

$$\|Q^* - Q^{\widehat{h}^*}\|_\infty \le \frac{4\gamma\varsigma^*_{\mathrm{QI}}}{(1-\gamma)^2} \tag{4.24}$$

$$\|Q^* - Q^{\hat{\widehat{h}}^*}\|_\infty \le \frac{2\gamma(2\varsigma^*_{\mathrm{QI}} + \gamma\varepsilon_{\mathrm{QI}})}{(1-\gamma)^2} \tag{4.25}$$

where $Q^{\widehat{h}^*}$ is the Q-function of a policy $\widehat{h}^*$ that is greedy in $F(\theta^*)$ (4.14), and $Q^{\hat{\widehat{h}}^*}$ is the Q-function of a policy $\hat{\widehat{h}}^*$ that is greedy in $F(\widehat{\theta}^*)$ (4.13).

[3]For simplicity, we assume that this minimum distance exists. If the minimum does not exist, then $\varsigma^*_{\mathrm{QI}}$ should be taken so that $\exists Q' \in \mathscr{F}_{F\circ P}$ with $\|Q' - Q^*\|_\infty \le \varsigma^*_{\mathrm{QI}}$.

Proof: The bound (4.22) was given in Section 3.4.4, and only relies on the properties of the fixed point θ^* and mappings F, P, and T, so it applies both to synchronous and asynchronous fuzzy Q-iteration.[4]

In order to obtain (4.23), a bound on $\|\widehat{\theta}^* - \theta^*\|_\infty$ is derived first. Let L be the number of iterations after which the algorithm stops, which is finite by Theorem 4.4. Therefore, $\widehat{\theta}^* = \theta_{L+1}$. We have:

$$\begin{aligned}\|\theta_L - \theta^*\|_\infty &\le \|\theta_{L+1} - \theta_L\|_\infty + \|\theta_{L+1} - \theta^*\|_\infty \\ &\le \varepsilon_{\mathrm{QI}} + \gamma\|\theta_L - \theta^*\|_\infty\end{aligned}$$

where the last step follows from the convergence condition $\|\theta_{L+1} - \theta_L\|_\infty \le \varepsilon_{\mathrm{QI}}$ and from the contracting nature of the updates (see also the proof of Theorem 4.4). From the last inequality, it follows that $\|\theta_L - \theta^*\|_\infty \le \frac{\varepsilon_{\mathrm{QI}}}{1-\gamma}$ and therefore that:

$$\|\theta_{L+1} - \theta^*\|_\infty \le \gamma\|\theta_L - \theta^*\|_\infty \le \frac{\gamma\varepsilon_{\mathrm{QI}}}{1-\gamma}$$

which is equivalent to:

$$\|\widehat{\theta}^* - \theta^*\|_\infty \le \frac{\gamma\varepsilon_{\mathrm{QI}}}{1-\gamma} \tag{4.26}$$

Using this inequality, the suboptimality of the Q-function $F(\widehat{\theta}^*)$ can be bounded with:

$$\begin{aligned}\|Q^* - F(\widehat{\theta}^*)\|_\infty &\le \|Q^* - F(\theta^*)\|_\infty + \|F(\theta^*) - F(\widehat{\theta}^*)\|_\infty \\ &\le \|Q^* - F(\theta^*)\|_\infty + \|\widehat{\theta}^* - \theta^*\|_\infty \\ &\le \frac{2\varsigma^*_{\mathrm{QI}}}{1-\gamma} + \frac{\gamma\varepsilon_{\mathrm{QI}}}{1-\gamma} \\ &\le \frac{2\varsigma^*_{\mathrm{QI}} + \gamma\varepsilon_{\mathrm{QI}}}{1-\gamma}\end{aligned}$$

thus obtaining (4.23), where the second step is true because F is a nonexpansion (which was shown in the proof of Theorem 4.1), and the third step follows from (4.22) and (4.26). The bound equally applies to synchronous and asynchronous fuzzy Q-iteration, because its derivation only relies on the fact that their updates are contractions with a factor γ.

The bounds (4.24) and (4.25), which characterize the suboptimality of the policies resulting from $F(\theta^*)$ and $F(\widehat{\theta}^*)$, follow from the equation (3.25), also given in Section 3.4.4. Recall that this equation relates the suboptimality of an arbitrary Q-function Q with the suboptimality of a policy h that is greedy in this Q-function:

$$\|Q^* - Q^h\|_\infty \le \frac{2\gamma}{(1-\gamma)}\|Q^* - Q\|_\infty \tag{4.27}$$

[4]The bound was given in Section 3.4.4 without a proof. The proof is not difficult to develop, and we refer the reader who wishes to understand how this can be done to (Tsitsiklis and Van Roy, 1996, Appendices), where a proof is given in the context of V-iteration. The same remark applies to (4.27).

To obtain (4.24) and (4.25), this inequality is applied to the Q-functions $F(\theta^*)$ and $F(\widehat{\theta}^*)$, using their suboptimality bounds (4.22) and (4.23). □

Examining (4.23) and (4.25), it can be seen that the suboptimality of the solution computed in a finite number of iterations is given by a sum of two terms. The second term depends linearly on the precision ε_{QI} with which the solution is computed, and is easy to control by setting ε_{QI} as close to 0 as needed. The first term in the sum depends linearly on ς^*_{QI}, which is in turn related to the accuracy of the fuzzy approximator, and is more difficult to control. The ς^*_{QI}-dependent term also contributes to the suboptimality of the asymptotic solutions (4.22), (4.24). Ideally, the optimal Q-function Q^* is a fixed point of $F \circ P$, in which case $\varsigma^*_{\text{QI}} = 0$ and fuzzy Q-iteration asymptotically converges to Q^*. For instance, Q^* is a fixed point of $F \circ P$ if it is exactly representable by the chosen approximator, i.e., if for all x, u:

$$Q^*(x,u) = \sum_{i=1}^{N} \phi_i(x) Q^*(x_i, u_j), \quad \text{where } j \in \arg\min_{j'} \|u - u_{j'}\|_2$$

Section 4.3.2 provides additional insight into the relationship between the suboptimality of the solution and the accuracy of the approximator.

4.3.2 Consistency

Next, we analyze the consistency of synchronous and asynchronous fuzzy Q-iteration. It is shown that the approximate solution $F(\theta^*)$ asymptotically converges to the optimal Q-function Q^*, as the largest distance between the cores of adjacent fuzzy sets and the largest distance between adjacent discrete actions both decrease to 0. An explicit relationship between the suboptimality of $F(\theta^*)$ and the accuracy of the approximator is derived.

The state resolution step δ_x is defined as the largest distance between any point in the state space and the nearest MF core. The action resolution step δ_u is defined similarly for the discrete actions. Formally:

$$\delta_x = \sup_{x \in X} \min_{i=1,\dots,N} \|x - x_i\|_2 \tag{4.28}$$

$$\delta_u = \sup_{u \in U} \min_{j=1,\dots,M} \|u - u_j\|_2 \tag{4.29}$$

where x_i is the core of the ith MF, and u_j is the jth discrete action. Smaller values of δ_x and δ_u indicate a higher resolution. The goal is to show that $\lim_{\delta_x \to 0,\, \delta_u \to 0} F(\theta^*) = Q^*$.

We assume that f and ρ are Lipschitz continuous, as formalized next.

Assumption 4.1 (Lipschitz continuity) *The dynamics f and the reward function ρ are Lipschitz continuous, i.e., there exist finite constants $L_f \geq 0$, $L_\rho \geq 0$ so that:*

$$\begin{aligned} \|f(x,u) - f(\bar{x},\bar{u})\|_2 &\leq L_f(\|x - \bar{x}\|_2 + \|u - \bar{u}\|_2) \\ |\rho(x,u) - \rho(\bar{x},\bar{u})| &\leq L_\rho(\|x - \bar{x}\|_2 + \|u - \bar{u}\|_2) \\ &\forall x, \bar{x} \in X, u, \bar{u} \in U \end{aligned}$$

We also require that the MFs are Lipschitz continuous.

Requirement 4.2 *Every MF ϕ_i is Lipschitz continuous, i.e., for every i there exists a finite constant $L_{\phi_i} \geq 0$ so that:*

$$\|\phi_i(x) - \phi_i(\bar{x})\|_2 \leq L_{\phi_i} \|x - \bar{x}\|_2, \quad \forall x, \bar{x} \in X$$

Finally, the MFs should be local and evenly distributed, in the following sense.

Requirement 4.3 *Every MF ϕ_i has a bounded support, which is contained in a ball with a radius proportional to δ_x. Formally, there exists a finite $\nu > 0$ so that:*

$$\{x \mid \phi_i(x) > 0\} \subset \{x \mid \|x - x_i\|_2 \leq \nu \delta_x\}, \quad \forall i$$

Furthermore, for every x, only a finite number of MFs are nonzero. Formally, there exists a finite $\kappa > 0$ so that:

$$|\{i \mid \phi_i(x) > 0\}| \leq \kappa, \quad \forall x$$

where $|\cdot|$ denotes set cardinality.

Lipschitz continuity conditions such as those of Assumption 4.1 are typically needed to prove the consistency of algorithms for approximate DP (e.g., Gonzalez and Rofman, 1985; Chow and Tsitsiklis, 1991). Moreover, note that Requirement 4.2 is not restrictive; for instance, triangular MFs (Example 4.1) and B-spline MFs are Lipschitz continuous.

Requirement 4.3 is satisfied in many cases of interest. For instance, it is satisfied by convex fuzzy sets with their cores distributed on an (equidistant or irregular) rectangular grid in the state space, such as the triangular partitions of Example 4.1. In such cases, every point x falls inside a hyperbox defined by the two adjacent cores that are closest to x_d on each axis d. Some points will fall on the boundary of several hyperboxes, in which case we can just pick any of these hyperboxes. Given Requirement 4.1 and because the fuzzy sets are convex, only the MFs with the cores in the corners of the hyperbox can take nonzero values in the chosen point. Since the number of corners is 2^D, where D is the dimension of X, we have:

$$|\{i \mid \phi_i(x) > 0\}| \leq 2^D$$

and a choice $\kappa = 2^D$ satisfies the second part of Requirement 4.3. Furthermore, along any axis of the state space, a given MF will be nonzero over an interval that spans at most two hyperboxes. From the definition of δ_x, the largest diagonal of any hyperbox is $2\delta_x$, and therefore:

$$\{x \mid \phi_i(x) > 0\} \subset \{x \mid \|x - x_i\|_2 \leq 4\delta_x\}$$

which means that a choice $\nu = 4$ satisfies the first part of Requirement 4.3.

The next lemma bounds the approximation error introduced by every iteration of the synchronous algorithm. Since we are ultimately interested in characterizing the convergence point θ^*, which is the same for both algorithms, the final consistency result (Theorem 4.6) applies to the asynchronous algorithm as well.

Lemma 4.2 (Bounded error) *Under Assumption 4.1 and if Requirements 4.2 and 4.3 are satisfied, there exists a constant* $\varepsilon_\delta \geq 0$, $\varepsilon_\delta = \mathrm{O}(\delta_x) + \mathrm{O}(\delta_u)$, *so that any sequence of Q-functions* $\widehat{Q}_0, \widehat{Q}_1, \widehat{Q}_2, \ldots$ *produced by synchronous fuzzy Q-iteration satisfies:*

$$\|\widehat{Q}_{\ell+1} - T(\widehat{Q}_\ell)\|_\infty \leq \varepsilon_\delta, \textit{ for any } \ell \geq 0$$

Proof: Since any Q-function $\widehat{Q}_\ell$ in a sequence produced by fuzzy Q-iteration is of the form $F(\theta_\ell)$ for some parameter vector θ_ℓ, it suffices to prove that any Q-function $\widehat{Q}$ of the form $F(\theta)$ for some θ satisfies:

$$\|(F \circ P \circ T)(\widehat{Q}) - T(\widehat{Q})\|_\infty \leq \varepsilon_\delta$$

For any pair (x,u):

$$\begin{aligned}
&\left|[(F \circ P \circ T)(\widehat{Q})](x,u) - [T(\widehat{Q})](x,u)\right| \\
&\quad = \left| \left(\sum_{i=1}^{N} \phi_i(x) [T(\widehat{Q})](x_i, u_j) \right) - [T(\widehat{Q})](x,u) \right| \\
&\qquad\qquad \text{(where } j \in \arg\min_{j'} \|u - u_{j'}\|_2 \text{)} \\
&\quad = \left| \left(\sum_{i=1}^{N} \phi_i(x) \left[\rho(x_i, u_j) + \gamma \max_{u'} \widehat{Q}(f(x_i, u_j), u') \right] \right) \right. \\
&\qquad\qquad \left. - \left[\rho(x,u) + \gamma \max_{u'} \widehat{Q}(f(x,u), u') \right] \right| \\
&\quad \leq \left| \left(\sum_{i=1}^{N} \phi_i(x) \rho(x_i, u_j) \right) - \rho(x,u) \right| \\
&\qquad\qquad + \gamma \left| \left(\sum_{i=1}^{N} \phi_i(x) \max_{u'} \widehat{Q}(f(x_i, u_j), u') \right) - \max_{u'} \widehat{Q}(f(x,u), u') \right|
\end{aligned} \tag{4.30}$$

Note that $\max_u \widehat{Q}(x,u)$ exists, because $\widehat{Q}$ can take at most M distinct values for any fixed x. This is because $\widehat{Q}$ is of the form $F(\theta)$ given in (4.3) for some θ, and is therefore constant in each set U_j, for $j = 1, \ldots, M$. The first term on the right-hand side of (4.30) is:

$$\begin{aligned}
\left| \sum_{i=1}^{N} \phi_i(x) \left[\rho(x_i, u_j) - \rho(x,u) \right] \right| &\leq \sum_{i=1}^{N} \phi_i(x) L_\rho (\|x_i - x\|_2 + \|u_j - u\|_2) \\
&\leq L_\rho \left[\|u_j - u\|_2 + \sum_{i=1}^{N} \phi_i(x) \|x_i - x\|_2 \right] \\
&\leq L_\rho (\delta_u + \kappa \nu \delta_x)
\end{aligned} \tag{4.31}$$

where the Lipschitz continuity of ρ was used, and the last step follows from the definition of δ_u and Requirement 4.3. The second term in the right-hand side of (4.30) is:

$$\begin{aligned}
&\gamma \left| \sum_{i=1}^{N} \phi_i(x) \left[\max_{u'} \widehat{Q}(f(x_i,u_j),u') - \max_{u'} \widehat{Q}(f(x,u),u') \right] \right| \\
&\qquad \le \gamma \sum_{i=1}^{N} \phi_i(x) \left| \max_{j'} \widehat{Q}(f(x_i,u_j),u_{j'}) - \max_{j'} \widehat{Q}(f(x,u),u_{j'}) \right| \\
&\qquad \le \gamma \sum_{i=1}^{N} \phi_i(x) \max_{j'} \left| \widehat{Q}(f(x_i,u_j),u_{j'}) - \widehat{Q}(f(x,u),u_{j'}) \right| \qquad (4.32)
\end{aligned}$$

The first step is true because $\widehat{Q}$ is constant in each set U_j, for $j = 1,\dots,M$. The second step is true because the difference between the maxima of two functions of the same variable is at most the maximum of the difference of the functions. Writing $\widehat{Q}$ explicitly as in (4.3), we have:

$$\begin{aligned}
\left| \widehat{Q}(f(x_i,u_j),u_{j'}) - \widehat{Q}(f(x,u),u_{j'}) \right| &= \left| \sum_{i'=1}^{N} \left[\phi_{i'}(f(x_i,u_j))\theta_{[i',j']} - \phi_{i'}(f(x,u))\theta_{[i',j']} \right] \right| \\
&\le \sum_{i'=1}^{N} \left| \phi_{i'}(f(x_i,u_j)) - \phi_{i'}(f(x,u)) \right| \left| \theta_{[i',j']} \right|
\end{aligned} \qquad (4.33)$$

Define $I' = \left\{ i' \mid \phi_{i'}(f(x_i,u_j)) \neq 0 \text{ or } \phi_{i'}(f(x,u)) \neq 0 \right\}$. Using Requirement 4.3, $|I'| \le 2\kappa$. Denote $L_\phi = \max_i L_{\phi_i}$ (where Requirement 4.2 is employed). Then, the right-hand side of (4.33) is equal to:

$$\begin{aligned}
\sum_{i' \in I'} \left| \phi_{i'}(f(x_i,u_j)) - \phi_{i'}(f(x,u)) \right| \left| \theta_{[i',j']} \right| &\le \sum_{i' \in I'} L_\phi L_f (\|x_i - x\|_2 + \|u_j - u\|_2) \|\theta\|_\infty \\
&\le 2\kappa L_\phi L_f (\|x_i - x\|_2 + \|u_j - u\|_2) \|\theta\|_\infty
\end{aligned} \qquad (4.34)$$

Using (4.33) and (4.34) in (4.32) yields:

$$\begin{aligned}
&\gamma \sum_{i=1}^{N} \phi_i(x) \max_{j'} \left| \widehat{Q}(f(x_i,u_j),u_{j'}) - \widehat{Q}(f(x,u),u_{j'}) \right| \\
&\qquad \le \gamma \sum_{i=1}^{N} \phi_i(x) \max_{j'} 2\kappa L_\phi L_f (\|x_i - x\|_2 + \|u_j - u\|_2) \|\theta\|_\infty \\
&\qquad \le 2\gamma\kappa L_\phi L_f \|\theta\|_\infty \left[\|u_j - u\|_2 + \sum_{i=1}^{N} \phi_i(x) \|x_i - x\|_2 \right] \\
&\qquad \le 2\gamma\kappa L_\phi L_f \|\theta\|_\infty (\delta_u + \kappa\nu\delta_x) \qquad (4.35)
\end{aligned}$$

where the last step follows from the definition of δ_u and Requirement 4.3. Finally, substituting (4.31) and (4.35) into (4.30) yields:

$$\left| [(F \circ P \circ T)(\widehat{Q})](x,u) - [T(\widehat{Q})](x,u) \right| \le (L_\rho + 2\gamma\kappa L_\phi L_f \|\theta\|_\infty)(\delta_u + \kappa\nu\delta_x) \quad (4.36)$$

Given a bounded initial parameter vector θ_0, all the parameter vectors considered by the algorithm are bounded, which can be shown as follows. By the Banach fixed point theorem, the optimal parameter vector θ^* (the unique fixed point of $P \circ T \circ F$) is finite. Also, we have $\|\theta_\ell - \theta^*\|_\infty \leq \gamma^\ell \|\theta_0 - \theta^*\|_\infty$ (see the proof of Theorem 4.4). Since $\|\theta_0 - \theta^*\|_\infty$ is bounded, all the other distances are bounded, and all the parameter vectors θ_ℓ are bounded. Let $B_\theta = \max_{\ell \geq 0} \|\theta_\ell\|_\infty$, which is bounded. Therefore, $\|\theta\|_\infty \leq B_\theta$ in (4.36), and the proof is complete with:

$$\varepsilon_\delta = (L_\rho + 2\gamma\kappa L_\phi L_f B_\theta)(\delta_u + \kappa\nu\delta_x) = \mathrm{O}(\delta_x) + \mathrm{O}(\delta_u)$$

□

Theorem 4.6 (Consistency) *Under Assumption 4.1 and if Requirements 4.2 and 4.3 are satisfied, synchronous and asynchronous fuzzy Q-iteration are consistent:*

$$\lim_{\delta_x \to 0, \delta_u \to 0} F(\theta^*) = Q^*$$

Furthermore, the suboptimality of the approximate Q-function satisfies:

$$\|F(\theta^*) - Q^*\|_\infty = \mathrm{O}(\delta_x) + \mathrm{O}(\delta_u)$$

Proof: First, it will be shown that $\|F(\theta^*) - Q^*\|_\infty \leq \frac{\varepsilon_\delta}{1-\gamma}$. Consider a sequence of Q-functions $\widehat{Q}_0, \widehat{Q}_1, \widehat{Q}_2, \ldots$ produced by synchronous fuzzy Q-iteration, and let us establish by induction a bound on $\|\widehat{Q}_\ell - T^\ell(\widehat{Q}_0)\|_\infty$ for $\ell \geq 1$. By Lemma 4.2, we have:

$$\|\widehat{Q}_1 - T(\widehat{Q}_0)\|_\infty \leq \varepsilon_\delta$$

Assume that for some $\ell \geq 1$, we have:

$$\|\widehat{Q}_\ell - T^\ell(\widehat{Q}_0)\|_\infty \leq \varepsilon_\delta(1 + \gamma + \cdots + \gamma^{\ell-1}) \tag{4.37}$$

Then, for $\ell + 1$:

$$\begin{aligned}
\|\widehat{Q}_{\ell+1} - T^{\ell+1}(\widehat{Q}_0)\|_\infty &\leq \|\widehat{Q}_{\ell+1} - T(\widehat{Q}_\ell)\|_\infty + \|T(\widehat{Q}_\ell) - T^{\ell+1}(\widehat{Q}_0)\|_\infty \\
&\leq \varepsilon_\delta + \gamma\|\widehat{Q}_\ell - T^\ell(\widehat{Q}_0)\|_\infty \\
&\leq \varepsilon_\delta + \gamma\varepsilon_\delta(1 + \gamma + \cdots + \gamma^{\ell-1}) \\
&\leq \varepsilon_\delta(1 + \gamma + \cdots + \gamma^\ell)
\end{aligned}$$

where in the second step we used Lemma 4.2 and the contraction property of T. The induction is therefore complete, and (4.37) is true for any $\ell \geq 1$.

The inequality (4.37) means that for any pair (x, u), we have:

$$\begin{aligned}
[T^\ell(\widehat{Q}_0)](x,u) - \varepsilon_\delta(1 + \gamma + \cdots + \gamma^{\ell-1}) &\leq \widehat{Q}_\ell(x,u) \\
&\leq [T^\ell(\widehat{Q}_0)](x,u) + \varepsilon_\delta(1 + \gamma + \cdots + \gamma^{\ell-1})
\end{aligned}$$

We take the limit of this pair of inequalities as $\ell \to \infty$, and use the facts that

$\lim_{\ell\to\infty} T^{\ell}(\widehat{Q}_0) = Q^*$ and $\lim_{\ell\to\infty} \widehat{Q}_\ell = F(\theta^*)$ (recall that $\widehat{Q}_\ell$ is the ℓth Q-function produced by synchronous fuzzy Q-iteration, which converges to $F(\theta^*)$) to obtain:

$$Q^*(x,u) - \frac{\varepsilon_\delta}{1-\gamma} \le [F(\theta^*)](x,u) \le Q^*(x,u) + \frac{\varepsilon_\delta}{1-\gamma}$$

for any (x,u), which means that $\|F(\theta^*) - Q^*\|_\infty \le \frac{\varepsilon_\delta}{1-\gamma}$. Note that a related bound for approximate V-iteration was proven along similar lines by Bertsekas and Tsitsiklis (1996, Sec. 6.5.3).

Now, using the explicit formula for ε_δ found in the proof of Lemma 4.2:

$$\begin{aligned}\lim_{\delta_x\to 0,\delta_u\to 0} \|F(\theta^*) - Q^*\|_\infty &= \lim_{\delta_x\to 0,\delta_u\to 0} \frac{\varepsilon_\delta}{1-\gamma} \\ &= \lim_{\delta_x\to 0,\delta_u\to 0} \frac{(L_\rho + 2\gamma\kappa L_\phi L_f B_\theta)(\delta_u + \kappa\nu\delta_x)}{1-\gamma} \\ &= 0\end{aligned}$$

and the first result of the theorem is proven. Furthermore, using the same lemma, $\frac{\varepsilon_\delta}{1-\gamma} = \mathrm{O}(\delta_x) + \mathrm{O}(\delta_u)$, which implies $\|F(\theta^*) - Q^*\|_\infty = \mathrm{O}(\delta_x) + \mathrm{O}(\delta_u)$, thus completing the proof. □

In addition to guaranteeing consistency, Theorem 4.6 also relates the suboptimality of the Q-function $F(\theta^*)$ to the accuracy of the fuzzy approximator. Using Theorem 4.5, the accuracy can be further related to the suboptimality of the policy $\widehat{h}^*$ greedy in $F(\theta^*)$, and to the suboptimality of the solution (Q-function $F(\widehat{\theta}^*)$ and corresponding policy $\hat{\widehat{h}}^*$) obtained after a finite number of iterations.

4.3.3 Computational complexity

In this section, the time and memory complexity of fuzzy Q-iteration are examined. It is easy to see that each iteration of the synchronous and asynchronous fuzzy Q-iteration (Algorithms 4.1 and 4.2) requires $\mathrm{O}(N^2M)$ time to run. Here, N is the number of MFs and M is the number of discrete actions, leading to a parameter vector of length NM. The complete algorithms consist of L iterations and thus require $\mathrm{O}(LN^2M)$ computation. Fuzzy Q-iteration requires $\mathrm{O}(NM)$ memory. The memory complexity is not proportional to L because, in practice, any $\theta_{\ell'}$ for which $\ell' < \ell$ can be discarded.

Example 4.3 Comparison with least-squares policy iteration. As an example, we compare the complexity of fuzzy Q-iteration with that of a representative algorithm from the approximate policy iteration class, namely least-squares policy iteration (LSPI) (Algorithm 3.11). We focus on the case in which both algorithms compute *the same number of parameters*. At each iteration, LSPI performs policy evaluation (with the least-squares temporal difference for Q-functions, Algorithm 3.8) and policy improvement. To approximate the Q-function, LSPI typically employs discrete-action approximators (Example 3.1), which consist of N state-dependent basis functions and M discrete actions, and have NM parameters. The time complexity of each policy evaluation is $O(N^3M^3)$ if "naive" matrix inversion is used to solve the linear system of size NM. More efficient algorithms than matrix inversion can be obtained, e.g., by incrementally computing the inverse, but the time complexity will still be larger than $O(N^2M^2)$. The memory complexity is $O(N^2M^2)$. Therefore, the asymptotic upper bounds on the time complexity per iteration and on the memory complexity are worse (larger) for LSPI than for fuzzy Q-iteration.

This comparison should be considered in light of some important differences between fuzzy Q-iteration and LSPI. The fact that both algorithms employ the same number of parameters means they employ *similarly*, but *not identically* powerful approximators: due to Requirements 4.1, 4.2, and 4.3, the class of approximators considered by fuzzy Q-iteration is smaller, and therefore less powerful. These requirements also enable fuzzy Q-iteration to perform more computationally efficient parameter updates, e.g., because the projection is reduced to an update (4.9). □

4.4 Optimizing the membership functions

The accuracy of the solution found by fuzzy Q-iteration crucially depends on the quality of the fuzzy approximator, which in turn is determined by the MFs and by the action discretization. We focus here on the problem of obtaining good MFs, and assume the action discretization is fixed. The MFs can be designed beforehand, in which case two possibilities arise. If prior knowledge about the shape of the optimal Q-function is available to design the MFs, then a moderate number of MFs may be sufficient to achieve a good approximator. However, such prior knowledge is often difficult to obtain without actually computing the optimal Q-function. When prior knowledge is not available, a large number of MFs must be defined to provide a good coverage and resolution over the entire state space, even in areas that will eventually be irrelevant to the policy.

In this section, we consider a different method, which does not require the MFs to be designed in advance. In this method, parameters encoding the location and shape of the MFs are optimized, while the number of MFs is kept constant. The goal is to obtain a set of MFs that are near optimal for the problem at hand. Since the MFs can be regarded as basis functions, this approach may be regarded as a basis function optimization technique, such as those introduced in Section 3.6.1. MF optimization is useful when prior knowledge about the shape of the optimal Q-function is not available and the number of MFs is limited.

4.4.1 A general approach to membership function optimization

Let the (normalized) MFs be parameterized by a vector $\xi \in \Xi$. Typically, the parameter vector ξ includes information about the location and shape of the MFs. Denote the MFs by $\phi_i(\cdot;\xi) : X \to \mathbb{R}$, $i = 1,\dots,N$, to highlight their dependence on ξ (where the dot stands for the argument x). The goal is to find a parameter vector that leads to good MFs. In the suboptimality bounds provided by Theorem 4.5, the quality of the approximator (and thereby the quality of the MFs) is indirectly represented by the minimum distance $\varsigma^*_{\mathrm{QI}}$ between Q^* and any fixed point of the mapping $F \circ P$. However, since Q^* is not available, $\varsigma^*_{\mathrm{QI}}$ cannot be directly computed nor used to evaluate the MFs.

Instead, we propose an score function (optimization criterion) that is directly related to the performance of the policy obtained. Specifically, we aim to find an optimal parameter vector ξ^* that maximizes the weighted sum of the returns from a finite set X_0 of representative states:

$$s(\xi) = \sum_{x_0 \in X_0} w(x_0) R^h(x_0) \tag{4.38}$$

The policy h is computed by running synchronous or asynchronous fuzzy Q-iteration to (near-)convergence with the MFs specified by the parameters ξ. The representative states are weighted by the function $w : X_0 \rightarrow (0,1]$. This score function was discussed before in the context of basis function optimization, see (3.55) in Section 3.6.1.

The infinite-horizon return from each representative state x_0 is estimated using a simulated trajectory of length K:

$$R^h(x_0) = \sum_{k=0}^{K} \gamma^k \rho(x_k, h(x_k)) \tag{4.39}$$

In this trajectory, $x_{k+1} = f(x_k, h(x_k))$ for $k \geq 0$. This simulation procedure is a variant of (3.64), specialized for the deterministic case considered in this chapter. By choosing the length K of the trajectory using (3.65), a desired precision $\varepsilon_{\mathrm{MC}}$ can be guaranteed for the return estimate.

The set X_0, together with the weight function w, determines the performance of the resulting policy. A good choice of X_0 and w will depend on the problem at hand. For instance, if the process only needs to be controlled starting from a known set of initial states, then X_0 should be equal to (or included in) this set. Initial states that are deemed more important can be assigned larger weights.

Because each policy is computed by running fuzzy Q-iteration with fixed MF parameters, this technique does not suffer from the convergence problems associated with the adaptation of the approximator while running the DP/RL algorithm (see Section 3.6.3).

This technique is not restricted to a particular optimization algorithm to search for ξ^*. However, the score (4.38) can generally be a nondifferentiable function of ξ, with multiple local optima, so a global, gradient-free optimization technique is preferable. In Section 4.4.3, an algorithm will be given to optimize the locations of triangular MFs using the cross-entropy (CE) method for optimization. First, the necessary background on CE optimization is outlined in the next section.

4.4.2 Cross-entropy optimization

This section provides a brief introduction to the CE method for optimization (Rubinstein and Kroese, 2004). In this introduction, the information presented and the notation employed are specialized to the application of CE optimization for finding MFs (to be used in fuzzy Q-iteration). For a detailed, general description of the CE method, see Appendix B.

Consider the following optimization problem:

$$\max_{\xi \in \Xi} s(\xi) \tag{4.40}$$

where $s : \Xi \rightarrow \mathbb{R}$ is the score function (optimization criterion) to maximize, and the parameters ξ take values in the domain Ξ. Denote the maximum by s^*. The CE method maintains a probability density with support Ξ. At each iteration, a number of samples are drawn from this density and the score values of these samples are computed. A (smaller) number of samples that have the best scores are kept, and the remaining samples are discarded. The probability density is then updated using the selected samples, such that at the next iteration the probability of drawing better samples is increased. The algorithm stops when the score of the worst selected sample no longer improves significantly.

Formally, a family of probability densities $\{p(\cdot; v)\}$ must be chosen, where the dot stands for the random variable ξ. This family has support Ξ and is parameterized by v. At each iteration τ of the CE algorithm, a number N_{CE} of samples are drawn from the density $p(\cdot; v_{\tau-1})$, their scores are computed, and the $(1 - \rho_{\text{CE}})$ quantile[5] λ_τ of the sample scores is determined, with $\rho_{\text{CE}} \in (0, 1)$. Then, a so-called associated stochastic problem is defined, which involves estimating the probability that the score of a sample drawn from $p(\cdot; v_{\tau-1})$ is at least λ_τ:

$$\mathrm{P}_{\xi \sim p(\cdot; v_{\tau-1})}(s(\xi) \geq \lambda_\tau) = \mathrm{E}_{\xi \sim p(\cdot; v_{\tau-1})}\{I(s(\xi) \geq \lambda_\tau)\} \tag{4.41}$$

where I is the indicator function, equal to 1 whenever its argument is true, and 0 otherwise.

The probability (4.41) can be estimated by importance sampling. For this problem, an importance sampling density is one that increases the probability of the interesting event $s(\xi) \geq \lambda_\tau$. An optimal importance sampling density in the family $\{p(\cdot; v)\}$, in the smallest cross-entropy (Kullback–Leibler divergence) sense, is given by a solution of:

$$\arg\max_{v} \mathrm{E}_{\xi \sim p(\cdot; v_{\tau-1})}\{I(s(\xi) \geq \lambda_\tau) \ln p(\xi; v)\} \tag{4.42}$$

An approximate solution v_τ of (4.42) is computed using:

$$v_\tau = v_\tau^{\ddagger}, \text{ where } v_\tau^{\ddagger} \in \arg\max_{v} \frac{1}{N_{\text{CE}}} \sum_{i_s=1}^{N_{\text{CE}}} I(s(\xi_{i_s}) \geq \lambda_\tau) \ln p(\xi_{i_s}; v) \tag{4.43}$$

---

[5] If the score values of the samples are ordered increasingly and indexed such that $s_1 \leq \cdots \leq s_{N_{\text{CE}}}$, then the $(1 - \rho_{\text{CE}})$ quantile is $\lambda_\tau = s_{\lceil (1-\rho_{\text{CE}}) N_{\text{CE}} \rceil}$.

Only the samples that satisfy $s(a_{i_s}) \geq \lambda_\tau$ contribute to this formula, since the contributions of the other samples are made to be zero by the product with the indicator function. In this sense, the updated density parameter only depends on these best samples, and the other samples are discarded.

CE optimization proceeds with the next iteration using the new density parameter v_τ (the probability (4.41) is never actually computed). The updated density aims at generating good samples with higher probability than the old density, thus bringing $\lambda_{\tau+1}$ closer to the optimum s^*. The goal is to eventually converge to a density that generates samples close to optimal value(s) of ξ with very high probability. The algorithm can be stopped when the $(1-\rho_{\text{CE}})$-quantile of the sample performance improves for $d_{\text{CE}} > 1$ consecutive iterations, but these improvements do not exceed ε_{CE}; alternatively, the algorithm stops when a maximum number of iterations $\tau_{\max}$ is reached. Then, the largest score among the samples generated in all the iterations is taken as the approximate solution of the optimization problem, and the corresponding sample as an approximate location of the optimum. Note that CE optimization can also use a so-called smoothing procedure to incrementally update the density parameters, but we do not employ such a procedure in this chapter (see Appendix B for details on this procedure).

4.4.3 Fuzzy Q-iteration with cross-entropy optimization of the membership functions

In this section, a complete algorithm is given for finding optimal MFs to be used in fuzzy Q-iteration. This algorithm employs the CE method to optimize the cores of triangular MFs (Example 4.1). Triangular MFs are chosen because they are the simplest MFs that ensure the convergence of fuzzy Q-iteration. CE optimization is chosen as an illustrative example of a global optimization technique that can be used for this problem. Many other optimization algorithms could be applied to optimize the MFs, e.g., genetic algorithms (Goldberg, 1989), tabu search (Glover and Laguna, 1997), pattern search (Torczon, 1997; Lewis and Torczon, 2000), etc.

Recall that the state space dimension is denoted by D. In this section, it is assumed that the state space is a hyperbox centered on the origin:

$$X = [-x_{\max,1}, x_{\max,1}] \times \cdots \times [-x_{\max,D}, x_{\max,D}]$$

where $x_{\max,d} \in (0, \infty)$, $d = 1, \ldots, D$. Separately for each state variable x_d, a triangular fuzzy partition is defined with core values $c_{d,1} < \cdots < c_{d,N_d}$, which give N_d triangular MFs. The product of each combination of single-dimensional MFs gives a pyramid-shaped D-dimensional MF in the fuzzy partition of X. The parameters to be optimized are the (scalar) free cores on each axis. The first and last core values on each axis are not free, but are always equal to the limits of the domain: $c_{d,1} = -x_{\max,d}$ and $c_{d,N_d} = x_{\max,d}$, hence, the number of free cores is only $N_\xi = \sum_{d=1}^{D}(N_d - 2)$. The parameter vector ξ can be obtained by collecting the free cores:

$$\xi = [c_{1,2}, \ldots, c_{1,N_1-1}, \ldots\ldots, c_{D,2}, \ldots, c_{D,N_D-1}]^{\mathrm{T}}$$

and has the domain:

$$\Xi = (-x_{\max,1}, x_{\max,1})^{N_1-2} \times \cdots \times (-x_{\max,D}, x_{\max,D})^{N_D-2}$$

The goal is to find a parameter vector ξ^* that maximizes the score function (4.38).

To apply CE optimization, we choose a family of densities with independent (univariate) Gaussian components for each of the N_ξ parameters.The Gaussian density for the ith parameter is determined by its mean η_i and standard deviation σ_i. Using Gaussian densities has the advantage that (4.43) has a closed-form solution (Rubinstein and Kroese, 2004), given by the mean and standard deviation of the best samples. By exploiting this property, simple update rules can be obtained for the density parameters, see, e.g., line 13 of the upcoming Algorithm 4.3. Note also that when the Gaussian family is used, the CE method can actually converge to a precise optimum location ξ^*, by letting each univariate Gaussian density converge to a degenerate (Dirac) distribution that assigns all the probability mass to the value ξ_i^*. This degenerate distribution is obtained for $\eta_i = \xi_i^*$ and $\sigma_i \to 0$.

Because the support of the chosen density is $\mathbb{R}^{N_\xi}$, which is larger than Ξ, samples that do not belong to Ξ are rejected and generated again. The density parameter vector v consists of the vector of means η and the vector of standard deviations σ, each of them containing N_ξ elements. The vectors η and σ are initialized using:

$$\eta_0 = 0, \quad \sigma_0 = [x_{\max,1}, \ldots, x_{\max,1}, \ldots\ldots, x_{\max,D}, \ldots, x_{\max,D}]^{\mathrm{T}}$$

where each bound $x_{\max,d}$ is replicated $N_d - 2$ times, for $d = 1, \ldots, D$. These values ensure that the samples cover the state space well in the first iteration of the CE method.

Algorithm 4.3 summarizes fuzzy Q-iteration with CE optimization of the MFs. Either the synchronous or the asynchronous variant of fuzzy Q-iteration could be used at line 7 of this algorithm, but the variant should not be changed during the CE procedure, since the two variants can produce different solutions for the same MFs and convergence threshold. At line 10 of Algorithm 4.3, the samples are sorted in an ascending order of their scores, to simplify the subsequent formulas. At line 13, the closed-form update of the Gaussian density parameters is employed. In these updates, the mathematical operators (e.g., division by a constant, square root) should be understood as working element-wise, separately for each element of the vectors considered. For a description of the stopping condition and parameters of the CE method, see again Section 4.4.2 or, for more details, Appendix B.

4.5 Experimental study

We dedicate the remainder of this chapter to an extensive experimental study of fuzzy Q-iteration. This study is organized in four parts, each focusing on different aspects relevant to the practical application of the algorithm. The first example illustrates the

ALGORITHM 4.3 Fuzzy Q-iteration with cross-entropy MF optimization.

Input: dynamics f, reward function ρ, discount factor γ,
set of discrete actions U_{d}, fuzzy Q-iteration convergence threshold $\varepsilon_{\mathrm{QI}}$,
representative states X_0, weight function w,
CE parameters $\rho_{\mathrm{CE}} \in (0,1)$, $N_{\mathrm{CE}} \geq 2$, $d_{\mathrm{CE}} \geq 2$, $\tau_{\max} \geq 2$, $\varepsilon_{\mathrm{CE}} \geq 0$

$\tau \leftarrow 0$
$\eta_0 \leftarrow 0$, $\sigma_0 \leftarrow [x_{\max,1}, \ldots, x_{\max,1}, \ldots \ldots, x_{\max,D}, \ldots, x_{\max,D}]^{\mathrm{T}}$
repeat
  $\tau \leftarrow \tau + 1$
  generate samples $\xi_1, \ldots, \xi_{N_{\mathrm{CE}}}$ from Gaussians given by $\eta_{\tau-1}$ and $\sigma_{\tau-1}$
  for $i_{\mathrm{s}} = 1, \ldots, N_{\mathrm{CE}}$ **do**
    run fuzzy Q-iteration with MFs $\phi_i(x; \xi_{i_{\mathrm{s}}})$, actions U_{d}, and threshold $\varepsilon_{\mathrm{QI}}$
    compute score $s(\xi_{i_{\mathrm{s}}})$ of resulting policy h, using (4.38)
  end for
  reorder and reindex samples s.t. $s_1 \leq \cdots \leq s_{N_{\mathrm{CE}}}$
  $\lambda_\tau \leftarrow s_{\lceil (1-\rho_{\mathrm{CE}}) N_{\mathrm{CE}} \rceil}$, the $(1-\rho_{\mathrm{CE}})$ quantile of the sample scores
  $i_\tau \leftarrow \lceil (1-\rho_{\mathrm{CE}}) N_{\mathrm{CE}} \rceil$, index of the first of the best samples
  $\eta_\tau \leftarrow \frac{1}{N_{\mathrm{CE}} - i_\tau + 1} \sum_{i_{\mathrm{s}}=i_\tau}^{N_{\mathrm{CE}}} \xi_{i_{\mathrm{s}}}$; $\sigma_\tau \leftarrow \sqrt{\frac{1}{N_{\mathrm{CE}} - i_\tau + 1} \sum_{i_{\mathrm{s}}=i_\tau}^{N_{\mathrm{CE}}} (\xi_{i_{\mathrm{s}}} - \eta_\tau)^2}$
until $(\tau > d_{\mathrm{CE}}$ **and** $|\lambda_{\tau-\tau'} - \lambda_{\tau-\tau'-1}| \leq \varepsilon_{\mathrm{CE}}$, for $\tau' = 0, \ldots, d_{\mathrm{CE}} - 1)$ **or** $\tau = \tau_{\max}$

Output: best sample $\widehat{\xi}^*$, its score, and corresponding fuzzy Q-iteration solution

convergence and consistency of fuzzy Q-iteration, using a DC motor problem. The second example employs a two-link manipulator to demonstrate the effects of action interpolation, and also to compare fuzzy Q-iteration with fitted Q-iteration. In the third example, the real-time control performance of fuzzy Q-iteration is illustrated using an inverted pendulum swing-up problem. For these three examples, the MFs are designed in advance. In the fourth and final example, the effects of optimizing the MFs (with the CE approach of Section 4.4) are studied in the classical car-on-the-hill benchmark.

4.5.1 DC motor: Convergence and consistency study

This section illustrates the practical impact of the convergence and consistency properties of fuzzy Q-iteration, using the DC motor problem introduced in Section 3.4.5. The DC motor system is chosen because its simplicity allows extensive simulations to be performed with reasonable computational costs. First, the convergence rates of synchronous and asynchronous fuzzy Q-iteration are empirically compared. Then, the change in solution quality as the approximation power increases is investigated, to illustrate how the consistency properties of fuzzy Q-iteration influence its behavior in practice. Recall that consistency was proven under the condition of Lipschitz continuous dynamics and rewards (Assumption 4.1). To examine the impact of violating this condition, we introduce discontinuities in the reward function, and repeat the consistency study.

DC motor problem

Consider the second-order discrete-time model of the DC motor:

$$f(x,u) = Ax + Bu$$
$$A = \begin{bmatrix} 1 & 0.0049 \\ 0 & 0.9540 \end{bmatrix}, \quad B = \begin{bmatrix} 0.0021 \\ 0.8505 \end{bmatrix} \tag{4.44}$$

where $x_1 = \alpha \in [-\pi, \pi]$ rad is the shaft angle, $x_2 = \dot{\alpha} \in [-16\pi, 16\pi]$ rad/s is the angular velocity, and $u \in [-10, 10]$ V is the control input (voltage). The state variables are restricted to their domains using saturation. The control goal is to stabilize the system around $x = 0$, and is described by the quadratic reward function:

$$\rho(x,u) = -x^{\mathrm{T}} Q_{\mathrm{rew}} x - R_{\mathrm{rew}} u^2$$
$$Q_{\mathrm{rew}} = \begin{bmatrix} 5 & 0 \\ 0 & 0.01 \end{bmatrix}, \quad R_{\mathrm{rew}} = 0.01 \tag{4.45}$$

with discount factor $\gamma = 0.95$. This reward function is shown in Figure 4.3.

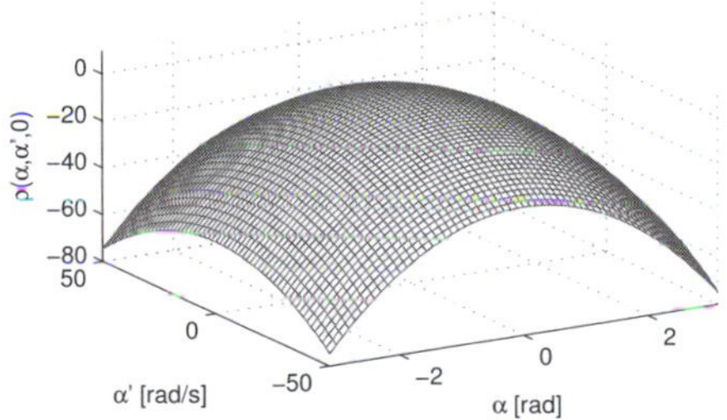

FIGURE 4.3
A state-dependent slice through the reward function (4.45), for $u = 0$. Reproduced with permission from (Buşoniu et al., 2008b), © 2008 IEEE.

Synchronous and asynchronous convergence

First, we compare the convergence rates of synchronous and asynchronous fuzzy Q-iteration. A triangular fuzzy partition with $N' = 41$ equidistant cores for each state variable is defined, leading to $N = 41^2$ fuzzy sets in the two-dimensional partition of X. The action space is discretized into 15 equidistant values. First, synchronous fuzzy Q-iteration is run with a very small threshold $\varepsilon_{\mathrm{QI}} = 10^{-8}$, to obtain an accurate approximation $\widehat{\theta}^*$ of the optimal parameter vector θ^*. Then, in order to compare how synchronous and asynchronous fuzzy Q-iteration approach $\widehat{\theta}^*$, the two algorithms are run until their parameter vectors are closer than 10^{-5} to $\widehat{\theta}^*$ in the infinity norm, i.e., until $\|\theta_\ell - \widehat{\theta}^*\|_\infty \leq 10^{-5}$. For these experiments, as well as throughout the remaining examples of this chapter, the parameter vector of fuzzy Q-iteration is initialized to zero.

Figure 4.4 presents the evolution of the distance between θ_ℓ and $\widehat{\theta}^*$ with the number of iterations ℓ, for both variants of fuzzy Q-iteration. The asynchronous algorithm approaches $\widehat{\theta}^*$ faster than the synchronous one, and gets within a 10^{-5} distance 20 iterations earlier (in 112 iterations, whereas the synchronous algorithm requires 132). Because the time complexity of one iteration is nearly the same for the two algorithms, this generally translates into computational savings for the asynchronous version.

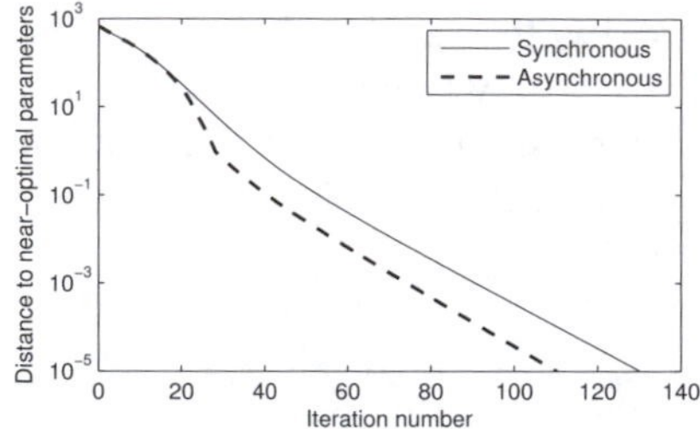

FIGURE 4.4 Convergence of synchronous and asynchronous fuzzy Q-iteration.

Figure 4.5 shows a representative fuzzy Q-iteration solution for the DC motor (specifically the solution corresponding to $\widehat{\theta}^*$). A state-dependent slice through the approximate Q-function is shown (obtained by setting the action u to 0), together with a greedy policy resulting from this Q-function and a controlled trajectory.

For the experiments in the remainder of this chapter, we will always employ synchronous fuzzy Q-iteration,[6] while referring to it simply as "fuzzy Q-iteration," for the sake of conciseness. Because small convergence thresholds $\varepsilon_{\mathrm{QI}}$ will be imposed, the parameter vectors obtained will always be near optimal, and therefore near to those that would be obtained by the asynchronous algorithm. Hence, the conclusions of the experiments also apply to asynchronous fuzzy Q-iteration.

Consistency and the effect of discontinuous rewards

Next, we investigate how the quality of the fuzzy Q-iteration solution changes as the approximation power increases, to illustrate the practical impact of the consistency properties of the algorithm. A triangular fuzzy partition with N' equidistant cores for each state variable is defined, leading to a total number of $N = N'^2$ fuzzy sets. The value of N' is gradually increased from 3 to 41. Similarly, the action is discretized

[6]The reason for this choice is implementation-specific. Namely, each synchronous iteration can be rewritten as a matrix multiplication, which in our MATLAB® implementation is executed using highly efficient low-level routines. The matrix implementation of synchronous fuzzy Q-iteration therefore runs much faster than the element-by-element updates of asynchronous fuzzy Q-iteration. If a specialized library for linear algebra were not available, the asynchronous algorithm would have the same cost per iteration as the synchronous one, and would be preferable because it can converge in fewer iterations, as predicted by the theory.

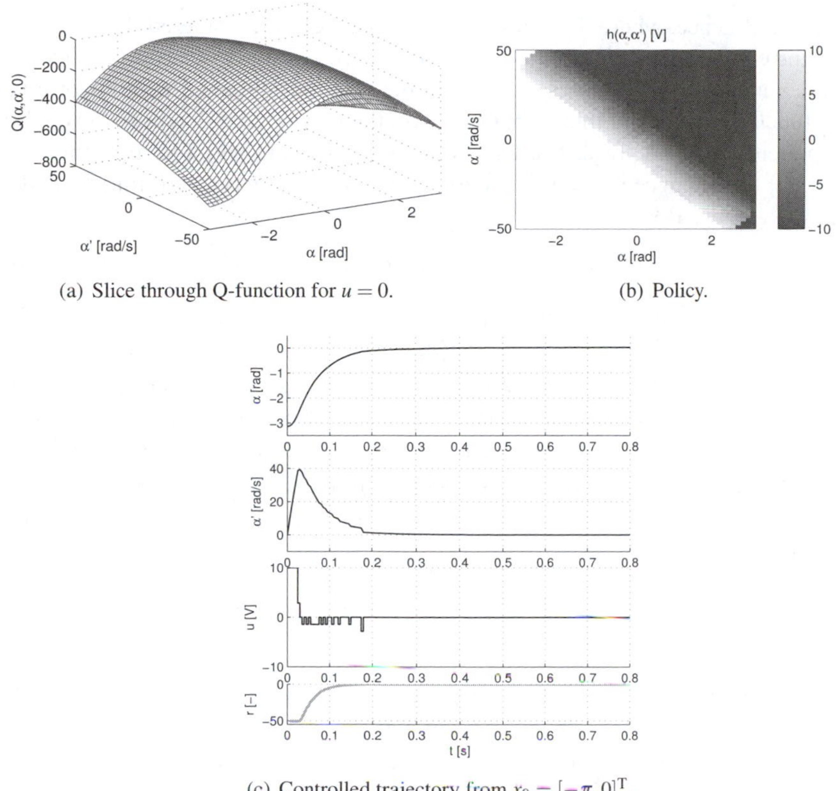

(a) Slice through Q-function for $u = 0$.

(b) Policy.

(c) Controlled trajectory from $x_0 = [-\pi, 0]^{\mathrm{T}}$.

FIGURE 4.5 A fuzzy Q-iteration solution for the DC motor.

into M equidistant values, with $M \in \{3, 5, \ldots, 15\}$ (only odd values are used because the 0 action is necessary to avoid chattering). The convergence threshold is set to $\varepsilon_{\mathrm{QI}} = 10^{-5}$ to ensure that the obtained parameter vector is close to the fixed point of the algorithm.

In a first set of experiments, fuzzy Q-iteration is run for each combination of N and M, with the original reward function (4.45). Recall that the consistency of fuzzy Q-iteration was proven under Lipschitz continuity assumptions on the dynamics and rewards (Assumption 4.1). Indeed, the transition function (4.44) of the DC motor is Lipschitz continuous with a Lipschitz constant $L_f \leq \max\{\|A\|_2, \|B\|_2\}$ (this bound for L_f holds for any system with linear dynamics), and the reward function (4.45) is also Lipschitz continuous, since it is smooth and has bounded support. So, for this first set of experiments, the consistency of fuzzy Q-iteration is guaranteed, and its solutions are expected to improve as N and M increase.

The aim of a second set of experiments is to study the practical effect of violating Lipschitz continuity, by adding discontinuities to the reward function. Discontinuous rewards are common practice due to the origins of reinforcement learning (RL) in artificial intelligence, where discrete-valued tasks are often considered. In our experiments, the choice of the discontinuous reward function cannot be arbitrary. Instead, to ensure a meaningful comparison between the solutions obtained with the original reward function (4.45) and those obtained with the new reward function, the quality of the policies must be preserved. One way to do this is to add a term of the form $\gamma\psi(f(x,u)) - \psi(x)$ to each reward $\rho(x,u)$, where $\psi : X \to \mathbb{R}$ is an arbitrary bounded function (Ng et al., 1999):

$$\rho'(x,u) = \rho(x,u) + \gamma\psi(f(x,u)) - \psi(x) \tag{4.46}$$

The quality of the policies is preserved by reward modifications of this form, in the sense that for any policy h, $Q^h_{\rho'} - Q^*_{\rho'} = Q^h_\rho - Q^*_\rho$, where Q_ρ denotes a Q-function under the reward function ρ, and $Q_{\rho'}$ a Q-function under ρ'. Indeed, it is easy to show by replacing ρ' in the expression (2.2) for the Q-function, that for any policy h, including any optimal policy, we have that $Q^h_{\rho'}(x,u) = Q^h_\rho(x,u) - \psi(x)\ \forall x,u$ (Ng et al., 1999). In particular, a policy is optimal for ρ' if and only if it is optimal for ρ.

We choose a discontinuous function ψ, which is positive only in a rectangular region around the origin:

$$\psi(x) = \begin{cases} 10 & \text{if } |x_1| \leq \pi/4 \text{ and } |x_2| \leq 4\pi \\ 0 & \text{otherwise} \end{cases} \tag{4.47}$$

With this form of ψ, the newly added term in (4.46) rewards transitions that take the state inside the rectangular region, and penalizes transitions that take it outside. A representative slice through the resulting reward function is presented in Figure 4.6 (compare with Figure 4.3). The additional positive rewards are visible as crests above the quadratic surface. The penalties are not visible in the figure, because their corresponding, downward-oriented crests are situated under the surface.

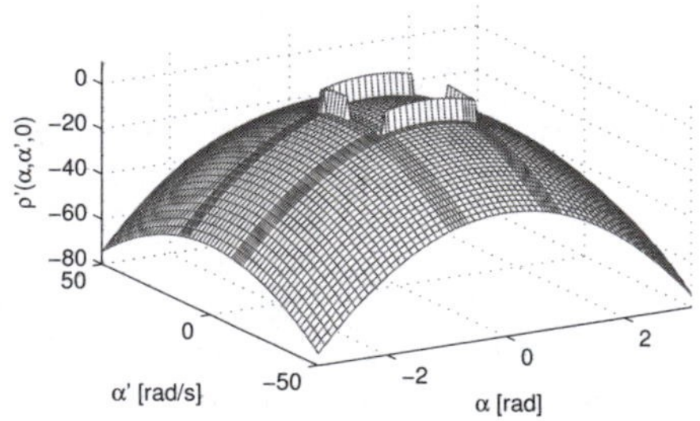

FIGURE 4.6
A slice through the modified reward function (4.46), for $u = 0$. Reproduced with permission from (Buşoniu et al., 2008b), © 2008 IEEE.

The performance (score) of the policies obtained with fuzzy Q-iteration is given in Figure 4.7. Each point in these graphs corresponds to the return of the policy obtained, averaged over the grid of initial states:

$$X_0 = \{-\pi, -5\pi/6, -4\pi/6, \ldots, \pi\} \times \{-16\pi, -14\pi, \ldots, 16\pi\} \tag{4.48}$$

The returns are evaluated using simulation (4.39), with a precision of $\varepsilon_{\mathrm{MC}} = 0.1$. While the reward functions used for Q-iteration are different, the performance evaluation is always done with the reward (4.45), to allow an easier comparison. As already explained, the change in the reward function preserves the quality of the policies, so comparing policies in this way is meaningful. The qualitative evolution of the performance is similar when evaluated with the modified reward function (4.46) with ψ as in (4.47).

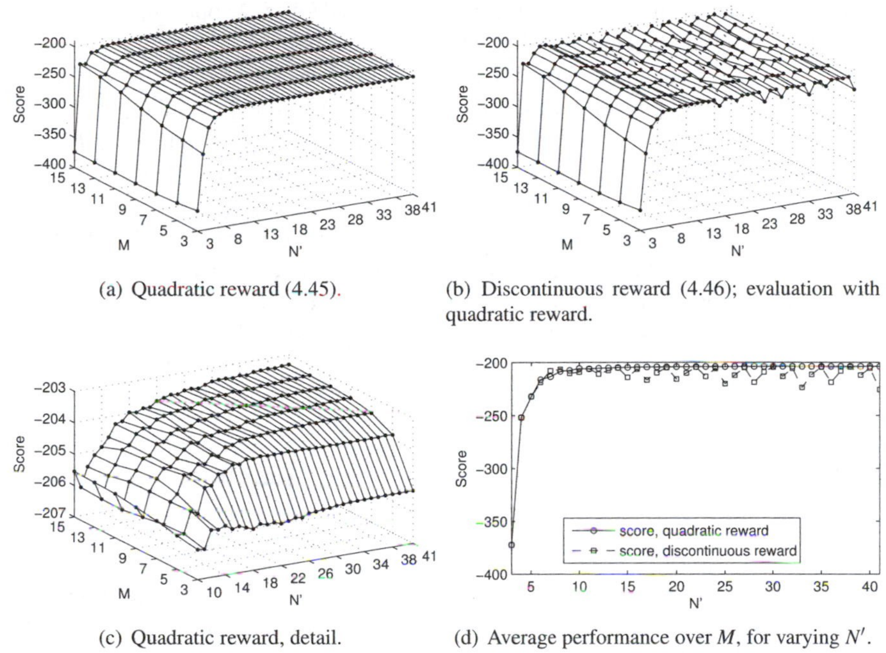

(a) Quadratic reward (4.45).

(b) Discontinuous reward (4.46); evaluation with quadratic reward.

(c) Quadratic reward, detail.

(d) Average performance over M, for varying N'.

FIGURE 4.7
The performance of fuzzy Q-iteration as a function of N and M, for quadratic and discontinuous rewards in the DC motor problem. Reproduced with permission from (Buşoniu et al., 2008b), © 2008 IEEE.

When the continuous reward is used, the performance of fuzzy Q-iteration is already near optimal for $N' = 20$ and is relatively smooth for $N' \geq 20$, see Figures 4.7(a) and 4.7(c). Also, the influence of the number of discrete actions is small for $N' \geq 4$. So, the consistency properties of fuzzy Q-iteration have a clear beneficial effect. However, when the reward is changed to the discontinuous function (4.46), thus violating the assumptions for consistency, the performance indeed varies significantly as N' increases, see Figure 4.7(b). For many values of N', the influence of M also becomes significant. Additionally, for many values of N' the performance is worse than with the continuous reward function, see Figure 4.7(d).

An interesting and somewhat counterintuitive fact is that the performance is not monotonic in N' and M. For a given value of N', the performance sometimes *de-*

creases as M increases. A similar effect occurs as M is kept fixed and N' varies. This effect is present with both reward functions, but is more pronounced in Figure 4.7(b) than in Figures 4.7(a) and 4.7(c). The magnitude of the changes decreases significantly as N' and M become large in Figures 4.7(a) and 4.7(c); this is not the case in Figure 4.7(b).

The negative effect of reward discontinuities on the consistency of the algorithm can be intuitively explained as follows. The discontinuous reward function (4.46) leads to discontinuities in the optimal Q-function. As the placement of the MFs changes with increasing N', the accuracy with which the fuzzy approximator captures these discontinuities changes as well. This accuracy depends less on the *number* of MFs, than on their *positions* (MFs should be ideally concentrated around the discontinuities). So, it may happen that for a certain, smaller value of N' the performance is better than for another, larger value. In contrast, the smoother optimal Q-function resulting from the continuous reward function (4.45) is easier to approximate using triangular MFs.

A similar behavior of fuzzy Q-iteration was observed in additional experiments with other discontinuous reward functions. In particular, adding more discontinuities similar to those in (4.46) does not significantly influence the evolution of the performance in Figure 4.7(b). Decreasing the magnitude of the discontinuities (e.g., replacing the value 10 by 1 in (4.46)) decreases the magnitude of the performance variations, but they are still present and they do not decrease as N and M increase.

4.5.2 Two-link manipulator: Effects of action interpolation, and comparison with fitted Q-iteration

In this section, fuzzy Q-iteration is applied to stabilize a two-link manipulator operating in a horizontal plane. Using this problem, the effects of employing the continuous-action, interpolated policy (4.15) are investigated, and fuzzy Q-iteration is compared with fitted Q-iteration (Algorithm 3.4 of Section 3.4.3). The two-link manipulator example also illustrates that fuzzy Q-iteration works well in problems having a higher dimensionality than the DC motor of Section 4.5.1; the two-link manipulator has four state variables and two action variables.

Two-link manipulator problem

The two-link manipulator, depicted in Figure 4.8, is described by the fourth-order, continuous-time nonlinear model:

$$M(\alpha)\ddot{\alpha} + C(\alpha, \dot{\alpha})\dot{\alpha} = \tau \tag{4.49}$$

where $\alpha = [\alpha_1, \alpha_2]^{\mathrm{T}}$ contains the angular positions of the two links, $\tau = [\tau_1, \tau_2]^{\mathrm{T}}$ contains the torques of the two motors, $M(\alpha)$ is the mass matrix, and $C(\alpha, \dot{\alpha})$ is the Coriolis and centrifugal forces matrix. The state signal contains the angles and angular velocities: $x = [\alpha_1, \dot{\alpha}_1, \alpha_2, \dot{\alpha}_2]^{\mathrm{T}}$, and the control signal is $u = \tau$. The angles α_1, α_2 vary in the interval $[-\pi, \pi)$ rad, and "wrap around" so that, e.g., a rotation of $3\pi/2$ for the first link corresponds to a value $\alpha_1 = -\pi/2$. The angular velocities $\dot{\alpha}_1, \dot{\alpha}_2$

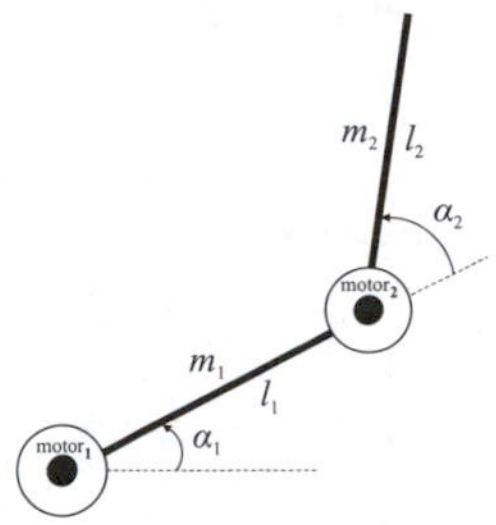

FIGURE 4.8 Schematic representation of the two-link manipulator.

are restricted to the interval $[-2\pi, 2\pi]$ rad/s using saturation, while the torques are constrained as follows: $\tau_1 \in [-1.5, 1.5]$ Nm, $\tau_2 \in [-1, 1]$ Nm. The discrete time step is set to $T_s = 0.05$ s, and the discrete-time dynamics f are obtained by numerically integrating (4.49) between consecutive time steps.

The matrices $M(\alpha)$ and $C(\alpha, \dot{\alpha})$ have the following form:

$$\begin{aligned} M(\alpha) &= \begin{bmatrix} P_1 + P_2 + 2P_3 \cos\alpha_2 & P_2 + P_3 \cos\alpha_2 \\ P_2 + P_3 \cos\alpha_2 & P_2 \end{bmatrix} \\ C(\alpha, \dot{\alpha}) &= \begin{bmatrix} b_1 - P_3 \dot{\alpha}_2 \sin\alpha_2 & -P_3(\dot{\alpha}_1 + \dot{\alpha}_2)\sin\alpha_2 \\ P_3 \dot{\alpha}_1 \sin\alpha_2 & b_2 \end{bmatrix} \end{aligned} \tag{4.50}$$

The meaning and values of the physical variables in the system are given in Table 4.1. Using these, the rest of the parameters in (4.50) can be computed as follows: $P_1 = m_1 c_1^2 + m_2 l_1^2 + I_1$, $P_2 = m_2 c_2^2 + I_2$, and $P_3 = m_2 l_1 c_2$.

TABLE 4.1 Parameters of the two-link manipulator.

Symbol	**Value**	**Units**	**Meaning**
$l_1; l_2$	0.4; 0.4	m	link lengths
$m_1; m_2$	1.25; 0.8	kg	link masses
$I_1; I_2$	0.066; 0.043	kg m^2	link inertias
$c_1; c_2$	0.2; 0.2	m	center of mass coordinates
$b_1; b_2$	0.08; 0.02	kg/s	damping in the joints

The control goal is the stabilization of the system around $\alpha = \dot{\alpha} = 0$, and is expressed by the quadratic reward function:

$$\rho(x, u) = -x^{\mathrm{T}} Q_{\mathrm{rew}} x, \quad \text{with } Q_{\mathrm{rew}} = \mathrm{diag}[1, 0.05, 1, 0.05] \tag{4.51}$$

The discount factor is set to $\gamma = 0.98$, which is large enough to allow rewards around the goal state to influence the values of states early in the trajectories, leading to an optimal policy that successfully stabilizes the manipulator.

Results of fuzzy Q-iteration, and effects of using interpolated actions

To apply fuzzy Q-iteration, triangular fuzzy partitions are defined for every state variable and then combined, as in Example 4.1. For the angles, a core is placed in the origin, and 6 logarithmically-spaced cores are placed on each side of the origin. For the velocities, a core is placed in the origin, and 3 logarithmically-spaced cores are used on each side of the origin. This leads to a total number of $(2 \cdot 6+1)^2 \cdot (2 \cdot 3+1)^2 = 8281$ MFs. The cores are spaced logarithmically to ensure a higher accuracy of the solution around the origin, while using only a limited number of MFs. This represents a mild form of prior knowledge about the importance of the state space region close to the origin. Each torque variable is discretized using 5 values: $\tau_1 \in \{-1.5, -0.36, 0, 0.36, 1.5\}$ and $\tau_2 \in \{-1, -0.24, 0, 0.24, 1\}$. These values are logarithmically spaced along the two axes of the action space. The convergence threshold is set to $\varepsilon_{\text{QI}} = 10^{-5}$.

With these values, fuzzy Q-iteration converged after 426 iterations. Figure 4.9 compares the discrete-action results with the corresponding continuous-action results. In particular, Figure 4.9(a) depicts the discrete-action policy given by (4.13), while Figure 4.9(b) depicts the interpolated, continuous-action policy computed with (4.15). The continuous-action policy is, of course, smoother than the discrete-action policy. Figures 4.9(c) and 4.9(d) show two representative trajectories of the two-link manipulator, controlled by the discrete-action and continuous-action policies, respectively. Both policies are able to stabilize the system after about 2 s. However, the discrete-action policy leads to more chattering of the control action and to a steady-state error for the angle of the second link, whereas the continuous-action policy alleviates these problems.

Compared to the DC motor, a larger number of triangular MFs and discrete actions are generally required to represent the Q-function for the manipulator problem, and the computational and memory demands of fuzzy Q-iteration increase accordingly. In fact, they increase exponentially with the number of state-action variables. For concreteness, assume that for a general problem with D state variables and C action variables, N' triangular MFs are defined along each state dimension, and each action dimension is discretized into M' actions. Then $N'^D M'^C$ parameters are required, leading to a time complexity per iteration of $\mathrm{O}(N'^{2D} M')$ and to a memory complexity of $\mathrm{O}(N'^D M')$ (see also Section 4.3.3). The DC motor therefore requires $N'^2 M'$ parameters, $\mathrm{O}(N^4 M')$ computations per iteration, and $\mathrm{O}(N^2 M')$ memory, whereas the manipulator requires $N'^4 M'^2$ parameters (i.e., $N'^2 M'$ times more than the DC motor), $\mathrm{O}(N^8 M'^2)$ computations per iteration, and $\mathrm{O}(N^4 M'^2)$ memory.

Comparison with fitted Q-iteration

Next, we compare the solution of fuzzy Q-iteration with a solution obtained by fitted Q-iteration with a nonparametric approximator (Algorithm 3.4). Even though fitted Q-iteration is a model-free, sample-based algorithm, it can easily be adapted to the model-based setting considered in this chapter by using the model to generate the samples. In order to make the comparison between the two algorithms more meaningful, fitted Q-iteration is supplied with the same state-action samples as those

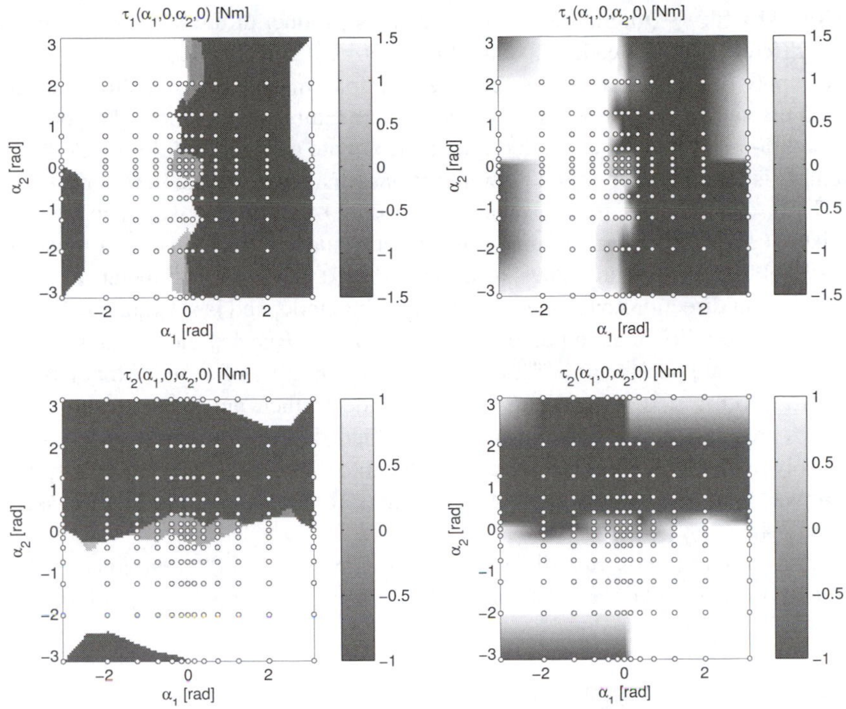

(a) A slice through the discrete-action policy, for $\dot{\alpha}_1 = \dot{\alpha}_2 = 0$ and parallel to the plane (α_1, α_2). The fuzzy cores for the angle variables are represented as small white disks with dark edges.

(b) A similarly obtained slice through the continuous-action policy.

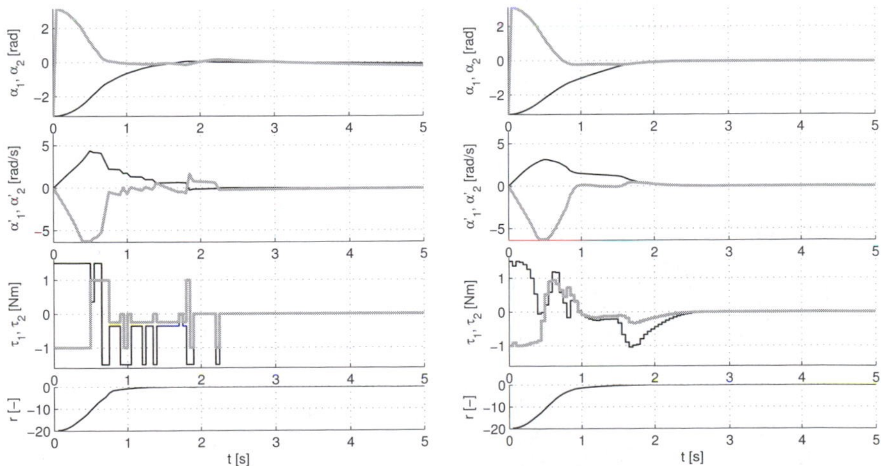

(c) A trajectory controlled by the discrete-action policy (thin black line – link 1, thick gray line – link 2). The initial state is $x_0 = [-\pi, 0, -\pi, 0]^T$.

(d) A trajectory from $x_0 = [-\pi, 0, -\pi, 0]^T$, controlled by the continuous-action policy.

FIGURE 4.9

Results of fuzzy Q-iteration for the two-link manipulator. The discrete-action results are shown on the left-hand side of the figure, and the continuous-action results on the right-hand side.

employed by fuzzy Q-iteration, namely the cross-product of the 8281 MF cores and the 25 discrete actions, leading to a total number of 207 025 samples.

To apply fitted Q-iteration, we choose a nonparametric approximator that combines a discretization of the action space with ensembles of extremely randomized trees (extra-trees) (Geurts et al., 2006) to approximate over the state space. A distinct ensemble is used for each of the discrete actions, in analogy to the fuzzy approximator. The discrete actions are the same as for fuzzy Q-iteration above. Each ensemble consists of $N_{\mathrm{tr}} = 50$ extremely randomized trees, and the tree construction parameters are set to their default values, as described next. The first parameter, K_{tr}, is the number of cut directions evaluated when splitting a node, and is set equal to 4, which is the dimensionality of the input to the regression trees (the 4-dimensional state variable). The second parameter, $n_{\mathrm{tr}}^{\mathrm{min}}$, is the minimum number of samples that has to be associated with a node in order to split that node further, and is set equal to 2, so the trees are fully developed. For a more detailed description of the ensembles of extremely randomized trees, see Appendix A. Note that a similar Q-function approximator was used in our application of fitted Q-iteration to the DC motor, discussed in Section 3.4.5.

Fitted Q-iteration is run for a predefined number of 400 iterations, and the Q-function found after the 400th iteration is considered satisfactory. Figure 4.10

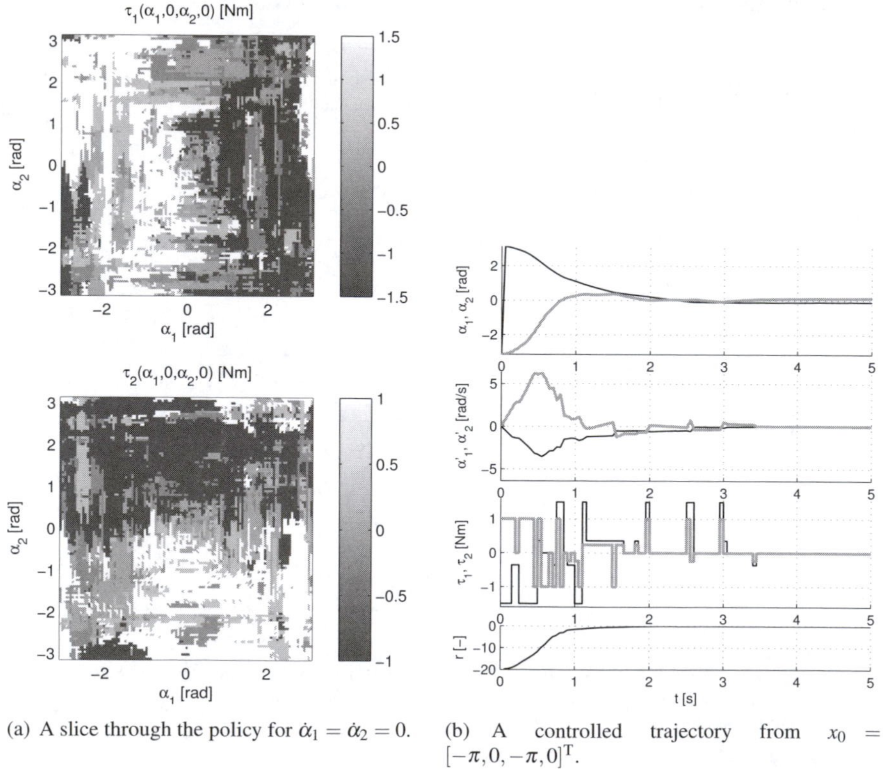

(a) A slice through the policy for $\dot{\alpha}_1 = \dot{\alpha}_2 = 0$.

(b) A controlled trajectory from $x_0 = [-\pi, 0, -\pi, 0]^{\mathrm{T}}$.

FIGURE 4.10 Results of fitted Q-iteration for the two-link manipulator.

presents a greedy policy resulting from this Q-function, together with a representative controlled trajectory. Although it roughly resembles the fuzzy Q-iteration policies of Figures 4.9(a) and 4.9(b), the fitted Q-iteration policy of Figure 4.10(a) contains spurious (and probably incorrect) actions for many states. The policy obtained by fitted Q-iteration stabilizes the system more poorly in Figure 4.10(b), than the solution of fuzzy Q-iteration in Figures 4.9(c) and 4.9(d). So, in this case, fuzzy Q-iteration with triangular MFs outperforms fitted Q-iteration with extra-trees approximation.

Note that instead of building a distinct ensemble of extra-trees for each of the discrete actions, fitted Q-iteration could also work with a single ensemble of trees that take continuous state-continuous action pairs as inputs. This might lead to a better performance, as it would allow the algorithm to identify structure along the action dimensions of the Q-functions. However, it would also make the results less comparable with those of fuzzy Q-iteration, which always requires action discretization, and for this reason we do not adopt this solution here.

4.5.3 Inverted pendulum: Real-time control

Next, fuzzy Q-iteration is used to swing up and to stabilize a real-life underactuated inverted pendulum. This application illustrates the performance of the fuzzy Q-iteration solutions in real-time control.

Inverted pendulum problem

The inverted pendulum is obtained by placing a mass off-center on a disk that rotates in a vertical plane and is driven by a DC motor (Figure 4.11).[7] Note that this DC motor is the same system which was modeled for use in simulations in Section 4.5.1, and earlier throughout the examples of Chapter 3. The control voltage is limited so that the motor does not provide enough power to push the pendulum up in a single rotation. Instead, the pendulum needs to be swung back and forth (destabilized) to gather energy, prior to being pushed up and stabilized. This creates a difficult, highly nonlinear control problem.

The continuous-time dynamics of the inverted pendulum are:

$$\ddot{\alpha} = \frac{1}{J}\left(mgl\sin(\alpha) - b\dot{\alpha} - \frac{K^2}{R}\dot{\alpha} + \frac{K}{R}u \right) \tag{4.52}$$

Table 4.2 shows the meanings and values of the parameters appearing in this equation. Note that some of these parameters (e.g., J and m) are rough estimates, and that the real system exhibits unmodeled dynamics such as static friction. The state signal consists of the angle and the angular velocity of the pendulum, i.e., $x = [\alpha, \dot{\alpha}]^{\mathrm{T}}$. The angle α "wraps around" in the interval $[-\pi, \pi)$ rad, where $\alpha = -\pi$ corresponds to pointing down and $\alpha = 0$ corresponds to pointing up. The velocity $\dot{\alpha}$ is restricted to

[7] This is different from the classical cart-pendulum system, in which the pendulum is attached to a cart and is indirectly actuated via the acceleration of the cart (e.g., Doya, 2000; Riedmiller et al., 2007). Here, the pendulum is actuated directly, and the system only has two state variables, as opposed to the cart-pendulum, which has four.

(a) The real inverted pendulum system.

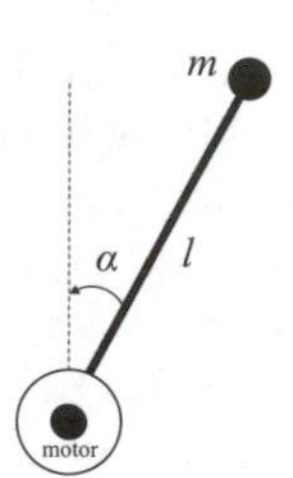

(b) A schematic representation.

FIGURE 4.11 The inverted pendulum.

the interval $[-15\pi, 15\pi]$ rad/s using saturation, and the control action (voltage) u is constrained to $[-3, 3]$ V. The sample time T_s is chosen to be 0.005 s, and the discrete-time dynamics f are obtained by numerically integrating (4.52) between consecutive time steps.

TABLE 4.2 Parameters of the inverted pendulum.

Symbol	Value	Units	Meaning
m	0.055	kg	mass
g	9.81	m/s^2	gravitational acceleration
l	0.042	m	distance from center of disk to mass
J	$1.91 \cdot 10^{-4}$	kg m^2	moment of inertia
b	$3 \cdot 10^{-6}$	Nms/rad	viscous damping
K	0.0536	Nm/A	torque constant
R	9.5	Ω	rotor resistance

The goal is to stabilize the pendulum in the unstable equilibrium $x = 0$ (pointing up). The following quadratic reward function is chosen to express this goal:

$$\begin{aligned} \rho(x,u) &= -x^{\mathrm{T}} Q_{\mathrm{rew}} x - R_{\mathrm{rew}} u^2 \\ Q_{\mathrm{rew}} &= \begin{bmatrix} 5 & 0 \\ 0 & 0.1 \end{bmatrix}, \quad R_{\mathrm{rew}} = 1 \end{aligned} \tag{4.53}$$

The discount factor is $\gamma = 0.98$. This discount factor is large so that rewards around the goal state (pointing up) influence the values of states early in the trajectories. This leads to an optimal policy that successfully swings up and stabilizes the pendulum.

Results of fuzzy Q-iteration

Triangular fuzzy partitions with 19 equidistant cores are defined for both state variables, and then combined as in Example 4.1. This relatively large number of MFs is chosen to ensure a good accuracy of the solution. The control action is discretized using 5 equidistant values, and the convergence threshold is set to $\varepsilon_{\mathrm{QI}} = 10^{-5}$.

With these settings, fuzzy Q-iteration converged after 659 iterations. Figure 4.12 shows the solution obtained, together with controlled trajectories (swing-ups) of the simulated and real-life pendulum, starting from the stable equilibrium $x_0 = [-\pi, 0]^{\mathrm{T}}$ (pointing down). In particular, Figure 4.12(c) is the trajectory of the simulation model (4.52), while Figure 4.12(d) is a trajectory of the real system. For the real system, only the angle is measured, and the angular velocity is estimated using a discrete difference, which results in a noisy signal. Even though the model is simplified and does not include effects such as measurement noise and static friction, the policy resulting from fuzzy Q-iteration performs well: it stabilizes the real system in about 1.5 s, around 0.25 s longer than in simulation. This discrepancy is due to the differences between the model and the real system. Note that, because only discrete actions are available, the control action chatters.

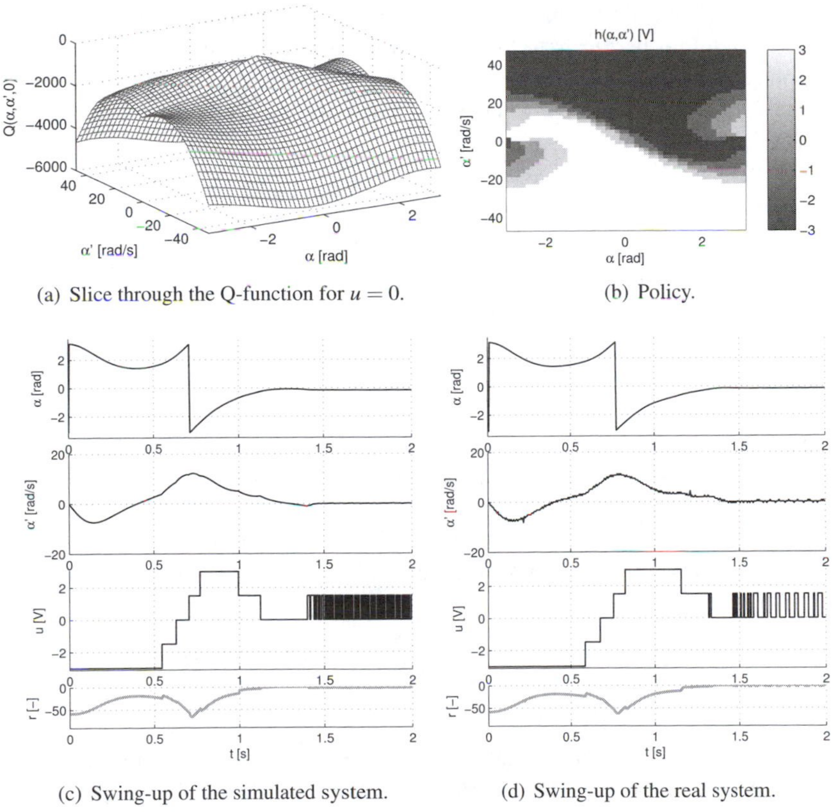

(a) Slice through the Q-function for $u = 0$.

(b) Policy.

(c) Swing-up of the simulated system.

(d) Swing-up of the real system.

FIGURE 4.12 Results of fuzzy Q-iteration for the inverted pendulum.

4.5.4 Car on the hill: Effects of membership function optimization

In this section, we study empirically the performance of fuzzy Q-iteration with CE optimization of the MFs (Algorithm 4.3). To this end, we apply fuzzy Q-iteration with optimized MFs to the car-on-the-hill problem (Moore and Atkeson, 1995), and compare the results with those of fuzzy Q-iteration with equidistant MFs.

Car-on-the-hill problem

The car on the hill is widely used as a benchmark in approximate DP/RL. It was first described by Moore and Atkeson (1995), and was used, e.g., by Munos and Moore (2002) as a primary benchmark for V-iteration with resolution refinement, and by Ernst et al. (2005) to validate fitted Q-iteration. In the car-on-the-hill problem, a point mass (the "car") must be driven past the top of a frictionless hill by applying a horizontal force, see Figure 4.13. For some initial states, the maximum available force is not sufficient to drive the car directly up the hill. Instead, it has to be driven up the opposite slope (left) and gather energy prior to accelerating towards the goal (right). This problem is roughly similar to the inverted pendulum of Section 4.5.3; there, the pendulum had to be swung back and forth to gather energy, which here corresponds to driving the car left and then right. An important difference is that the pendulum had to be stabilized, whereas the car only has to be driven past the top, which is easier to do.

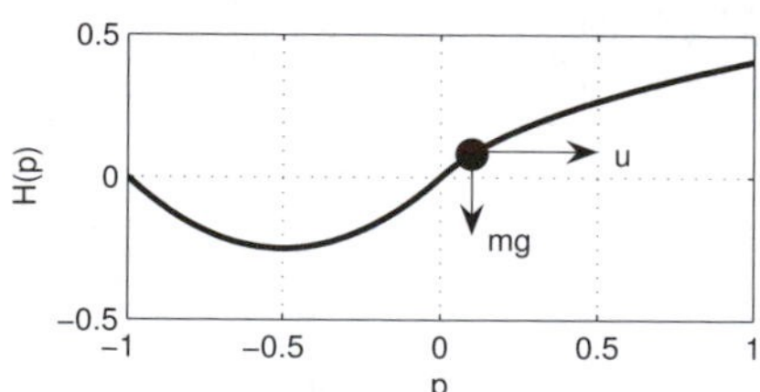

FIGURE 4.13
The car on the hill. The "car" is represented as a black bullet, and its goal is to drive out of the figure to the right.

The continuous-time dynamics of the car are (Moore and Atkeson, 1995; Ernst et al., 2005):

$$\ddot{p} = \frac{1}{1 + \left(\frac{\mathrm{d}H(p)}{\mathrm{d}p}\right)^2} \left(u - g\,\frac{\mathrm{d}H(p)}{\mathrm{d}p} - \dot{p}^2\,\frac{\mathrm{d}H(p)}{\mathrm{d}p}\,\frac{\mathrm{d}^2 H(p)}{\mathrm{d}^2 p} \right) \tag{4.54}$$

where $p \in [-1,1]$ m is the horizontal position of the car, $\dot{p} \in [-3,3]$ m/s is its velocity, $u \in [-4,4]$ N is the horizontal force applied, $g = 9.81$ m/s^2 is the gravitational acceleration, and H denotes the shape of the hill, which is given by:

$$H(p) = \begin{cases} p^2 + p & \text{if } p < 0 \\ \frac{p}{\sqrt{1+5p^2}} & \text{if } p \geq 0 \end{cases}$$

Furthermore, a unity mass of the car is assumed. The discrete time step is set to $T_s = 0.1$ s, and the discrete-time dynamics f are obtained by numerically integrating (4.54) between consecutive time steps.

The state signal consists of the position and velocity of the car, $x = [p, \dot{p}]^T$, while the control action u is the applied force. The state space is $X = [-1, 1] \times [-3, 3]$ plus a terminal state (see below), and the action space is $U = [-4, 4]$. Whenever the position or velocity exceed the bounds, the car reaches the terminal state, from which it can no longer escape, and the trial terminates. Throughout the remainder of this example, the action space is discretized into $U_d = \{-4, 4\}$. These two values are sufficient to obtain a good solution, given that the car does not have to be stabilized, but only driven past the top of the hill, which only requires it to be fully accelerated towards the left and right.

The goal is to drive past the top of the hill to the right with a speed within the allowed limits. Reaching a terminal state in any other way is considered a failure. The reward function chosen to express this goal is:

$$r_{k+1} = \rho(x_k, u_k) = \begin{cases} -1 & \text{if } x_{1,k+1} < -1 \text{ or } |x_{2,k+1}| > 3 \\ 1 & \text{if } x_{1,k+1} > 1 \text{ and } |x_{2,k+1}| \leq 3 \\ 0 & \text{otherwise} \end{cases} \tag{4.55}$$

The discount factor is $\gamma = 0.95$.

This reward function is represented in Figure 4.14(a). It is chosen to be discontinuous to provide a challenging problem for the MF optimization algorithm, by making the Q-function difficult to approximate. To illustrate this difficulty, Figure 4.14(b) depicts an approximately optimal Q-function. This Q-function was obtained with fuzzy Q-iteration using a very fine fuzzy partition, which contains 401×301 MFs. (Even though the consistency of fuzzy Q-iteration is not guaranteed since the reward is discontinuous, this fine partition should at least lead to a rough approximation of the optimal Q-function.) Clearly, the large number of discontinuities appearing in this Q-function make it difficult to approximate.

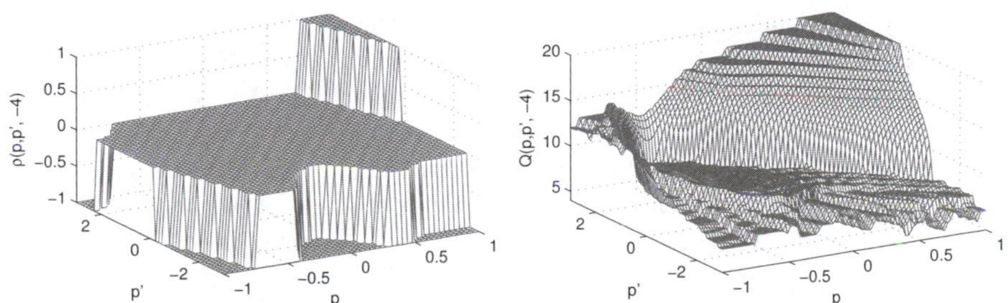

(a) A slice through ρ for $u = -4$.

(b) A slice through an approximately optimal Q-function, also for $u = -4$.

FIGURE 4.14
Reward function and an approximately optimal Q-function for the car on the hill.

Results of fuzzy Q-iteration with MF optimization

To apply fuzzy Q-iteration with CE optimization of the MFs, triangular fuzzy partitions are defined for both state variables and then combined as in Example 4.1. The number of MFs is chosen to be the same for each of the two variables, and is denoted by N'. This number is gradually increased from 3 to 20.[8] A given value of N' corresponds to a total number of $N = N'^2$ MFs in the fuzzy partition of X. To compute the score function (optimization criterion) (4.38), the following equidistant grid of representative states is chosen:

$$X_0 = \{-1, -0.75, -0.5, \ldots, 1\} \times \{-3, -2, -1, \ldots, 3\}$$

and each point is weighted by $\frac{1}{|X_0|}$. Since the representative states are uniformly distributed and equally weighted, the algorithm is expected to lead to a uniformly good performance across the state space. The parameters of the CE optimization method are set to typical values, as follows: $N_{\text{CE}} = 5 \cdot 2 \cdot N_\xi$, $\rho_{\text{CE}} = 0.05$, and $d_{\text{CE}} = 5$. The number of samples N_{CE} is set to be 5 times the number of parameters needed to describe the probability density used in CE optimization. Recall that one mean and one standard deviation are needed to describe the Gaussian density for each of the N_ξ MF parameters. In turn, $N_\xi = (N' - 2)^2$, because there are $N' - 2$ free cores to optimize along each of the two axes of X. Additionally, the maximum number of CE iterations is set to $\tau_{\max} = 50$, and the same value 10^{-3} is used as admissible error ε_{MC} in the return estimation, as fuzzy Q-iteration convergence threshold ε_{QI}, and as CE convergence threshold ε_{CE}. With these settings, 10 independent runs are performed for each value of N'.

Figure 4.15 compares the results obtained using *optimized* MFs with those obtained using the same number of *equidistant* MFs. In particular, Figure 4.15(a) shows the mean score across the 10 independent runs of the MF optimization algorithm, together with 95% confidence intervals on this mean. This figure also includes the performance with equidistant MFs, and the best possible performance that can be obtained with the two discrete actions.[9] The optimized MFs reliably provide a better performance than the same number of equidistant MFs. For $N' \geq 12$, they lead to a nearly optimal performance. As also observed in the consistency study of Section 4.5.1, the discontinuous reward function leads to unpredictable variations of the performance as the number of equidistant MFs is increased. Optimizing the MFs recovers a more predictable performance increase, because the MFs are adjusted to better represent the discontinuities of the Q-function. Figure 4.15(b) shows the computational cost of fuzzy Q-iteration with MF optimization and with equidistant MFs.

[8] The experiments stop at 20 MFs to limit the computation time per experiment in the order of hours. To run the experiments, we used MATLAB 7 on a PC with an Intel T2400 1.83 GHz CPU and 2 GB RAM.

[9] This optimal performance is obtained using the following brute-force procedure. All the possible sequences of actions of a sufficient length K are generated, and the system is controlled in an open-loop fashion with all these sequences, starting from every state x_0 in X_0. For a given state x_0, the largest discounted return obtained in this way is optimal under the action discretization U_{d}. The length K is sufficient if, from any initial state in X_0, an optimal trajectory leads to a terminal state after at most K steps.

The performance gained by optimizing the MFs comes at a large computational cost, several orders of magnitude higher than the cost incurred by the equidistant MFs.

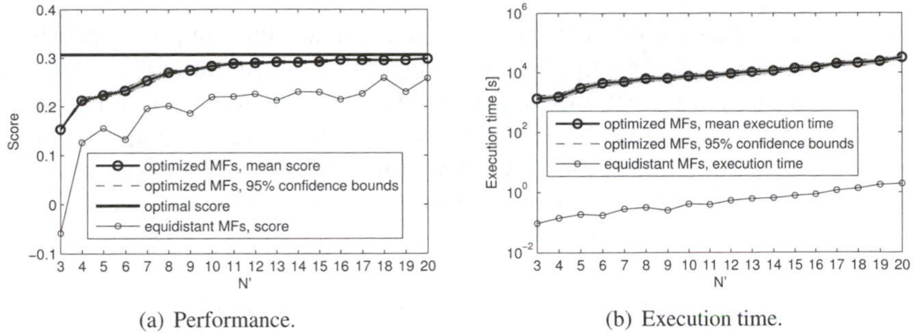

(a) Performance. (b) Execution time.

FIGURE 4.15
Comparison between fuzzy Q-iteration with optimized and equidistant MFs for the car on the hill.

Figure 4.16 presents a representative set of final, optimized MFs. In this figure, the number of MFs on each axis is $N' = 10$. To better understand this placement of the MFs, see again the approximately optimal Q-function of Figure 4.14(b). It is impossible to capture all the discontinuities of this Q-function with only 10 MFs on each axis. Instead, the MF optimization algorithm concentrates most of the MFs in the region of the state space where $p \approx -0.8$. In this region the car, having accumulated sufficient energy, has to stop moving left and accelerate toward the right; this is a critical control decision. Therefore, this placement of MFs illustrates that, when the number of MFs is insufficient to accurately represent the Q-function over the entire state space, the optimization algorithm focuses the approximator on the regions that *are most important for the performance*. The MFs on the velocity axis $\dot{p}$ are concentrated towards large values, possibly in order to represent more accurately the top-left region of the Q-function, which is most irregular in the neighborhood of $p = -0.8$.

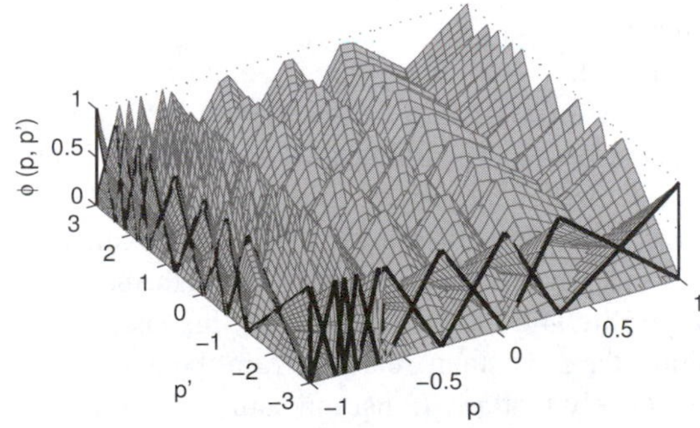

FIGURE 4.16 A representative set of optimized MFs for the car on the hill.

4.6 Summary and discussion

In this chapter, we have considered fuzzy Q-iteration, an algorithm for approximate value iteration that represents Q-functions using a fuzzy partition of the state space and a discretization of the action space. Fuzzy Q-iteration was shown to be convergent to a near-optimal solution, and consistent under continuity assumptions on the dynamics and the reward function. A version of the algorithm where parameters are updated in an asynchronous fashion was proven to converge at least as fast as the synchronous variant. As an alternative to designing the MFs in advance, we have developed a method to optimize the parameters of a constant number of MFs. A detailed experimental study of fuzzy Q-iteration was also performed, which led to the following important conclusions: discontinuous reward functions can harm the performance of fuzzy Q-iteration; in certain problems, fuzzy Q-iteration can outperform fitted Q-iteration with nonparametric approximation; and MF optimization is beneficial for performance but computationally intensive.

While fuzzy Q-iteration has been presented in this chapter as an algorithm for solving problems with deterministic dynamics, it can also be extended to the stochastic case. Consider, for instance, asynchronous fuzzy Q-iteration, given in Algorithm 4.2. In the stochastic case, the parameter update at line 5 of this algorithm would become:

$$\theta_{[i,j]} \leftarrow \mathrm{E}_{x' \sim \tilde{f}(x_i,u_j,\cdot)} \left\{ \tilde{\rho}(x_i,u_j,x') + \gamma \max_{j'} \sum_{i'=1}^{N} \phi_{i'}(x') \theta_{[i',j']} \right\}$$

where x' is sampled using the probability density function $\tilde{f}(x_i,u_j,\cdot)$ of the next state x', given x_i and u_j. In general, the expectation in this update cannot be computed exactly, but must be estimated from a finite number of samples. In this case, our analysis does not apply directly, but the finite-sample results outlined in Section 3.4.4 may help to analyze the effect of the finite-sampling errors. Moreover, in some special cases, e.g., when there is a finite number of possible successor states, the expectation can be computed exactly, in which case the theoretical analysis of this chapter applies after some minor changes.

Model-free, RL algorithms with fuzzy approximation can be derived similarly to the RL algorithms of Section 3.4.2. For instance, the fuzzy approximator can easily be employed in gradient-based Q-learning (Algorithm 3.3), leading to a fuzzy Q-learning algorithm. The theoretical properties of such model-free algorithms can be investigated using the framework of nonexpansive approximators (Section 3.4.4). Another possibility is to learn a model of the MDP from the data (transition samples), and then apply fuzzy Q-iteration to this model. To this end, it suffices to learn, for all the discrete actions, the next states reached from the MF cores and the resulting rewards. Since it is unlikely that any transition samples will be located precisely at the MF cores, the algorithm must learn from samples located nearby, which requires smoothness assumptions on the dynamics and reward function (such as the Lipschitz continuity already assumed for the consistency analysis).

To improve the scalability of the fuzzy approximator to high-dimensional problems, MFs that lead to subexponential complexity should be used (e.g., Gaussians), in combination with techniques to find good MFs automatically, such as the MF optimization technique of Section 4.4. If the computational demands of MF optimization become prohibitive, other approaches for finding the MFs must be explored, such as resolution refinement (Section 3.6.2). Furthermore, action-space approximators more powerful than discretization could be studied, e.g., using a fuzzy partition of the action space. Such approximators may naturally lead to continuous-action policies.

The extensive analysis and experimentation presented in this chapter serve to strengthen the knowledge about approximate value iteration developed in Chapter 3. Continuing along similar lines, the next two chapters consider in detail, respectively, algorithms for approximate policy iteration and for approximate policy search.

Bibliographical notes

This chapter integrates and extends the authors' earlier work on fuzzy Q-iteration (Buşoniu et al., 2008c, 2007, 2008b,d). For the theoretical analysis, certain limiting assumptions made in some of this work were removed, such as an originally discrete action space in (Buşoniu et al., 2008c, 2007), and a restrictive bound on the Lipschitz constant of the process dynamics in (Buşoniu et al., 2008b). The MF optimization approach of Section 4.4 was proposed in (Buşoniu et al., 2008d).

Fuzzy approximators have typically been used in model-free (RL) techniques such as Q-learning (Horiuchi et al., 1996; Jouffe, 1998; Glorennec, 2000) and actor-critic algorithms (Berenji and Vengerov, 2003; Lin, 2003). A method that shares important similarities with our MF optimization approach was proposed by Menache et al. (2005). They applied the CE method to optimize the locations and shapes of a constant number of basis functions for approximate policy evaluation, using the Bellman error as an optimization criterion.

5

Approximate policy iteration for online learning and continuous-action control

This chapter considers a model-free, least-squares algorithm for approximate policy iteration. An online variant of this algorithm is developed, and some important issues that appear in online reinforcement learning are emphasized along the way. Additionally, a procedure to integrate prior knowledge about the policy in this online variant is described, and a continuous-action approximator for the offline variant is introduced. These developments are experimentally evaluated for several control problems.

5.1 Introduction

Whereas Chapter 4 focused on an algorithm for approximate value iteration, the present chapter concerns the second major class of techniques for approximate DP/RL: approximate policy iteration (PI). In PI, policies are evaluated by constructing their value functions, which are then used to find new, improved policies. Approximate PI was reviewed in Section 3.5, and this chapter builds and expands on the foundation given there. In particular, the least-squares policy iteration (LSPI) algorithm (Lagoudakis and Parr, 2003a) is selected, and three extensions to it are introduced: an online variant, an approach to integrate prior knowledge in this variant, and a continuous-action approximator for (offline) LSPI.

The first topic of this chapter is therefore the development of an online variant of LSPI. In online reinforcement learning (RL), a solution is learned from data collected by interacting with the controlled system. A suitable online algorithm, in the first place, must quickly provide a good performance, instead of only at the end of the learning process, as is the case in offline RL. Second, it must explore novel action choices, even at the risk of a temporarily reduced performance, in order to avoid local optima and to eventually achieve a (near-)optimal performance. LSPI is originally offline: it improves the policy only after an accurate Q-function has been computed from a large batch of samples. In order to transform it into a good online algorithm, the two requirements above must be satisfied. To quickly obtain a good performance, policy improvements are performed once every few transitions, before an accurate evaluation of the current policy can be completed. Such policy improvements are sometimes called "optimistic" (Sutton, 1988; Bertsekas, 2007). To satisfy the exploration requirement, online LSPI must sometimes try actions different from those given by the current policy. Online LSPI can be combined with many exploration procedures, and in this chapter, the classical, so-called ε-greedy exploration (Sutton and Barto, 1998) is applied: at every step, an exploratory, uniformly distributed random action is applied with probability ε, and the action given by the current policy is applied with probability $1-\varepsilon$.

RL is usually envisioned as working in a purely model-free fashion, without any prior knowledge about the problem. However, using prior knowledge can be highly beneficial if it is available. In the second topic of this chapter, we illustrate how prior knowledge about the policy can be exploited to increase the learning rate of online LSPI. In particular, policies that are monotonic in the state variables are considered. Such policies are suitable for controlling, e.g., (nearly) linear systems, or systems that are (nearly) linear and have monotonic input nonlinearities (such as saturation or dead-zone nonlinearities). A speedup of the learning process is then expected, because the online LSPI algorithm restricts its focus to the class of monotonic policies, and no longer invests valuable learning time in trying other, unsuitable policies.

A third important development in this chapter is a continuous-action Q-function approximator for offline LSPI, which combines state-dependent basis functions with orthogonal polynomial approximation over the action space. Continuous actions are useful in many classes of control problems. For instance, when a system must be stabilized around an unstable equilibrium, any discrete-action policy will lead to undesirable chattering of the control action.

These developments are empirically studied in three problems that were also employed in earlier chapters: the inverted pendulum, the two-link manipulator, and the DC motor. In particular, online LSPI is evaluated for inverted pendulum swing-up (for which it is compared with offline LSPI and with another online PI algorithm, and real-time learning results are given), as well as for two-link manipulator stabilization. The effects of using prior knowledge in online LSPI are then investigated for DC motor stabilization. We finally return to the inverted pendulum to examine the effects of continuous-action, polynomial approximation.

The remainder of this chapter starts by briefly revisiting LSPI, in Section 5.2. Then, online LSPI is developed in Section 5.3, the procedure to integrate prior knowledge about the policy is presented in Section 5.4, and the continuous-action, polynomial Q-function approximator is explained in Section 5.5. Section 5.6 provides the empirical evaluation of these three techniques, and Section 5.7 closes the chapter with a summary and discussion.

5.2 A recapitulation of least-squares policy iteration

LSPI is an offline algorithm for approximate policy iteration that evaluates policies using the least-squares temporal difference for Q-functions (LSTD-Q) and performs exact policy improvements. LSTD-Q was described in detail in Section 3.5.2 and was presented in a procedural form in Algorithm 3.8, while LSPI was discussed in Section 3.5.5 and summarized in Algorithm 3.11. Here, we only provide a summary of these results, and make some additional remarks regarding the practical implementation of the algorithm.

In LSPI, Q-functions are approximated using a linear parametrization:

$$\widehat{Q}(x,u) = \phi^{\mathrm{T}}(x,u)\theta$$

where $\phi(x,u) = [\phi_1(x,u), \ldots, \phi_n(x,u)]^{\mathrm{T}}$ is a vector of n basis functions (BFs), and $\theta \in \mathbb{R}^n$ is a parameter vector. To find the approximate Q-function of the current policy, the parameter vector is computed from a batch of transition samples, using LSTD-Q. Then, an improved, greedy policy in this Q-function is determined, the approximate Q-function of this improved policy is found, and so on.

Algorithm 5.1 presents LSPI integrated with an explicit description of the LSTD-Q policy evaluation step. This explicit form makes it easier to compare offline LSPI with the online variant that will be introduced later.

ALGORITHM 5.1 Offline least-squares policy iteration.

Input: discount factor γ,
BFs $\phi_1, \ldots, \phi_n : X \times U \to \mathbb{R}$, samples $\{(x_{l_s}, u_{l_s}, x'_{l_s}, r_{l_s}) \,|\, l_s = 1, \ldots, n_s\}$
1: initialize policy h_0
2: **repeat** at every iteration $\ell = 0, 1, 2, \ldots$
3: $\quad \Gamma_0 \leftarrow 0$, $\Lambda_0 \leftarrow 0$, $z_0 \leftarrow 0$ $\quad$ ▷ start LSTD-Q policy evaluation
4: $\quad$ **for** $l_s = 1, \ldots, n_s$ **do**
5: $\quad\quad \Gamma_{l_s} \leftarrow \Gamma_{l_s-1} + \phi(x_{l_s}, u_{l_s})\phi^{\mathrm{T}}(x_{l_s}, u_{l_s})$
6: $\quad\quad \Lambda_{l_s} \leftarrow \Lambda_{l_s-1} + \phi(x_{l_s}, u_{l_s})\phi^{\mathrm{T}}(x'_{l_s}, h(x'_{l_s}))$
7: $\quad\quad z_{l_s} \leftarrow z_{l_s-1} + \phi(x_{l_s}, u_{l_s}) r_{l_s}$
8: $\quad$ **end for**
9: $\quad$ solve $\frac{1}{n_s}\Gamma_{n_s}\theta_\ell = \gamma \frac{1}{n_s}\Lambda_{n_s}\theta_\ell + \frac{1}{n_s} z_{n_s}$ $\quad$ ▷ finalize policy evaluation
10: $\quad h_{\ell+1}(x) \leftarrow u$ where $u \in \arg\max_{\bar{u}} \phi^{\mathrm{T}}(x,\bar{u})\theta_\ell, \quad \forall x$ $\quad$ ▷ policy improvement
11: **until** $h_{\ell+1}$ is satisfactory
Output: $\widehat{h}^* = h_{\ell+1}$

The parameter θ_ℓ obtained by LSTD-Q at line 9 of Algorithm 5.1 leads to an approximate Q-function $\widehat{Q}_\ell(x,u) = \phi^{\mathrm{T}}(x,u)\theta_\ell$, which has a precise formal meaning, as explained next. The linear system at line 9 approximates the projected Bellman equation given in matrix form by (3.38), and repeated here for easy reference:

$$\Gamma\theta^{h_\ell} = \gamma\Lambda\theta^{h_\ell} + z \tag{5.1}$$

The approximation is obtained by replacing the matrices Γ, Λ, and the vector z with estimates derived from the samples. The matrix equation (5.1) is in turn equivalent to the original projected Bellman equation (3.34):

$$\widehat{Q}^{h_\ell} = (P^w \circ T^h)(\widehat{Q}^{h_\ell})$$

where $\widehat{Q}^{h_\ell}(x,u) = \phi^{\mathrm{T}}(x,u)\theta^{h_\ell}$, and the mapping P^w performs a weighted least-squares projection onto the space spanned by the BFs. The weight function is identical to the distribution of the state-action samples used in LSTD-Q. The Q-function $\widehat{Q}_\ell$ obtained by LSTD-Q is thus an estimate of the solution $\widehat{Q}^{h_\ell}$ to the projected Bellman equation.

Note that, because θ_ℓ appears on both sides of the equation at line 9, this equation can be simplified to:

$$\frac{1}{n_\mathrm{s}}(\Gamma_{n_\mathrm{s}} - \gamma\Lambda_{n_\mathrm{s}})\theta_\ell = \frac{1}{n_\mathrm{s}} z_{n_\mathrm{s}}$$

and therefore the matrices Γ and Λ do not have to be estimated separately. Instead, the combined matrix $\Gamma - \gamma\Lambda$ can be estimated as a single object, thereby reducing the memory demands of LSPI.

At line 10 of Algorithm 5.1, an improved policy is found that is greedy in the approximate Q-function $\widehat{Q}_\ell$. In practice, improved policies do not have to be explicitly computed and stored. Instead, for any given state x, improved (greedy) actions can be computed on-demand, by using:

$$h_{\ell+1}(x) = u, \quad \text{where } u \in \arg\max_{\bar{u}} \phi^{\mathrm{T}}(x,\bar{u})\theta_\ell \tag{5.2}$$

The maximization in this equation must be solved efficiently, because at every policy evaluation, a greedy action has to be computed for each of the n_s samples (see line 6 of Algorithm 5.1). Efficient maximization is possible when a suitable approximator is chosen (i.e., when suitable BFs are defined). For instance, with a discrete-action approximator of the type introduced in Example 3.1, the maximization can be solved by enumeration over the set of discrete actions.

As long as the policy evaluation error is bounded, LSPI eventually produces policies with a bounded suboptimality (see Section 3.5.6). It is not, however, guaranteed to converge to a fixed policy – although it often does in practice. For instance, the value function parameters might converge to limit cycles, so that every point on the cycle yields a near-optimal policy.

5.3 Online least-squares policy iteration

LSPI is an offline RL algorithm: it employs data collected in advance to learn a policy that should perform well at the end of the learning process. However, one of the main goals of RL is to develop algorithms that learn online, by interacting with the controlled system. Therefore, in this section we extend LSPI to online learning.

A good online algorithm must satisfy two requirements. First, by *exploiting* the data collected by interaction, it must quickly provide a good performance, instead of only at the end of the learning process. Second, it must also eventually achieve a (near-)optimal performance, without getting stuck in a local optimum. To this end, actions different from those indicated by the current policy must be *explored*, even at the risk of a temporarily reduced performance. Hence, this second requirement is partly in conflict with the first, and the combination of the two is traditionally called the exploration-exploitation trade-off in the RL literature (Thrun, 1992; Kaelbling, 1993; Sutton and Barto, 1998). One way to formalize this trade-off is to use the notion of regret, which roughly speaking is the cumulative difference between the optimal returns and the returns actually obtained over the learning process (see, e.g., Auer et al., 2002; Audibert et al., 2007; Auer et al., 2009; Bubeck et al., 2009). Minimizing the regret leads to fast learning and efficient exploration, by requiring that the performance becomes near optimal (which ensures exploration is applied), and that this happens as soon as possible (which ensures that learning is fast and is delayed by exploration only as much as necessary).

To ensure that our online variant of LSPI learns quickly (thereby satisfying the first requirement above), policy improvements must be performed once every few transitions, before an accurate evaluation of the current policy can be completed. This is a crucial difference with offline LSPI, which improves the policy only after an accurate Q-function has been obtained by running LSTD-Q on a large batch of samples. In the extreme case, online LSPI improves the policy after every transition, and applies the improved policy to obtain a new transition sample. Then, another policy improvement takes place, and the cycle repeats. Such a variant of PI is called fully optimistic (Sutton, 1988; Bertsekas, 2007). In general, online LSPI improves the policy once every several (but not too many) transitions; this variant is called partially optimistic.

To satisfy the second requirement, online LSPI must explore, by trying other actions than those given by the current policy. Without exploration, only the actions dictated by the current policy would be performed in every state, and samples of the other actions in that state would not be available. This would lead to a poor estimation of the Q-values of these other actions, and the resulting Q-function would not be reliable for policy improvement. Furthermore, exploration helps obtain data from regions of the state space that would not be reached using only the greedy policy. In this chapter, the classical, ε-greedy exploration (Sutton and Barto, 1998) is used: at every step k, a uniform random exploratory action is applied with probability $\varepsilon_k \in [0,1]$, and the greedy (maximizing) action with probability $1-\varepsilon_k$. Typically, ε_k decreases over time (as k increases), so that the algorithm increasingly exploits the current policy, as this policy (expectedly) approaches the optimal one. Other exploration procedures are possible, see, e.g., (Li et al., 2009) for a comparison in the context of LSPI with online sample collection.

Algorithm 5.2 presents online LSPI with ε-greedy exploration. The differences with offline LSPI are clearly visible in a comparison with Algorithm 5.1. In particular, online LSPI collects its own samples by interacting with the system (line 6), during which it employs exploration (line 5). Also, instead of waiting until many samples have been processed to perform policy improvements, online LSPI solves for the Q-function parameters and improves the policy at short intervals, using the currently available values of Γ, Λ, and z (lines 11–12).

Online LSPI uses two new, essential parameters that are not present in offline LSPI: the number of transitions $K_\theta > 0$ between consecutive policy improvements,

ALGORITHM 5.2 Online least-squares policy iteration with ε-greedy exploration.

Input: discount factor γ,
BFs $\phi_1, \dots, \phi_n : X \times U \to \mathbb{R}$,
policy improvement interval K_θ, exploration schedule $\{\varepsilon_k\}_{k=0}^{\infty}$,
a small constant $\beta_\Gamma > 0$

$\ell \leftarrow 0$, initialize policy h_0
$\Gamma_0 \leftarrow \beta_\Gamma I_{n \times n}$, $\Lambda_0 \leftarrow 0$, $z_0 \leftarrow 0$
measure initial state x_0
for every time step $k = 0, 1, 2, \dots$ **do**
  $u_k \leftarrow \begin{cases} h_\ell(x_k) & \text{with probability } 1 - \varepsilon_k \text{ (exploit)} \\ \text{a uniform random action in } U & \text{with probability } \varepsilon_k \text{ (explore)} \end{cases}$
  apply u_k, measure next state x_{k+1} and reward r_{k+1}
  $\Gamma_{k+1} \leftarrow \Gamma_k + \phi(x_k, u_k)\phi^{\mathrm{T}}(x_k, u_k)$
  $\Lambda_{k+1} \leftarrow \Lambda_k + \phi(x_k, u_k)\phi^{\mathrm{T}}(x_{k+1}, h_\ell(x_{k+1}))$
  $z_{k+1} \leftarrow z_k + \phi(x_k, u_k)\, r_{k+1}$
  if $k = (\ell + 1)K_\theta$ **then**
    solve $\frac{1}{k+1}\Gamma_{k+1}\theta_\ell = \frac{1}{k+1}\Lambda_{k+1}\theta_\ell + \frac{1}{k+1}z_{k+1}$ ▷ finalize policy evaluation
    $h_{\ell+1}(x) \leftarrow \arg\max_u \phi^{\mathrm{T}}(x, u)\theta_\ell, \quad \forall x$ ▷ policy improvement
    $\ell \leftarrow \ell + 1$
  end if
end for

and the exploration schedule $\{\varepsilon_k\}_{k=0}^{\infty}$. When $K_\theta = 1$, the policy is updated after every sample and online LSPI is fully optimistic. When $K_\theta > 1$, the algorithm is partially optimistic. Note that the number K_θ should not be chosen too large, and a significant amount of exploration is recommended, i.e., ε_k should not approach 0 too fast. In this chapter, the exploration probability is initially set to a value ε_0, and decays exponentially[1] once every second with a decay rate of $\varepsilon_{\mathrm{d}} \in (0, 1)$:[2]

$$\varepsilon_k = \varepsilon_0\, \varepsilon_{\mathrm{d}}^{\lfloor k T_{\mathrm{s}} \rfloor} \tag{5.3}$$

where T_{s} is the sampling time of the controlled system, and $\lfloor \cdot \rfloor$ denotes the largest integer smaller than or equal to the argument (floor). Like in the offline case, improved policies do not have to be explicitly computed; instead, improved actions can be computed on-demand. To ensure its invertibility, Γ is initialized to a small multiple $\beta_\Gamma > 0$ of the identity matrix.

Offline LSPI rebuilds Γ, Λ, and z from scratch before every policy improvement.

[1] An exponential decay does not asymptotically lead to infinite exploration, which is required by some online RL algorithms (Section 2.3.2). Nevertheless, for an experiment having a finite duration, ε_{d} can be chosen large enough to provide any desired amount of exploration.

[2] The exploration probability ε_k decays once every second, instead of once every time step (sampling period), in order to ensure that exploration schedules are comparable even among systems with different sampling times. Of course, a very similar effect can be obtained by decaying ε_k once every time step, with a larger ε_{d} when $T_{\mathrm{s}} < 1$, or a smaller ε_{d} when $T_{\mathrm{s}} > 1$.

Online LSPI cannot do this, because the few samples that arrive before the next policy improvement are not sufficient to construct informative new estimates of Γ, Λ, and z. Instead, these estimates are continuously updated. The underlying assumption is that the Q-functions of subsequent policies are similar, which means that the previous values of Γ, Λ, and z are also representative of the improved policy.

An alternative would be to store the samples and use them to rebuild Γ, Λ, and z before each policy improvement. This would incur larger computational costs, which would also increase with the number of samples observed, and might therefore make the algorithm too slow for online real-time learning after many samples have been observed. Such difficulties often appear when batch RL algorithms like LSPI must be applied in real time, and some general ways to address them are discussed, e.g., by Ernst et al. (2006a, Section 5). In Algorithm 5.2, we ensure that the computational and memory demands are independent of the number of samples observed, by exploiting optimistic policy updates and reusing Γ, Λ, and z.

More specifically, the time complexity per step of online LSPI is $O(n^3)$. The cost is the largest at time steps where policy improvements are performed, because this involves solving the linear system at line 11. This cost can be reduced by using computationally efficient methods to solve the system, but will still be larger than $O(n^2)$. The memory required to store Γ, Λ, and z is $O(n^2)$. Like offline LSPI, the online algorithm can estimate the combined matrix $\Gamma - \gamma\Lambda$ instead of Γ and Λ separately, thereby reducing its memory requirements.

Before closing this section, we discuss the separation of the learning process into distinct trials. As previously explained in Chapter 2, trials arise naturally in problems with terminal states, in which a trial is defined as a trajectory starting from some initial state and ending in a terminal state. A terminal state, once reached, can no longer be exited. So, the system must be reset in some fashion to an initial state, thereby starting a new trial. For instance, many robot manipulators have safeguards that stop the robot's motion if its pose gets outside the operating range (reaches a terminal state), after which human intervention is required to reposition the robot (start a new trial). If the problem does not have terminal states, it is possible to learn from a single, long trial. However, even in this case, it may still be beneficial for learning to terminate trials artificially. For instance, when learning a stabilizing control law, if the system has been successfully stabilized and exploration is insufficient to drive the state away from the equilibrium, there is little more to be learned from that trial, and a new trial starting from a new state will be more useful. In the sequel, we denote by T_{trial} the duration of such an artificially terminated trial.

5.4 Online LSPI with prior knowledge

RL is typically envisioned as working without any prior knowledge about the controlled system or about the optimal solution. However, in practice, a certain amount of prior knowledge is often available, and using this prior knowledge can be very beneficial. Prior knowledge can refer, e.g., to the policy, to the value function, or to the system dynamics. We focus on using prior knowledge about the optimal policy, or more generally about good policies that are not necessarily optimal. This focus is motivated by the fact that it is often easier to obtain knowledge about the policy than about the value function.

A general way of specifying knowledge about the policy is by defining constraints. For instance, one might know, and therefore require, that the policy is (globally or piecewise) monotonic in the state variables, or inequality constraints might be available on the state and action variables. The main benefit of constraining policies is a speedup of the learning process, which is expected because the algorithm restricts its focus to the constrained class of policies, and no longer invests valuable learning time in trying other, unsuitable policies. This speedup is especially relevant in online learning, although it may help reduce the computational demands of offline learning as well.

The original (online or offline) LSPI does not explicitly represent policies, but computes them on demand by using (5.2). Therefore, the policy is implicitly defined by the Q-function. In principle, it is possible to use the constraints on the policy in order to derive corresponding constraints on the Q-function. However, this derivation is very hard to perform in general, due to the complex relationship between a policy and its Q-function. A simpler solution is to represent the policy *explicitly* (and, in general, approximately), and to enforce the constraints in the policy improvement step. This is the solution adopted in the sequel.

In the remainder of this section, we develop an online LSPI algorithm for explicitly parameterized, globally monotonic policies. Such a policy is monotonic with respect to any state variable, if the other state variables are held constant. Monotonic policies are suitable for controlling important classes of systems. For instance, a monotonic policy is appropriate for a (nearly) linear system, or a nonlinear system in a neighborhood of an equilibrium where it is nearly linear. This is because linear policies, which work well for controlling linear systems, are monotonic. Monotonic policies also work well for some linear systems with monotonic input nonlinearities (such as saturation or dead-zone nonlinearities). In such cases, the policy may be strongly nonlinear, but still monotonic. Of course, in general, the global monotonicity requirement is restrictive. It can be made less restrictive, e.g., by requiring that the policy is monotonic only over a subregion of the state space, such as in a neighborhood of an equilibrium. Multiple monotonicity regions can also be considered.

5.4.1 Online LSPI with policy approximation

Policy iteration with explicit policy approximation was described in Section 3.5.5. Here, we specialize this discussion for online LSPI. Consider a linearly parameterized policy representation of the form (3.12), repeated here:

$$\widehat{h}(x) = \varphi^{\mathrm{T}}(x)\vartheta \tag{5.4}$$

where $\varphi(x) = [\varphi_1(x), \dots, \varphi_{\mathcal{N}}(x)]^{\mathrm{T}}$ is a vector containing $\mathcal{N}$ state-dependent BFs, and ϑ is the policy parameter vector. A scalar action is assumed, but the parametrization can easily be extended to multiple action variables. When no prior knowledge about the policy is available, approximate policy improvement can be performed by solving the unconstrained linear least-squares problem (3.47), which for linearly parameterized Q-functions and scalar actions becomes:

$$\begin{gathered}\vartheta_{\ell+1} = \vartheta^{\ddagger}, \text{ where } \vartheta^{\ddagger} \in \arg\min_{\vartheta} \sum_{i_s=1}^{\mathcal{N}_s} \left(\varphi^{\mathrm{T}}(x_{i_s})\vartheta - u_{i_s}\right)^2 \\ \text{and } u_{i_s} \in \arg\max_{u} \phi^{\mathrm{T}}(x_{i_s}, u)\theta_{\ell}\end{gathered} \tag{5.5}$$

Here, the parameter vector $\vartheta_{\ell+1}$ leads to an improved policy, and $\{x_1, \ldots, x_{\mathcal{N}_s}\}$ is a set of samples to be used for policy improvement.

To obtain online LSPI with parameterized policies, the exact policy improvement at line 12 of Algorithm 5.2 is replaced by (5.5). Moreover, the parameterized policy (5.4) is used to choose actions at line 5, and in the updates of Λ at line 8.

An added benefit of the approximate policy (5.4) is that it produces continuous actions. Note however that, if a discrete-action Q-function approximator is employed, the continuous actions given by the policy must be quantized during learning into actions belonging to a discrete set U_{d}. In this case, the policy evaluation step actually estimates the Q-function of a *quantized* version of the policy. The quantization function $q_{\mathrm{d}} : U \to U_{\mathrm{d}}$ is used, given by:

$$q_{\mathrm{d}}(u) = u^{\ddagger}, \quad \text{where } u^{\ddagger} \in \arg\min_{u_j \in U_{\mathrm{d}}} \left|u - u_j\right| \tag{5.6}$$

5.4.2 Online LSPI with monotonic policies

Consider a problem with a D-dimensional state space $X \subseteq \mathbb{R}^D$. In this section, it is assumed that X is a hyperbox:

$$X = [x_{\min,1}, x_{\max,1}] \times \cdots \times [x_{\min,D}, x_{\max,D}]$$

where $x_{\min,d} \in \mathbb{R}$, $x_{\max,d} \in \mathbb{R}$, and $x_{\min,d} < x_{\max,d}$, for $d = 1, \ldots, D$.

We say that a policy h is monotonic along the dth dimension of the state space if and only if, for any pair $x \in X, \bar{x} \in X$ of states that fulfill:

$$\begin{aligned} x_d &\leq \bar{x}_d \\ x_{d'} &= \bar{x}_{d'} \quad \forall d' \neq d \end{aligned}$$

the policy satisfies:

$$\delta_{\mathrm{mon},d} \cdot h(x) \leq \delta_{\mathrm{mon},d} \cdot h(\bar{x}) \tag{5.7}$$

where the scalar $\delta_{\mathrm{mon},d} \in \{-1, 1\}$ specifies the monotonicity direction: if $\delta_{\mathrm{mon},d}$ is -1 then h is decreasing along dimension d, whereas if $\delta_{\mathrm{mon},d}$ is 1 then h is increasing. We say that a policy is (fully) monotonic if it is monotonic along every dimension of the state space. In this case, the monotonicity directions are collected in a vector $\delta_{\mathrm{mon}} = [\delta_{\mathrm{mon},1}, \ldots, \delta_{\mathrm{mon},D}]^{\mathrm{T}} \in \{-1, 1\}^D$.

In this chapter, policies are approximated using axis-aligned, normalized radial basis functions (RBFs) (see Example 3.1) that are distributed on a grid with $\mathcal{N}_1 \times \cdots \times \mathcal{N}_D$ elements. The grid spacing is equidistant along each dimension, and all the RBFs have identical widths. Before examining how (5.7) can be satisfied when using such RBFs, a notation is required to relate the D-dimensional position of an RBF on the grid with its scalar index in the vector φ. Consider the RBF located at indices i_d along every dimension $d = 1, \ldots, D$, where $i_d \in \{1, \ldots, \mathcal{N}_d\}$. The position of this RBF is therefore described by the D-dimensional index $(i_1, \ldots, i_D)$. We introduce the notation $[i_1, \ldots, i_D]$ for the corresponding scalar index of the RBF in φ, which is computed as follows:

$$[i_1, \ldots, i_D] = i_1 + (i_2 - 1)\mathcal{N}_1 + (i_3 - 1)\mathcal{N}_1\mathcal{N}_2 + \cdots\cdots + (i_D - 1)\mathcal{N}_1\mathcal{N}_2\cdots\mathcal{N}_{D-1}$$

This formula can be understood more easily as a generalization of the two-dimensional case, for which $[i_1, i_2] = i_1 + (i_2 - 1)\mathcal{N}_1$. In this two-dimensional case, the grid (matrix) of RBFs has $\mathcal{N}_1 \times \mathcal{N}_2$ elements, and the vector φ is obtained by first taking the left-most column of the grid (matrix), which contains $\mathcal{N}_1$ elements, then appending the second column from the left (also with $\mathcal{N}_1$ elements), and so on.

Thus, the RBF at position $(i_1, \ldots, i_D)$ on the grid sits at index $i = [i_1, \ldots, i_D]$ in the vector φ, and is multiplied by the policy parameter $\vartheta_i = \vartheta_{[i_1,\ldots,i_D]}$ when the approximate policy (5.4) is computed. The D-dimensional center of this RBF is $[c_{1,i_1}, \ldots, c_{D,i_D}]^{\mathrm{T}}$, where c_{d,i_d} denotes the i_dth grid coordinate along dimension d. Without any loss of generality, the coordinates are assumed to increase monotonically along each dimension d:

$$c_{d,1} < \cdots < c_{d,\mathcal{N}_d}$$

Furthermore, we impose that the first and last grid elements are placed at the limits of the domain: $c_{d,1} = x_{\min,d}$ and $c_{d,\mathcal{N}_d} = x_{\max,d}$.

With these conditions, and also because the normalized RBFs are equidistant and identically shaped, we conjecture that in order to satisfy (5.7) it suffices to properly order the parameters corresponding to each sequence of RBFs along all the grid lines, and in every dimension of the state space.[3] An example of this ordering relationship, for a 3×3 grid of RBFs, is:

$$\begin{array}{ccccc} \vartheta_{[1,1]} & \leq & \vartheta_{[1,2]} & \leq & \vartheta_{[1,3]} \\ \geq & & \geq & & \geq \\ \vartheta_{[2,1]} & \leq & \vartheta_{[2,2]} & \leq & \vartheta_{[2,3]} \\ \geq & & \geq & & \geq \\ \vartheta_{[3,1]} & \leq & \vartheta_{[3,2]} & \leq & \vartheta_{[3,3]} \end{array} \tag{5.8}$$

In this case, the policy is decreasing along the first dimension of X (vertically in the equation), and increasing along the second dimension (horizontally in the equation).

[3]To our knowledge, this monotonicity property has not yet been formally proven; however, it can be verified empirically for many RBF configurations.

For a general grid in D dimensions, the monotonicity conditions can be written:

$$\begin{aligned}
&\delta_{\text{mon},1} \cdot \vartheta_{[1,i_2,i_3,\ldots,i_D]} \le \delta_{\text{mon},1} \cdot \vartheta_{[2,i_2,i_3,\ldots,i_D]} \le \cdots \le \delta_{\text{mon},1} \cdot \vartheta_{[\mathscr{N}_1,i_2,\ldots,i_D]} \\
&\qquad\qquad\qquad\qquad\qquad\qquad \text{for all } i_2, i_3, \ldots, i_D, \\
&\delta_{\text{mon},2} \cdot \vartheta_{[i_1,1,i_3,\ldots,i_D]} \le \delta_{\text{mon},2} \cdot \vartheta_{[i_1,2,i_3,\ldots,i_D]} \le \cdots \le \delta_{\text{mon},2} \cdot \vartheta_{[i_1,\mathscr{N}_2,i_3,\ldots,i_D]} \\
&\qquad\qquad\qquad\qquad\qquad\qquad \text{for all } i_1, i_3, \ldots, i_D, \\
&\qquad \ldots \qquad\qquad \ldots \qquad\qquad \ldots \\
&\delta_{\text{mon},D} \cdot \vartheta_{[i_1,i_2,i_3,\ldots,1]} \le \delta_{\text{mon},D} \cdot \vartheta_{[i_1,i_2,i_3,\ldots,2]} \le \cdots \le \delta_{\text{mon},D} \cdot \vartheta_{[i_1,i_2,i_3,\ldots,\mathscr{N}_D]} \\
&\qquad\qquad\qquad\qquad\qquad\qquad \text{for all } i_1, i_2, \ldots, i_{D-1}
\end{aligned} \tag{5.9}$$

The total number of inequalities in this equation is:

$$\sum_{d=1}^{D} \left((\mathscr{N}_d - 1) \prod_{d'=1,\ d' \neq d}^{D} \mathscr{N}_{d'} \right)$$

The monotonicity of the policy is enforced in online LSPI by replacing the unconstrained policy improvement with the constrained least-squares problem:

$$\begin{aligned}
\vartheta_{\ell+1} = \vartheta^{\ddagger}, \text{ where } \vartheta^{\ddagger} \in \underset{\vartheta \text{ satisfying (5.9)}}{\arg\min} \sum_{i_s=1}^{\mathscr{N}_s} \left(\varphi^{\mathrm{T}}(x_{i_s})\vartheta - u_{i_s} \right)^2 \\
\text{and } u_{i_s} \in \arg\max_{u} \phi^{\mathrm{T}}(x_{i_s}, u)\theta_{\ell}
\end{aligned} \tag{5.10}$$

and then using the approximate, monotonic policy $\widehat{h}_{\ell+1}(x) = \varphi^{\mathrm{T}}(x)\vartheta_{\ell+1}$. The problem (5.10) is solved using quadratic programming (see, e.g., Nocedal and Wright, 2006).

For clarity, Algorithm 5.3 summarizes online LSPI incorporating monotonic policies, a general linear parametrization of the Q-function, and ε-greedy exploration. If a discrete-action Q-function approximator is used in this algorithm, the approximate action must additionally be quantized with (5.6) at lines 5 and 8.

This framework can easily be generalized to multiple action variables, in which case a distinct policy parameter vector can be used for every action variable, and the monotonicity constraints can be enforced separately, on each of these parameter vectors. This also means that different monotonicity directions can be imposed for different action variables.

5.5 LSPI with continuous-action, polynomial approximation

Most versions of LSPI from the literature employ discrete actions (Lagoudakis et al., 2002; Lagoudakis and Parr, 2003a; Mahadevan and Maggioni, 2007). Usually, a number N of BFs are defined over the state space only, and are replicated for each

ALGORITHM 5.3 Online least-squares policy iteration with monotonic policies.

Input: discount factor γ,
Q-function BFs $\phi_1, \ldots, \phi_n : X \times U \to \mathbb{R}$, policy BFs $\varphi_1, \ldots, \varphi_{\mathcal{N}} : X \to \mathbb{R}$
policy improvement interval K_θ, exploration schedule $\{\varepsilon_k\}_{k=0}^{\infty}$,
a small constant $\beta_\Gamma > 0$

$\ell \leftarrow 0$, initialize policy parameter ϑ_0
$\Gamma_0 \leftarrow \beta_\Gamma I_{n\times n}$, $\Lambda_0 \leftarrow 0$, $z_0 \leftarrow 0$
measure initial state x_0
for every time step $k = 0, 1, 2, \ldots$ **do**
$$u_k \leftarrow \begin{cases} \varphi^{\mathrm{T}}(x_k)\vartheta_\ell & \text{with probability } 1-\varepsilon_k \text{ (exploit)} \\ \text{a uniform random action in } U & \text{with probability } \varepsilon_k \text{ (explore)} \end{cases}$$
  apply u_k, measure next state x_{k+1} and reward r_{k+1}
  $\Gamma_{k+1} \leftarrow \Gamma_k + \phi(x_k, u_k)\phi^{\mathrm{T}}(x_k, u_k)$
  $\Lambda_{k+1} \leftarrow \Lambda_k + \phi(x_k, u_k)\phi^{\mathrm{T}}(x_{k+1}, \varphi^{\mathrm{T}}(x_{k+1})\vartheta_\ell)$
  $z_{k+1} \leftarrow z_k + \phi(x_k, u_k)\, r_{k+1}$
  if $k = (\ell+1)K_\theta$ **then**
    solve $\frac{1}{k+1}\Gamma_{k+1}\theta_\ell = \frac{1}{k+1}\Lambda_{k+1}\theta_\ell + \frac{1}{k+1}z_{k+1}$
    $$\vartheta_{\ell+1} \leftarrow \vartheta^{\ddagger} \in \mathop{\arg\min}_{\vartheta \text{ satisfying (5.9)}} \sum_{i_s=1}^{\mathcal{N}_s} \left(\varphi^{\mathrm{T}}(x_{i_s})\vartheta - u_{i_s}\right)^2, u_{i_s} \in \arg\max_u \phi^{\mathrm{T}}(x_{i_s}, u)\theta_\ell$$
    $\ell \leftarrow \ell + 1$
  end if
end for

of the M discrete actions, leading to a total number $n = NM$ of BFs and parameters. Such approximators were discussed in Example 3.1. However, there exist important classes of control problems in which continuous actions are required. For instance, when a system must be stabilized around an unstable equilibrium, any discrete-action policy will lead to undesirable chattering of the control action and to limit cycles.

Therefore, in this section we introduce a continuous-action Q-function approximator for LSPI. This approximator works for problems with scalar control actions. It uses state-dependent BFs and orthogonal polynomials of the action variable, thus separating approximation over the state space from approximation over the action space. Note that because the action that maximizes the Q-function for a given state is not restricted to discrete values in (5.2), this approximator produces continuous-action policies. A polynomial approximator is chosen because it allows one to efficiently solve maximization problems over the action variable (and thus perform policy improvements), by computing the roots of the polynomial's derivative. Moreover, orthogonal polynomials are preferred to plain polynomials, because they lead to numerically better conditioned regression problems at the policy improvement step.

Similarly to the case of discrete-action approximators, a set of N state-dependent BFs is defined: $\bar{\phi}_i : X \to \mathbb{R}$, $i = 1, \ldots, N$. Only scalar control actions u are considered, bounded to an interval $U = [u_{\mathrm{L}}, u_{\mathrm{H}}]$. To approximate over the action dimension of the state-action space, Chebyshev polynomials of the first kind are chosen as an illustrative example of orthogonal polynomials, but many other types of orthogonal polynomials can alternatively be used. Chebyshev polynomials of the first kind are defined by the recurrence relation:

$$\begin{aligned}\psi_0(\bar{u}) &= 1\\ \psi_1(\bar{u}) &= \bar{u}\\ \psi_{j+1}(\bar{u}) &= 2\bar{u}\psi_j(\bar{u}) - \psi_{j-1}(\bar{u})\end{aligned}$$

They are orthogonal to each other on the interval $[-1,1]$ relative to the weight function $1/\sqrt{1-\bar{u}^2}$, i.e., they satisfy:

$$\int_{-1}^{1} \frac{1}{\sqrt{1-\bar{u}^2}} \psi_j(\bar{u})\psi_{j'}(\bar{u})d\bar{u} = 0, \quad j = 0,1,2,\dots, \; j' = 0,1,2,\dots, \; j' \neq j$$

In order to take advantage of the orthogonality property, the action space U must be scaled and translated into the interval $[-1,1]$. This is simply accomplished using the affine transformation:

$$\bar{u} = -1 + 2\frac{u - u_{\mathrm{L}}}{u_{\mathrm{H}} - u_{\mathrm{L}}} \tag{5.11}$$

The approximate Q-values for an orthogonal polynomial approximator of degree M_{p} are computed as follows:

$$\widehat{Q}(x,u) = \sum_{j=0}^{M_{\mathrm{p}}} \psi_j(\bar{u}) \sum_{i=1}^{N} \bar{\phi}_i(x)\theta_{[i,j+1]} \tag{5.12}$$

This can be written as $\widehat{Q}(x,u) = \phi^{\mathrm{T}}(x,\bar{u})\theta$ for the state-action BF vector:

$$\begin{aligned}\phi(x,\bar{u}) = [&\bar{\phi}_1(x)\psi_0(\bar{u}),\dots,\bar{\phi}_N(x)\psi_0(\bar{u}),\\ &\bar{\phi}_1(x)\psi_1(\bar{u}),\dots,\bar{\phi}_N(x)\psi_1(\bar{u}),\\ &\dots,\\ &\bar{\phi}_1(x)\psi_{M_{\mathrm{p}}}(\bar{u}),\dots,\bar{\phi}_N(x)\psi_{M_{\mathrm{p}}}(\bar{u})]^{\mathrm{T}}\end{aligned} \tag{5.13}$$

The total number of state-action BFs (and therefore the total number of parameters) is $n = N(M_{\mathrm{p}}+1)$. So, given the same number N of state BFs, a polynomial approximator of degree M_{p} has the same number of parameters as a discrete-action approximator with $M = M_{\mathrm{p}}+1$ discrete actions.

To find the greedy action (5.2) for a given state x, the approximate Q-function (5.12) for that value of x is first computed, which yields a polynomial in $\bar{u}$. Then, the roots of the derivative of this polynomial that lie in the interval $(-1,1)$ are found, the approximate Q-values are computed for each root and also for -1 and 1, and the action that corresponds to the largest Q-value is chosen. This action is then translated back into $U = [u_{\mathrm{L}}, u_{\mathrm{H}}]$:

$$\begin{aligned}&h(x) = u_{\mathrm{L}} + (u_{\mathrm{H}} - u_{\mathrm{L}})\frac{1 + \arg\max_{\bar{u}\in\bar{U}(x)} \phi^{\mathrm{T}}(x,\bar{u})\theta}{2}, \quad \text{where:}\\ &\bar{U}(x) = \{-1,1\} \cup \left\{\bar{u}' \in (-1,1) \;\middle|\; \frac{\mathrm{d}\psi(\bar{u}')}{\mathrm{d}\bar{u}} = 0, \text{where } \psi(\bar{u}) = \phi^{\mathrm{T}}(x,\bar{u})\theta\right\}\end{aligned} \tag{5.14}$$

In this equation, $\frac{\mathrm{d}\psi(\bar{u}')}{\mathrm{d}\bar{u}}$ denotes the derivative $\frac{\mathrm{d}\psi(\bar{u})}{\mathrm{d}\bar{u}}$ evaluated at $\bar{u}'$. In some cases, the polynomial will attain its maximum inside the interval $(-1,1)$, but in other cases, it may not, which is why the boundaries $\{-1,1\}$ must also be tested. Efficient algorithms can be used to compute the polynomial roots with high accuracy.[4] Therefore, the proposed parametrization allows the maximization problems in the policy improvement (5.2) to be solved efficiently and with high accuracy, and is well-suited to the use with LSPI.

Polynomial approximation can be used in offline, as well as in online LSPI. In this chapter, we will evaluate it for the offline case.

5.6 Experimental study

In this section, we experimentally evaluate the extensions to the LSPI algorithm that were introduced above. First, an extensive empirical study of online LSPI is performed in Sections 5.6.1 and 5.6.2, using respectively an inverted pendulum and a robotic manipulator problem. Then, in Section 5.6.3, the benefits of using prior knowledge in online LSPI are investigated, for a DC motor example. Finally, in Section 5.6.4, we return to the inverted pendulum problem to examine the effects of continuous-action polynomial approximation.

5.6.1 Online LSPI for the inverted pendulum

The inverted pendulum swing-up problem is challenging and highly nonlinear, but low-dimensional, which means extensive experiments can be performed with reasonable computational costs. Using this problem, we study the effects of the exploration decay rate, of the policy improvement interval, and of the trial length on the performance of online LSPI. Then, we compare the final performance of online LSPI with the performance of offline LSPI; and we compare online LSPI with an online PI algorithm that uses, instead of LSTD-Q, the least-squares policy evaluation for Q-functions (LSPE-Q) (Algorithm 3.9). Finally, we provide real-time learning results for the inverted pendulum system.

Inverted pendulum problem

The inverted pendulum problem was introduced in Section 4.5.3, and is described here only briefly. The pendulum consists of a weight attached to a rod that is actuated by a DC motor and rotates in a vertical plane (see Figure 4.11 on page 158). The goal is to stabilize the pendulum in the pointing up position. Due to the limited torque of the DC motor, from certain states (e.g., pointing down) the pendulum cannot be

[4]In our implementation, the roots are computed as the eigenvalues of the companion matrix of the polynomial $\frac{\mathrm{d}\psi(\bar{u})}{\mathrm{d}\bar{u}}$, using the "roots" function of MATLAB® (see, e.g., Edelman and Murakami, 1995).

pushed up in a single rotation, but must be swung back and forth to gather energy prior to being pushed up and stabilized.

A continuous-time model of the pendulum dynamics is:

$$\ddot{\alpha} = \frac{1}{J}\left(mgl\sin(\alpha) - b\dot{\alpha} - \frac{K^2}{R}\dot{\alpha} + \frac{K}{R}u\right)$$

where $J = 1.91 \cdot 10^{-4}\,\text{kgm}^2$, $m = 0.055\,\text{kg}$, $g = 9.81\,\text{m/s}^2$, $l = 0.042\,\text{m}$, $b = 3 \cdot 10^{-6}\,\text{Nms/rad}$, $K = 0.0536\,\text{Nm/A}$, $R = 9.5\,\Omega$. The angle α "wraps around" in the interval $[-\pi, \pi)$ rad, so that, e.g., a rotation of $3\pi/2$ corresponds to a value $\alpha = -\pi/2$. When $\alpha = 0$, the pendulum is pointing up. The velocity $\dot{\alpha}$ is restricted to $[-15\pi, 15\pi]$ rad/s, using saturation, and the control action u is constrained to $[-3, 3]$ V. The state vector is $x = [\alpha, \dot{\alpha}]^{\mathrm{T}}$. The sampling time is $T_{\mathrm{s}} = 0.005$ s, and the discrete-time transitions are obtained by numerically integrating the continuous-time dynamics. The stabilization goal is expressed by the reward function:

$$\rho(x,u) = -x^{\mathrm{T}} Q_{\mathrm{rew}} x - R_{\mathrm{rew}} u^2$$

$$Q_{\mathrm{rew}} = \begin{bmatrix} 5 & 0 \\ 0 & 0.1 \end{bmatrix}, \quad R_{\mathrm{rew}} = 1$$

with discount factor $\gamma = 0.98$. This discount factor is large so that rewards around the goal state (pointing up) influence the values of states early in the trajectories. This leads to an optimal policy that successfully swings up and stabilizes the pendulum.

A near-optimal solution (Q-function and policy) for this problem is given in Figure 5.1. This solution was computed with fuzzy Q-iteration (Chapter 4) using a fine grid of membership functions in the state space, and a fine discretization of the action space.

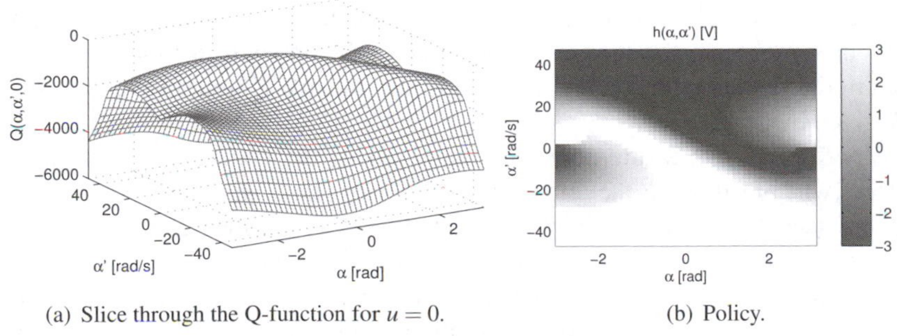

(a) Slice through the Q-function for $u = 0$. (b) Policy.

FIGURE 5.1 A near-optimal solution for the inverted pendulum.

Approximator and performance criteria

To approximate the Q-function, a discrete-action approximator of the type introduced in Example 3.1 of Chapter 3 is employed. Recall that for such an approximator, N state-dependent BFs $\bar{\phi}_1, \ldots, \bar{\phi}_N : X \to \mathbb{R}$ are defined and replicated for every action in a discretized set $U_\mathrm{d} = \{u_1, \ldots, u_M\}$. Approximate Q-values can be computed for any state-discrete action pair with:

$$\widehat{Q}(x, u_j) = \phi^\mathrm{T}(x, u_j)\,\theta$$

where, in the state-action BF vector $\phi(x, u_j)$, the BFs not corresponding to the current discrete action are taken to be equal to 0:

$$\phi(x, u_j) = [\underbrace{0, \ldots, 0}_{u_1}, \ldots, \underbrace{0, \bar{\phi}_1(x), \ldots, \bar{\phi}_N(x)}_{u_j}, 0, \ldots, \underbrace{0, \ldots, 0}_{u_M}]^\mathrm{T} \in \mathbb{R}^{NM} \tag{5.15}$$

An equidistant 11×11 grid of normalized Gaussian RBFs (3.6) is used to approximate the Q-function over the state space, and the action space is discretized into 3 discrete values: $U_\mathrm{d} = \{-3, 0, 3\}$. This leads to a total number of $n = 11^2 \cdot 3 = 363$ state-action BFs. The RBFs are axis-aligned and identical in shape, and their width b_d along each dimension d is equal to ${b'_d}^2/2$, where b'_d is the distance between adjacent RBFs along that dimension (the grid step). These RBFs yield a smooth interpolation of the Q-function over the state space. Recalling that the angle spans a domain of size 2π and that the angular velocity spans 30π, we obtain $b'_1 = \frac{2\pi}{11-1} \approx 0.63$ and $b'_2 = \frac{30\pi}{11-1} \approx 9.42$, which lead to $b_1 \approx 0.20$ and $b_2 \approx 44.41$.

We first define a performance measure that evaluates the quality of the policy computed online for a set of representative initial states spanning the entire state space. In particular, *after* each online LSPI experiment is completed, snapshots of the current policy at increasing moments of time are evaluated by estimating, with precision $\varepsilon_\mathrm{MC} = 0.1$, their average return (score) over the grid of initial states:

$$X_0 = \{-\pi, -\pi/2, 0, \pi/2\} \times \{-10\pi, -3\pi, -\pi, 0, \pi, 3\pi, 10\pi\} \tag{5.16}$$

During performance evaluation, learning and exploration are turned off. This produces a curve recording the control performance of the policy over time. The return for each initial state is estimated by simulating only the first K steps of the trajectory, with K given by (2.41).

The performance measure above is computed in the absence of exploration. Therefore, when evaluating the effects of exploration, an additional measure is required that does take these effects into account. To obtain this measure, the system is periodically reset *during* the learning process to the initial state $x_0 = [-\pi, 0]^\mathrm{T}$ (pointing down), and the empirical return obtained along a learning trial starting from this state is recorded, *without* turning off exploration and learning. Repeating this evaluation multiple times over the learning process gives a curve indicating the evolution of the return obtained while using exploration.

The two performance measures are respectively called "score" and "return with exploration" in the sequel. They are not directly comparable, first because they handle exploration differently, and second because the score evaluates the performance over X_0, whereas the return with exploration only considers the single state x_0. Nevertheless, if two experiments employ the same exploration schedule, comparing their score also gives a good idea about how they compare qualitatively in terms of return with exploration. Because of this, and also to preserve consistency with the performance criteria used in other chapters of this book, we will rely on the score as a primary performance measure, and we will use the return with exploration only when the effect of different exploration schedules must be assessed. Note that, because the reward function is negative, both performance measures will be negative, and smaller absolute values for these measures correspond to better performance.

Effects of the tuning parameters

In this section, we study the effects of varying the tuning parameters of online LSPI. In particular, we change the exploration decay rate ε_d, the number of transitions between consecutive policy improvements K_θ, and the trial length T_{trial}. Each online experiment is run for 600 s, and is divided into trials having the length T_{trial}. The decaying exploration schedule (5.3) is used, with the initial exploration probability $\varepsilon_0 = 1$, so that a fully random policy is used at first. The parameter β_Γ is set to 0.001. Furthermore, 20 independent runs of each experiment are performed, in order to obtain statistically significant results.

To study the influence of ε_d, the following values are chosen, $\varepsilon_d = 0.8913, 0.9550, 0.9772, 0.9886, 0.9924$, and 0.9962, so that ε_k becomes 0.1 after respectively 20, 50, 100, 200, 300, and 600 s. Larger values of ε_d correspond to more exploration. The policy is improved once every $K_\theta = 10$ transitions, and the trial length is $T_{trial} = 1.5$ s, which is sufficient to swing up and stabilize the inverted pendulum. The initial state of each trial is drawn from a uniform random distribution over X. Figure 5.2 shows how the *score* (average return over X_0) of the policies learned by online LSPI evolves. In particular, the curves in Figure 5.2(a) represent the mean performance across the 20 runs, while Figure 5.2(b) additionally shows 95% confidence intervals, but only considers the extreme values of ε_d, in order to avoid cluttering. The score converges in around 120 s of simulated time. The final score improves with more exploration, and the difference between the score with large and small exploration schedules is statistically significant, as illustrated in Figure 5.2(b). These results are not surprising, since the considerations in Section 5.3 already indicated that online LSPI requires significant exploration.

However, too much exploration may decrease the control performance obtained during learning. Since exploration is turned off while computing the score, this effect is not visible in Figure 5.2. To examine the effects of exploration, the experiments above are repeated, but this time resetting the system to the initial state $x_0 = [-\pi, 0]^T$ in one out of every 10 trials (i.e., once every 15 s), and recording the *return with exploration* during every such trial. Figure 5.3 presents the results, from which it appears that too large an exploration schedule negatively affects the rate of improvement of this performance measure. The differences between small and large exploration schedules are statistically significant, as illustrated in Figure 5.3(b). This means that, when selecting the exploration schedule, a trade-off between the score of the policy and the return with exploration must be resolved. For this example, an acceptable compromise between the two performance measures is obtained for $\varepsilon_d = 0.9886$, corresponding to $\varepsilon_k = 0.1$ after 200 s. We therefore choose this explo-

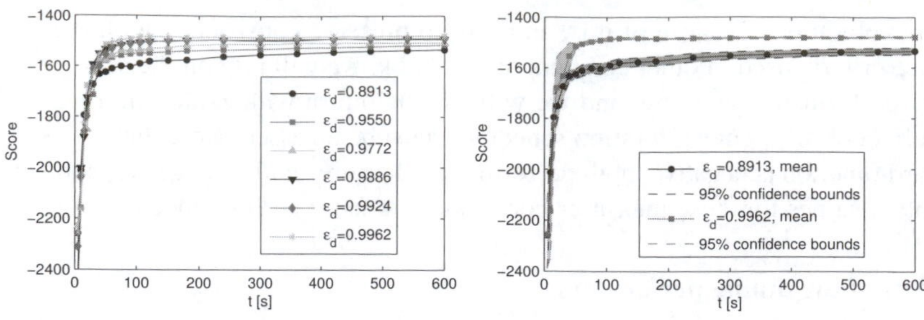

(a) Mean score for all the experiments.

(b) Mean score with 95% confidence intervals, for the smallest and largest values of ε_{d}.

FIGURE 5.2
Score of online LSPI for varying ε_{d} in the inverted pendulum problem. The marker locations indicate the moments in time when the policies were evaluated.

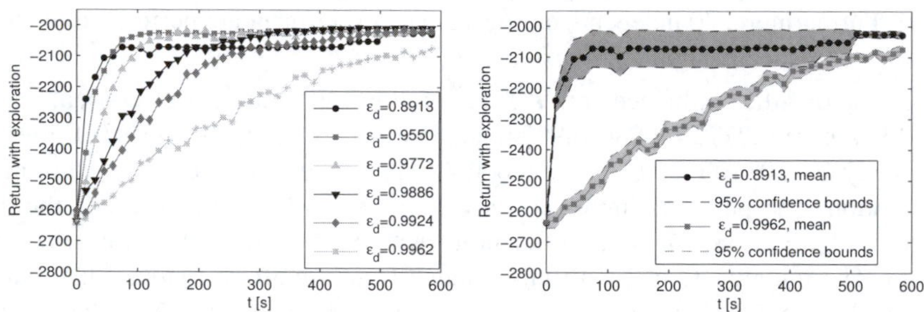

(a) Mean return with exploration for all the experiments.

(b) Mean and 95% confidence intervals for the return with exploration obtained using the smallest and largest values of ε_{d}.

FIGURE 5.3
Return with exploration obtained by online LSPI for varying ε_{d}, in the inverted pendulum problem.

ration schedule for all the upcoming simulation experiments with online LSPI for the inverted pendulum.

To study the influence of the number K_θ of transitions between policy improvements, the following values are used: $K_\theta = 1, 10, 100, 1000$, and 5000. When $K_\theta = 1$, the algorithm is fully optimistic: the policy is improved after every sample. The trial length is $T_{\text{trial}} = 1.5\,$s, and the initial state of each trial is drawn from a uniform random distribution over X. As already mentioned, the exploration decay rate is $\varepsilon_{\mathrm{d}} = 0.9886$. Figure 5.4 shows how the performance of the policies learned by online LSPI evolves.[5] In Figure 5.4(a) all the values of K_θ lead to a similar performance except the cases in which the policy is updated very rarely. For instance,

[5]In this figure, the performance is measured using the score, which is representative because all the experiments use the same exploration schedule. For similar reasons, the score is employed as a performance measure throughout the remainder of this chapter.

when $K_\theta = 5000$, the performance is worse, and the difference with the performance of smaller K_θ is statistically significant, as illustrated in Figure 5.4(b). This indicates that policy improvements should not be performed too rarely in online LSPI.

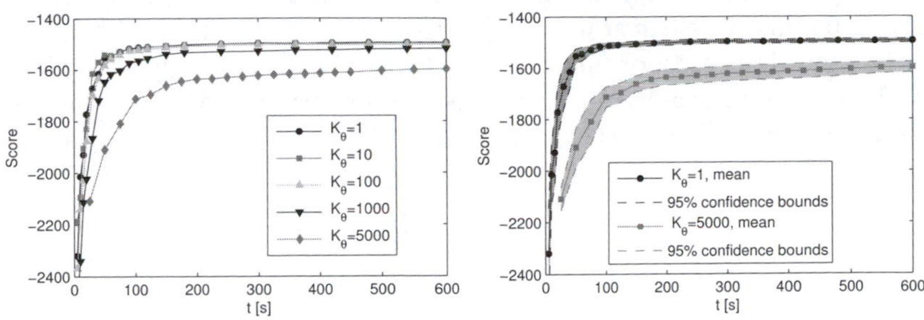

(a) Mean score for all the experiments.

(b) Mean score with 95% confidence intervals, for the smallest and largest values of K_θ.

FIGURE 5.4
Performance of online LSPI for varying K_θ in the inverted pendulum problem.

To study the effects of the trial length, the values $T_{\text{trial}} = 0.75, 1.5, 3, 6$, and 12 s are used, corresponding to, respectively, 800, 400, 200, 100, and 50 learning trials in the 600 s of learning. The initial state of each trial is drawn from a uniform random distribution over X, the policy is improved once every $K_\theta = 10$ samples, and the exploration decay rate is $\varepsilon_d = 0.9886$. These settings gave a good performance in the experiments above. Figure 5.5 reports the performance of online LSPI. Long trials (6 and 12 s) are detrimental to the learning rate, as well as to the final performance. Short trials are beneficial for the performance because the more frequent re-initializations to random states provide more information to the learning algorithm. This difference between the performance with short and long trials is statistically significant, see Figure 5.5(b).

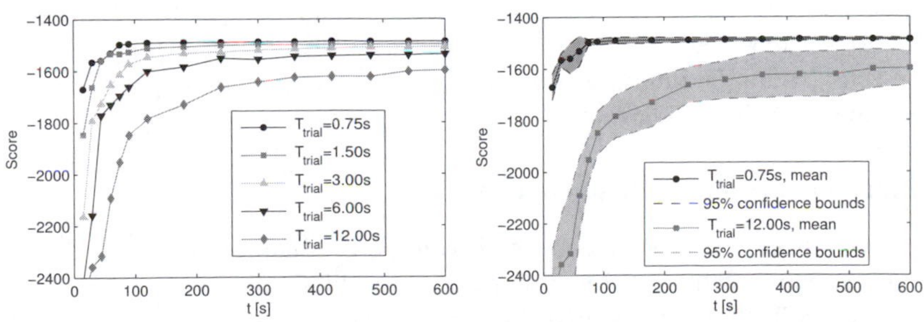

(a) Mean score for all the experiments.

(b) Mean score with 95% confidence intervals, for the smallest and largest values of T_{trial}.

FIGURE 5.5
Performance of online LSPI for varying T_{trial} in the inverted pendulum problem.

Figure 5.6 shows the mean execution time for varying ε_{d}, K_θ, and T_{trial}, taken across the 20 independent runs of each experiment.[6] The 95% confidence intervals are too small to be visible at the scale of the figure, so they are left out. The execution time is larger for smaller K_θ, because the most computationally expensive operation is solving the linear system at line 11 of Algorithm 5.2, which must be done once every K_θ steps. The execution time does not change significantly with the exploration schedule or with the trial length, since choosing random actions and re-initializing the state are computationally cheap operations.

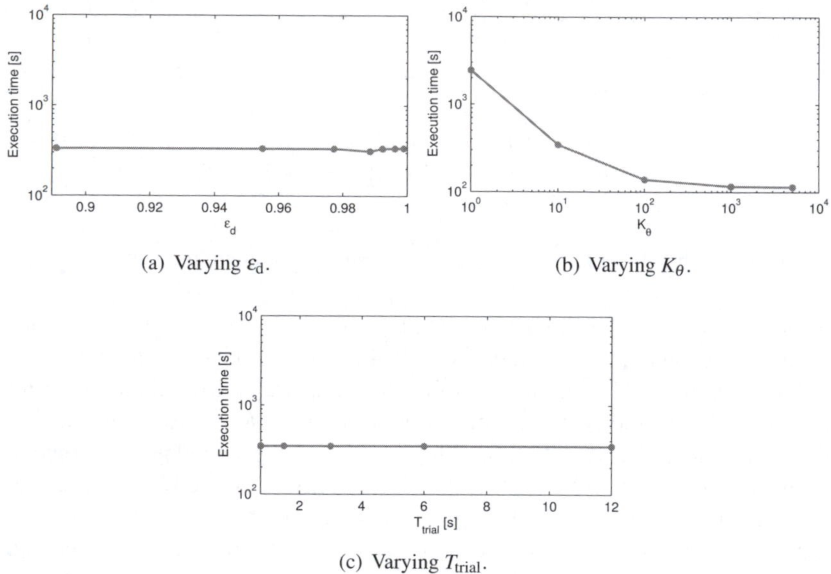

(a) Varying ε_{d}.

(b) Varying K_θ.

(c) Varying T_{trial}.

FIGURE 5.6 Mean execution time of online LSPI for the inverted pendulum.

Note that for fully optimistic updates, $K_\theta = 1$, the execution time is around 2430 s, longer than the length of 600 s for the simulated experiment, and therefore online LSPI cannot be run in real time for this value of K_θ. Some possible ways to address this problem will be discussed in Section 5.6.2.

Comparison of online LSPI and offline LSPI

In this section, online LSPI is compared with the original, offline LSPI algorithm. The online experiments described above are reused for this comparison. To apply offline LSPI, the same approximator is employed as in the online case. While the online algorithm generates its own samples during learning, a number of $n_{\mathrm{s}} = 20000$ pregenerated random samples are used for offline LSPI, uniformly distributed throughout the state-discrete action space $X \times U_{\mathrm{d}}$. Offline LSPI is run 20 times with

[6] All the execution times reported in this chapter were recorded while running the algorithms in MATLAB 7 on a PC with an Intel Core 2 Duo E6550 2.33 GHz CPU and with 3 GB RAM.

independent sets of samples. Table 5.1 compares the score (average return over X_0) of the policies found offline with that of the *final* policies found at the end of the online experiments, as well as the execution time of offline and online LSPI. Two representative online experiments from the study of ε_d are selected for comparison: the experiment with the best mean performance, and the experiment with the worst mean performance. Representative experiments from the study of K_θ and T_{trial} are selected in a similar way.

TABLE 5.1 Comparison of offline and online LSPI (mean; 95% confidence interval).

Experiment	**Performance (score)**	**Execution time** [s]
Offline	-1496.8; $[-1503.6, -1490.0]$	82.7; [79.6, 85.8]
$\varepsilon_d = 0.9962$ (best)	-1479.0; $[-1482.3, -1475.7]$	335.9; [332.8, 339.1]
$\varepsilon_d = 0.8913$ (worst)	-1534.0; $[-1546.9, -1521.1]$	333.6; [331.6, 335.7]
$K_\theta = 1$ (best)	-1494.3; $[-1501.5, -1487.2]$	2429.9; [2426.2, 2433.5]
$K_\theta = 5000$ (worst)	-1597.8; $[-1618.1, -1577.4]$	114.0; [113.7, 114.2]
$T_{trial} = 0.75$ s (best)	-1486.8; $[-1492.4, -1481.2]$	346.3; [345.1, 347.5]
$T_{trial} = 12$ s (worst)	-1598.5; $[-1664.2, -1532.8]$	346.6; [345.5, 347.7]

The table indicates that the final performance of online LSPI is comparable with the performance of its offline counterpart. On the other hand, online LSPI is more computationally expensive, because it performs more policy improvements. The offline algorithm employs 20000 samples, whereas online LSPI processes the same number of samples in 100 s of simulated time, and 120000 samples during the entire learning process. Nevertheless, as indicated, e.g., in Figure 5.2(a), for reasonable parameter settings, the score of the policy found by online LSPI is already good after 120 s, i.e., after processing 24000 samples. Therefore, online LSPI is also comparable with offline LSPI in the number of samples that are sufficient to find a good policy. Note that online LSPI processes samples only once, whereas the offline algorithm loops through the samples once at every iteration.

Figure 5.7 presents a representative final solution of online LSPI, computed with $K_\theta = 10$, $\varepsilon_d = 0.9886$, and $T_{trial} = 1.5$ s, in comparison to a representative solution of the offline algorithm. The two policies, shown in Figures 5.7(a) and 5.7(b), have a similar large-scale structure but differ in some regions. The Q-function found online (Figure 5.7(c)) resembles the near-optimal Q-function of Figure 5.1(a) more closely than the Q-function found offline (Figure 5.7(d)), which has an extra peak in the origin of the state space. The swing-up trajectories, shown in Figures 5.7(e) and 5.7(f), are similar, but the online solution leads to more chattering.

Comparison of online LSPI and online PI with LSPE-Q

In this section, we consider an online PI algorithm that evaluates policies with LSPE-Q (Algorithm 3.9), rather than with LSTD-Q, as online LSPI does. Recall that LSPE-Q updates the matrices Γ, Λ, and the vector z in the same way as LSTD-Q. However, unlike LSTD-Q, which finds the parameter vector θ by solving a one-shot linear

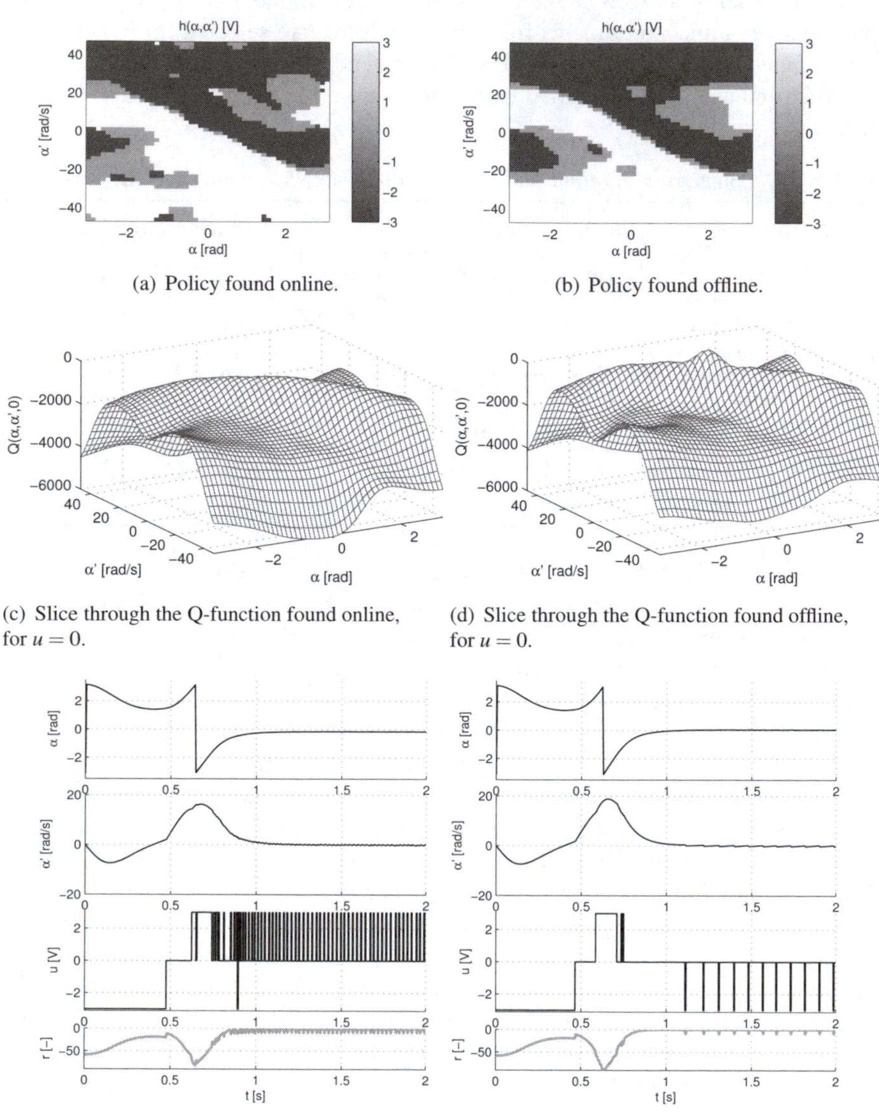

(a) Policy found online.

(b) Policy found offline.

(c) Slice through the Q-function found online, for $u = 0$.

(d) Slice through the Q-function found offline, for $u = 0$.

(e) Swing-up using the policy found online.

(f) Swing-up using the policy found offline.

FIGURE 5.7
Representative solutions found with online LSPI (left) and with offline LSPI (right) for the inverted pendulum.

problem, LSPE-Q updates the parameter vector incrementally, after every sample. In the online context, the update at step k has the form:

$$\begin{aligned}\theta_{k+1} = \theta_k + \alpha_{\text{LSPE}}(\theta^{\ddagger}_{k+1} - \theta_k), \text{ where:}\\ \frac{1}{k+1}\Gamma_{k+1}\theta^{\ddagger}_{k+1} = \gamma\frac{1}{k+1}\Lambda_{k+1}\theta_k + \frac{1}{k+1}z_{k+1}\end{aligned} \tag{5.17}$$

where α_{LSPE} is a step size parameter. The policy is improved optimistically, once every K_θ transitions. Algorithm 5.4 shows online PI with LSPE-Q, in a variant that employs ε-greedy exploration; this variant will be used in the sequel.

ALGORITHM 5.4 Online policy iteration with LSPE-Q and ε-greedy exploration.

Input: discount factor γ,
BFs $\phi_1, \dots, \phi_n : X \times U \to \mathbb{R}$,
policy improvement interval K_θ, exploration schedule $\{\varepsilon_k\}_{k=0}^{\infty}$,
step size $\alpha_{\text{LSPE}} > 0$, a small constant $\beta_\Gamma > 0$

$\ell \leftarrow 0$, initialize policy h_0, initialize parameters θ_0
$\Gamma_0 \leftarrow \beta_\Gamma I_{n\times n}$, $\Lambda_0 \leftarrow 0$, $z_0 \leftarrow 0$
measure initial state x_0
for every time step $k = 0, 1, 2, \dots$ **do**
  $u_k \leftarrow \begin{cases} h_\ell(x_k) & \text{with probability } 1-\varepsilon_k \text{ (exploit)} \\ \text{a uniform random action in } U & \text{with probability } \varepsilon_k \text{ (explore)} \end{cases}$
  apply u_k, measure next state x_{k+1} and reward r_{k+1}
  $\Gamma_{k+1} \leftarrow \Gamma_k + \phi(x_k,u_k)\phi^{\mathrm{T}}(x_k,u_k)$
  $\Lambda_{k+1} \leftarrow \Lambda_k + \phi(x_k,u_k)\phi^{\mathrm{T}}(x_{k+1},h_\ell(x_{k+1}))$
  $z_{k+1} \leftarrow z_k + \phi(x_k,u_k)\,r_{k+1}$
  $\theta_{k+1} \leftarrow \theta_k + \alpha_{\text{LSPE}}(\theta^{\ddagger}_{k+1} - \theta_k)$, where $\frac{1}{k+1}\Gamma_{k+1}\theta^{\ddagger}_{k+1} = \gamma\frac{1}{k+1}\Lambda_{k+1}\theta_k + \frac{1}{k+1}z_{k+1}$
  if $k = (\ell+1)K_\theta$ **then**
    $h_{\ell+1}(x) \leftarrow \arg\max_u \phi^{\mathrm{T}}(x,u)\theta_{k+1}, \quad \forall x$
    $\ell \leftarrow \ell + 1$
  end if
end for

The fully optimistic variant of online PI with LSPE-Q (for $K_\theta = 1$) was studied, e.g., by Jung and Polani (2007a). Bertsekas (2007) and Jung and Polani (2007a) conjectured that LSPE-Q is more promising for online PI than LSTD-Q, due to its incremental nature.

Next, we apply online PI with LSPE-Q to the swing-up problem. We do not study the influence of all the parameters, but only focus on parameter K_θ, by running a set of experiments that parallels the study of K_θ for online LSPI. The approximator, exploration schedule, and trial length are the same as in that experiment, and the same K_θ values are used. The matrix Γ is initialized to $0.001 \cdot I_{n\times n}$. Online PI with LSPE-Q has an additional step size parameter, α_{LSPE}, which was not present in online LSPI. In order to choose α_{LSPE}, preliminary experiments were performed for each value of K_θ, using several values of α_{LSPE}: 0.001, 0.01, 0.1, and 1. In these experiments, the following values of α_{LSPE} performed reasonably: 0.001, 0.01, 0.001, 0.01, and 0.1, for, respectively, $K_\theta = 1, 10, 100, 1000$, and 5000. With these values of α_{LSPE}, 20 independent runs are performed for every K_θ.

Figure 5.8 presents the performance of online PI with LSPE-Q across these 20 runs; compare with Figure 5.4. A reliably improving performance is only obtained for $K_\theta = 1$ and $\alpha_{\text{LSPE}} = 0.001$. In this experiment, due to the very small step size, learning is slower than for online LSPI in Figure 5.4. For all the other experiments, online PI with LSPE-Q is less reliable than online LSPI: there is a larger variation in performance across the 20 runs, as illustrated by the larger 95% confidence intervals for $K_\theta = 5000$ in Figure 5.8(b); the experiments for which confidence intervals are not shown have a similar character. To explain why this is the case, recall from Section 3.5.2 that in order to guarantee the convergence of LSPE-Q, it is required that state-action samples are generated according to their steady-state probabilities under the current policy. In online PI, the policy is changed often and many exploratory actions are taken, which severely violates this requirement, destabilizing the update (5.17). While online LSPI is also affected by imprecision in the values of Γ, Λ, and z, it may be more stable because it only uses them to compute one-shot solutions, rather than updating the parameter vector recursively, like online PI with LSPE-Q. Even though a very small step size may recover a stable performance improvement for online PI with LSPE-Q (as illustrated for $K_\theta = 1$ with $\alpha_{\text{LSPE}} = 0.001$), this is not guaranteed (as illustrated for $K_\theta = 100$, which uses the same value of α_{LSPE} but nevertheless remains unstable).

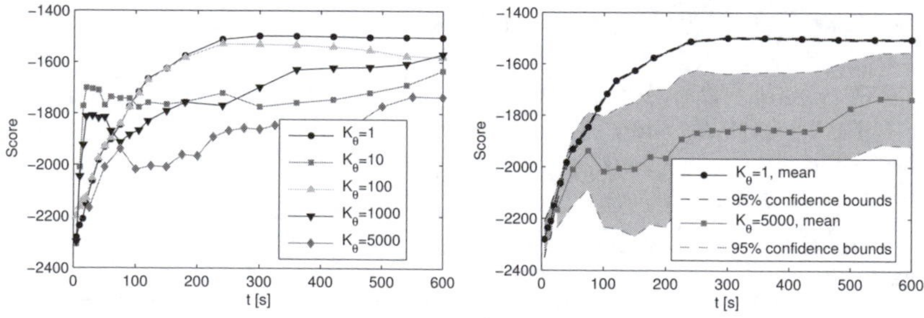

(a) Mean score for all the experiments.

(b) Mean score with 95% confidence intervals, for the extreme values of K_θ.

FIGURE 5.8
Performance of online PI with LSPE-Q for varying K_θ in the inverted pendulum problem.

Figure 5.9 presents the mean execution time of online PI with LSPE-Q, and repeats the execution time of online LSPI from Figure 5.6(b), for an easy comparison. Online PI with LSPE-Q solves a linear system at every step, so, for $K_\theta > 1$, it is more computationally expensive than online LSPI, which solves a linear system only before policy improvements.

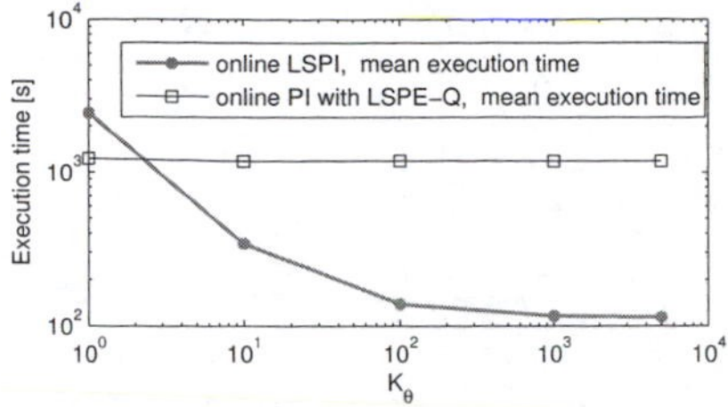

FIGURE 5.9
Execution time of online PI with LSPE-Q for varying K_θ, compared with the execution time of online LSPI, in the inverted pendulum problem.

Online LSPI for the real inverted pendulum

Next, online LSPI is used to control the inverted pendulum system in real time, rather than in simulation as in the earlier sections. To make the problem slightly easier for the learning controller, the sampling time is increased to $T_s = 0.02$ s (from 0.005 s), and the maximum available control is increased to 3.2 V (from 3 V); even so, the pendulum must still be swung back and forth to gather energy before it can be turned upright. The same approximator is used as in the simulation experiments, and online LSPI is run for 300 s, divided into 2 s long trials. Half of the trials start in the stable equilibrium (pointing down), and half in a random initial state obtained by applying a sequence of random actions. The initial exploration probability is $\varepsilon_0 = 1$ and decays with $\varepsilon_d = 0.9848$, which leads to a final exploration probability of 0.01. Policy improvements are performed only after each trial, because solving the linear system at line 11 of Algorithm 5.2 at arbitrary time steps may take longer than the sampling time.

Figure 5.10(a) presents a subsequence of learning trials, containing 1 out of each 10 trials. All of these trials are among those starting with the pendulum pointing down. These trajectories *include* the effects of exploration. Figure 5.10(b) shows a swing-up of the pendulum with the final policy and without exploration. The controller successfully learns how to swing up and stabilize the pendulum, giving a good performance roughly 120 s into learning. This is similar to the learning rate observed in the simulation experiments. Figure 5.10(c) shows the final policy obtained, indicating also some of the state samples collected during learning; compare with the near-optimal policy of Figure 5.1(b) on page 181. For small velocities (around the zero coordinate on the vertical axis), the policy found online resembles the near-optimal policy, but it is incorrect for large velocities. This is because, using the procedure described above to re-initialize the state in the beginning of each trial, the samples are concentrated in the low-velocity areas of the state space. The performance would improve if a (suboptimal) controller were available to re-initialize the state to arbitrary values.

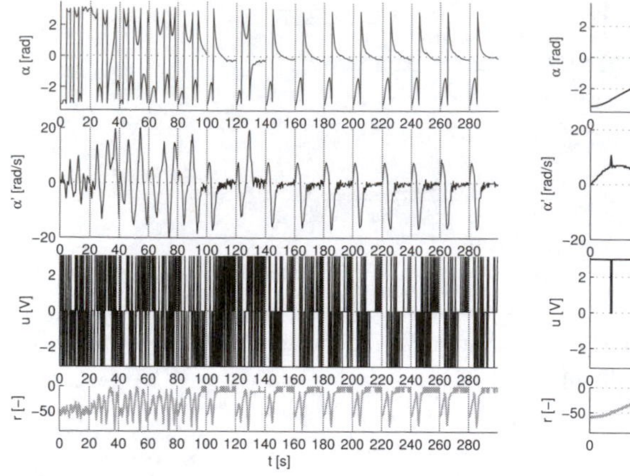

(a) A subsequence of learning trials. Each trial is 2 s long, and only 1 out of every 10 trials is shown. The starting time of each trial is given on the horizontal axis, and trials are separated by vertical lines.

(b) A swing-up using the final policy.

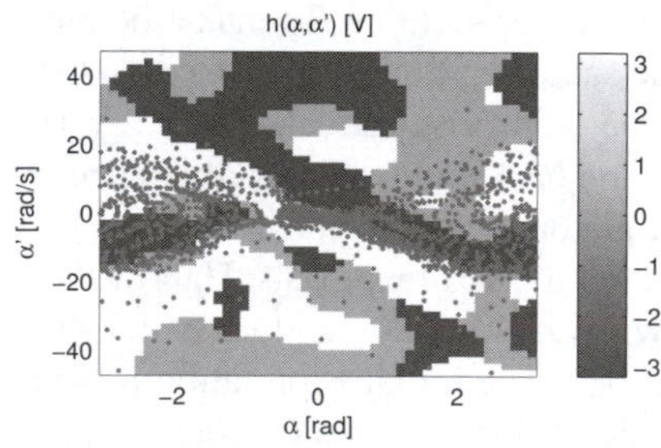

(c) Policy, with some of the state samples observed during learning indicated as gray dots.

FIGURE 5.10 Real-time results of online LSPI for the inverted pendulum.

5.6.2 Online LSPI for the two-link manipulator

This section examines the performance of online LSPI for a system with higher dimensionality than that of the inverted pendulum: a two-link robotic manipulator operating in a horizontal plane.

Two-link manipulator problem

Since the two-link manipulator problem was already described in Section 4.5.2, it is only briefly recapitulated here. The two-link manipulator, has 4 state variables, 2 action variables, and the following continuous-time dynamics:

$$M(\alpha)\ddot{\alpha} + C(\alpha, \dot{\alpha})\dot{\alpha} = \tau$$

where $\alpha = [\alpha_1, \alpha_2]^\mathrm{T}$ contains the angular positions of the two links, $\tau = [\tau_1, \tau_2]^\mathrm{T}$ contains the torques of the two motors, $M(\alpha)$ is the mass matrix, and $C(\alpha, \dot{\alpha})$ is the Coriolis and centrifugal forces matrix. For the values of these matrices, see Section 4.5.2, page 153; see also the schematic representation of the manipulator in Figure 4.8. The state signal contains the angles and angular velocities: $x = [\alpha_1, \dot{\alpha}_1, \alpha_2, \dot{\alpha}_2]^\mathrm{T}$, and the control signal is $u = \tau$. The angles α_1, α_2 "wrap around" in the interval $[-\pi, \pi)$ rad, and the angular velocities $\dot{\alpha}_1, \dot{\alpha}_2$ are restricted to the interval $[-2\pi, 2\pi]$ rad/s using saturation. The torques are constrained as follows: $\tau_1 \in [-1.5, 1.5]$ Nm, $\tau_2 \in [-1, 1]$ Nm. The discrete time step is set to $T_\mathrm{s} = 0.05$ s, and the discrete-time dynamics f are obtained by numerically integrating (4.49) between consecutive time steps.

The goal is to stabilize the system around $\alpha = \dot{\alpha} = 0$, and is expressed by the quadratic reward function:

$$\rho(x, u) = -x^\mathrm{T} Q_\mathrm{rew} x, \quad \text{with } Q_\mathrm{rew} = \mathrm{diag}[1, 0.05, 1, 0.05]$$

The discount factor is set to $\gamma = 0.98$.

Approximator, parameter settings, and performance criterion

Like for the inverted pendulum, the Q-function approximator combines state-dependent RBFs with discretized actions. An equidistant grid of $5 \times 5 \times 5 \times 5$ identically shaped axis-aligned RBFs is defined over the four-dimensional state space. The discretized actions are $[\tau_1, \tau_2]^\mathrm{T} \in \{-1.5, 0, 1.5\} \times \{-1, 0, 1\}$. This leads to a total number of $5^4 \cdot 9 = 5625$ state-action BFs. The learning experiment has a duration of 7200 s, and is divided into trials that have a length of 10 s (which is sufficient to stabilize the system) and start from uniformly distributed random initial states. The policy is improved once every $K_\theta = 50$ transitions, and β_Γ is set to 0.001. The initial exploration rate is $\varepsilon_0 = 1$ and decays with $\varepsilon_\mathrm{d} = 0.999041$, leading to an exploration probability of 0.001 at the end of the experiment.

The performance of the policies computed online is evaluated using the average return (score) over a set of initial states containing a regular grid of link angles:

$$X_0 = \{-\pi, -2\pi/3, -\pi/3, \ldots, \pi\} \times \{0\} \times \{-\pi, -2\pi/3, -\pi/3, \ldots, \pi\} \times \{0\}$$

The returns are estimated with a precision of $\varepsilon_\mathrm{MC} = 0.1$.

Results of online LSPI

Figure 5.11 shows the performance of online LSPI across 10 independent runs. The algorithm first reaches a near-final performance after 1200 s, during which 24000 samples are collected; a similar number of samples was required for the two-dimensional inverted pendulum. So, the learning rate of online LSPI scales up well to the higher-dimensional manipulator problem.

Figure 5.12 presents a policy found by online LSPI, together with a representative trajectory that is controlled by this policy; compare this solution, e.g., with the fuzzy Q-iteration solution shown in Figure 4.9 on page 155, in Chapter 4. The

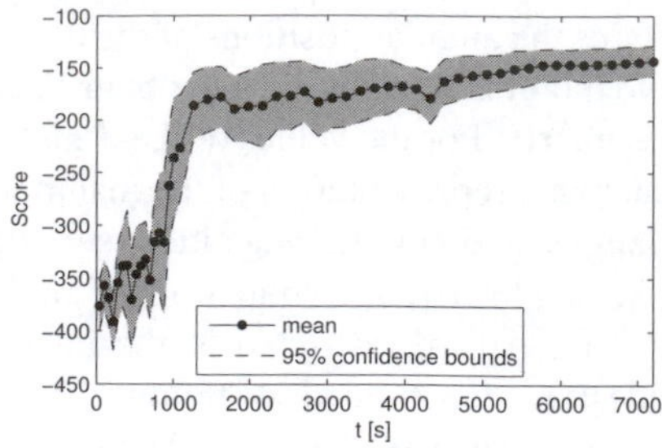

FIGURE 5.11 Performance of online LSPI for the robotic manipulator.

large-scale structure of the policy in Figure 5.12(a) roughly resembles that of the fuzzy Q-iteration policy, but the details are different and likely suboptimal. In Figure 5.12(a), the system is stabilized after 2 s, but the trajectory exhibits chattering of the control actions and, as a result, oscillation of the states. In comparison, the fuzzy Q-iteration trajectories of Figure 4.9 do not exhibit as much chattering.

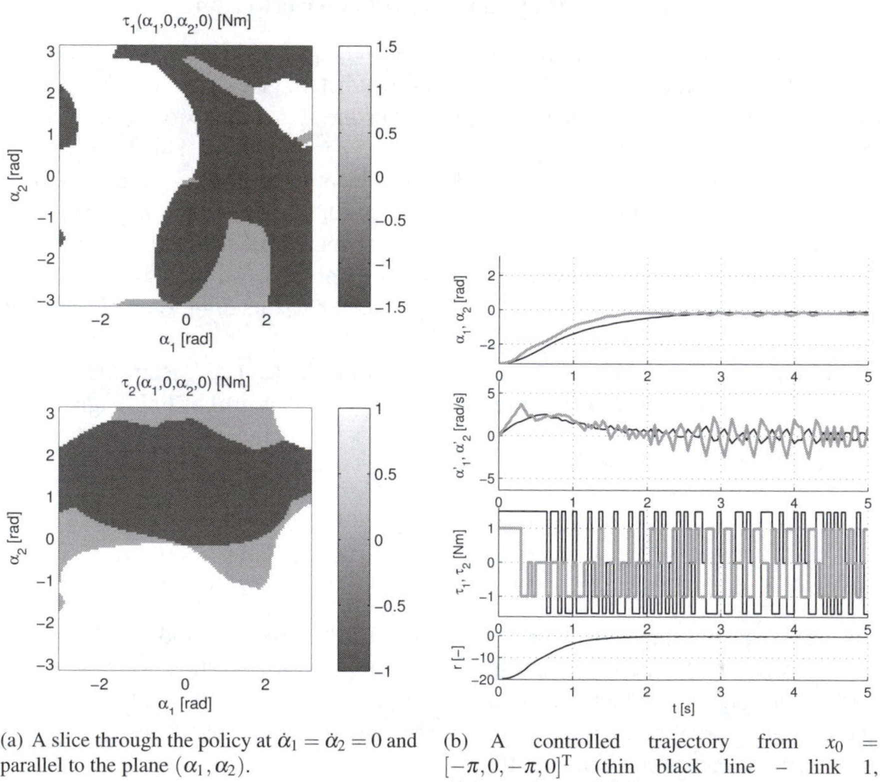

(a) A slice through the policy at $\dot{\alpha}_1 = \dot{\alpha}_2 = 0$ and parallel to the plane (α_1, α_2).

(b) A controlled trajectory from $x_0 = [-\pi, 0, -\pi, 0]^{\mathrm{T}}$ (thin black line – link 1, thick gray line – link 2).

FIGURE 5.12 Solution found by online LSPI for the two-link manipulator.

The poorer performance of online LSPI is caused mainly by the limitations of the chosen approximator. While increasing the number of RBFs or discrete actions would help, it may also lead to excessive computational costs.[7] Recall that online LSPI has a time complexity of $O(n^3) = O(N^3 M^3)$ per (policy improvement) step, where $N = 5^4$ is the number of state-dependent BFs and $M = 9$ the number of discrete actions. The actual execution time of online LSPI was around 20 hours per run, longer than the 1200 s interval of simulated time, so the experiment cannot be reproduced in real time. This illustrates the difficulty of using generic, uniform-resolution approximators in high-dimensional problems. Additionally, having more RBFs means that more parameters must be determined, which in turn requires more data. This can be problematic when data is costly.

A good way to avoid these difficulties is to determine automatically a small number of BFs well suited to the problem at hand. While we do not study this possibility here, we have reviewed in Section 3.6 some approaches for automatically finding BFs to be used in least-squares algorithms for policy evaluation. Most of approaches these work in a batch, offline setting (Menache et al., 2005; Mahadevan and Maggioni, 2007; Xu et al., 2007; Kolter and Ng, 2009), and would need to be modified to work online. Approaches that work on a sample-by-sample basis are more readily adaptable to online LSPI (Engel et al., 2005; Jung and Polani, 2007a). Another possibility to reduce the computational demands is to employ incremental updates of the parameters, rather than solving linear systems of equations (Geramifard et al., 2006, 2007), but this approach does not reduce the data demands of the algorithm.

For completeness, we also attempted to apply offline LSPI and online PI with LSPE-Q to the manipulator problem. Offline LSPI failed to converge even when provided with 10^5 samples, while online PI with LSPE-Q exceeded the memory resources of our machine[8] and therefore could not be run at all. Recall from Section 5.3 that, in practice, online LSPI only needs to store a single $n \times n$ estimate of $\Gamma - \gamma\Lambda$, whereas online PI with LSPE-Q must estimate Γ and Λ separately.

5.6.3 Online LSPI with prior knowledge for the DC motor

In this section, we investigate the effects of using prior knowledge about the policy in online LSPI. We consider prior knowledge about the monotonicity of the policy, as discussed in Section 5.4, and perform an experimental study on the DC motor problem, which was introduced in Section 3.4.5 and used again in Section 4.5.1.

DC motor problem

The DC motor is described by the discrete-time model:

[7] By comparison, fuzzy Q-iteration can use a more accurate approximator without incurring excessive computational costs, because it only needs to store and update vectors of parameters, whereas LSPI needs to store matrices and solve linear systems of equations.

[8] Our machine was equipped with 3 GB of RAM and was configured to use 2 GB of swap space.

$$f(x,u) = Ax + Bu$$
$$A = \begin{bmatrix} 1 & 0.0049 \\ 0 & 0.9540 \end{bmatrix}, \quad B = \begin{bmatrix} 0.0021 \\ 0.8505 \end{bmatrix}$$

where $x_1 = \alpha \in [-\pi, \pi]$ rad is the shaft angle, $x_2 = \dot{\alpha} \in [-16\pi, 16\pi]$ rad/s is the angular velocity, and $u \in [-10, 10]$ V is the control input (voltage). The state variables are restricted to their domains using saturation. The goal is to stabilize the system around $x = 0$, and is described by the quadratic reward function:

$$\rho(x,u) = -x^{\mathrm{T}} Q_{\mathrm{rew}} x - R_{\mathrm{rew}} u^2$$
$$Q_{\mathrm{rew}} = \begin{bmatrix} 5 & 0 \\ 0 & 0.01 \end{bmatrix}, \quad R_{\mathrm{rew}} = 0.01$$

with discount factor $\gamma = 0.95$.

Because the dynamics are linear and the reward function is quadratic, the optimal policy would be a linear state feedback if the constraints on the state and action variables were disregarded (Bertsekas, 2007, Section 3.2). The optimal feedback gain can be computed using an extension of linear quadratic control to the discounted case, as explained in Footnote 16 of Section 3.7.3. The resulting feedback gain for the DC motor is $[-12.92, -0.68]^{\mathrm{T}}$. By additionally restricting the control input to the admissible range $[-10, 10]$ using saturation, the following policy is obtained:

$$h(x) = \mathrm{sat}\left\{[-12.92, -0.68]^{\mathrm{T}} \cdot x, -10, 10\right\} \tag{5.18}$$

which is monotonically decreasing along both axes of the state space. This monotonicity property will be used in the sequel to accelerate the learning rate of online LSPI. Note that the actual values of the linear state feedback gains are not required to derive this prior knowledge, but only their sign must be known. The policy (5.18) is shown in Figure 5.13(a).

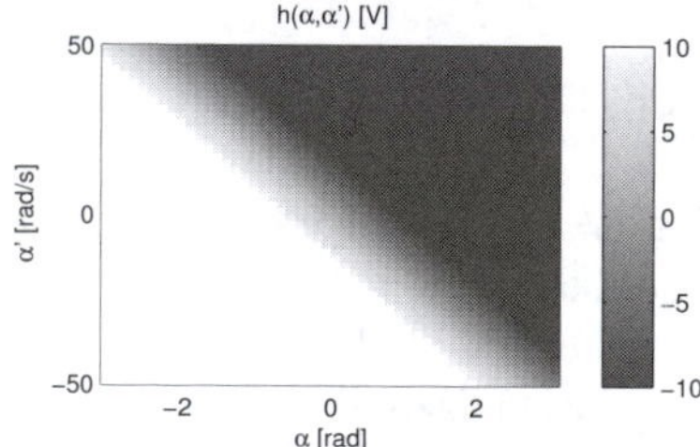

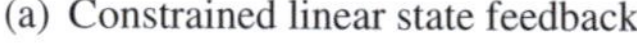
(a) Constrained linear state feedback.

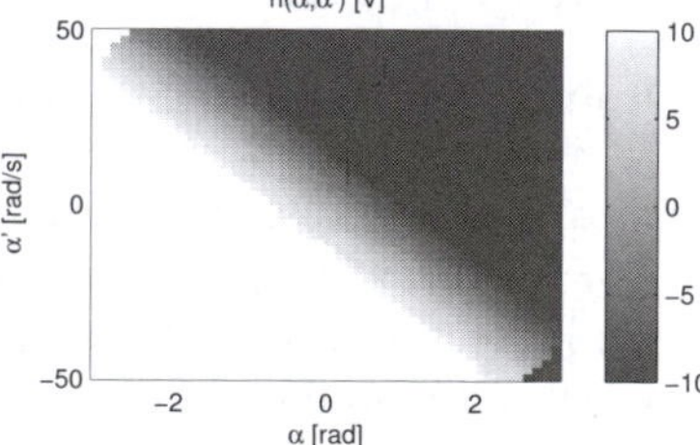

(b) Near-optimal policy found by fuzzy Q-iteration.

FIGURE 5.13 Near-optimal policies for the DC motor.

For comparison, Figure 5.13(b) presents a near-optimal policy, computed by fuzzy Q-iteration with an accurate approximator (this policy is repeated from Figure 3.5(b)). Over a large region of the state space, this policy is linear, and therefore monotonic. The only nonlinear, nonmonotonic regions appear in the top-left and bottom-right corners of the figure, and are probably due to the constraints on the state variables. So, the class of monotonic policies to which online LSPI will be restricted does indeed contain near-optimal solutions.

Approximator, parameter settings, and performance criterion

To apply online LSPI with monotonic policies to the DC motor, the Q-function is approximated using state-dependent RBFs and discretized actions, as in the inverted pendulum and manipulator examples above. The RBFs are axis-aligned, their centers are arranged on an equidistant 9×9 grid in the state space, and their width b_d along each dimension d is equal to ${b'_d}^2/2$, where b'_d is the distance between adjacent RBFs along that dimension. These RBFs lead to a smooth interpolation over the state space. Since the domains of the state variables are $[-\pi, \pi]$ for the angle and $[-16\pi, 16\pi]$ for the angular velocity, we obtain $b'_1 = \frac{2\pi}{9-1} \approx 0.79$ and $b'_2 = \frac{32\pi}{9-1} \approx 12.57$, leading to $b_1 \approx 0.31$ and $b_2 \approx 78.96$. The discretized action space is $U_\mathrm{d} = \{-10, 0, 10\}$, leading to a total number of $9^2 \cdot 3 = 243$ state-action BFs.

As explained in Section 5.4, we employ a linear policy parametrization (5.4) and enforce the monotonicity constraints in the policy improvements (5.10). The policy RBFs are identical to the Q-function RBFs, so the policy has 81 parameters. An added benefit of this parametrization is that it produces continuous actions. Nevertheless, as explained in Section 5.4.1, these actions must be discretized during learning, because the Q-function approximator can only employ discrete actions. To perform the policy improvements (5.10), 1000 uniformly distributed, random state samples are generated. Since these samples do not include information about the dynamics or the rewards, a model is not required to generate them.

The learning experiment has a length of 600 s, and is divided into 1.5 s long trials with uniformly distributed random initial states. The policy improvement interval is $K_\theta = 100$, and the exploration schedule starts from $\varepsilon_0 = 1$ and decays with a rate of $\varepsilon_\mathrm{d} = 0.9886$, leading to $\varepsilon = 0.1$ at $t = 200$ s. Policies are evaluated by estimating with precision $\varepsilon_\mathrm{MC} = 0.1$ their average return (score) over the grid of initial states:

$$X_0 = \{-\pi, -\pi/2, 0, \pi/2, \pi\} \times \{-10\pi, -5\pi, -2\pi, -\pi, 0, \pi, 2\pi, 5\pi, 10\pi\}$$

Results of online LSPI with prior knowledge, and comparison to online LSPI without prior knowledge

Figure 5.14 shows the learning performance of online LSPI with prior knowledge about the monotonicity of the policy, in comparison to the performance of the original online LSPI algorithm, which does not use prior knowledge. Mean values across 40 independent runs are reported, together with 95% confidence intervals on these means. Using prior knowledge leads to much faster and more reliable learning: the score reliably converges in around 50 s of simulation time. In contrast, online LSPI without prior knowledge requires more than 300 s of simulation time to reach a near-optimal performance, and has a larger variation in performance across the 40 runs, which can be seen in the wide 95% confidence intervals.

The mean execution time of online LSPI with prior knowledge is 1034.2 s, with

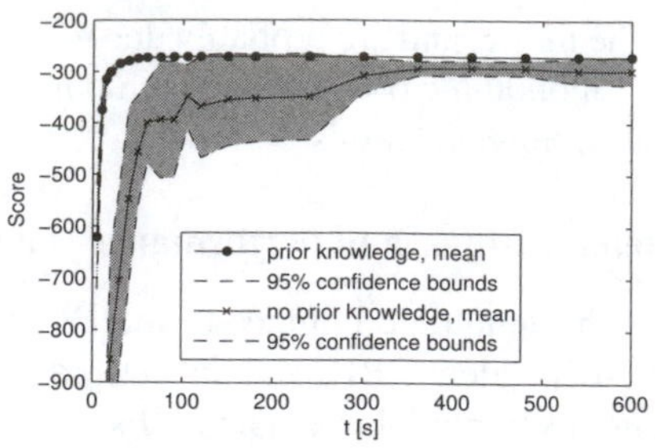

FIGURE 5.14
Comparison between online LSPI with prior knowledge and the original online LSPI algorithm, in the DC motor problem.

a 95% confidence interval of $[1019.6, 1048.7]$ s. For the original online LSPI algorithm, the mean execution time is 87.6 s with a confidence interval of $[84.0, 91.3]$ s. The execution time is larger for online LSPI with prior knowledge, because the constrained policy improvements (5.10) are more computationally demanding than the original policy improvements (5.2). Note that the execution time of online LSPI with prior knowledge is larger than the duration of the simulation (600 s), so the algorithm cannot be applied in real time.

Figure 5.15 compares a representative solution obtained using prior knowledge with one obtained by the original online LSPI algorithm. The policy obtained without using prior knowledge (Figure 5.15(b)) violates monotonicity in several areas. In the controlled trajectory of Figure 5.15(e), the control performance of the monotonic policy is better, mainly because it outputs continuous actions. Unfortunately, this advantage cannot be exploited *during* learning, when the actions must be discretized to make them suitable for the Q-function approximator. (Recall also that the same action discretization is employed in both the experiment without prior knowledge and in the experiment employing prior knowledge.)

5.6.4 LSPI with continuous-action approximation for the inverted pendulum

In this fourth and final example, we return to the inverted pendulum problem, and use it to evaluate the continuous-action approximators introduced in Section 5.5. To this end, offline LSPI with continuous-action approximation is applied to the inverted pendulum, and the results are compared to those obtained by offline LSPI with discrete actions.

Approximator, parameter settings, and performance criterion

Both the continuous-action and the discrete-action representations of the Q-function employ a set of state-dependent RBFs. These RBFs are identical to those used in the online LSPI experiments of Section 5.6.1. Namely, they are axis-aligned, identically

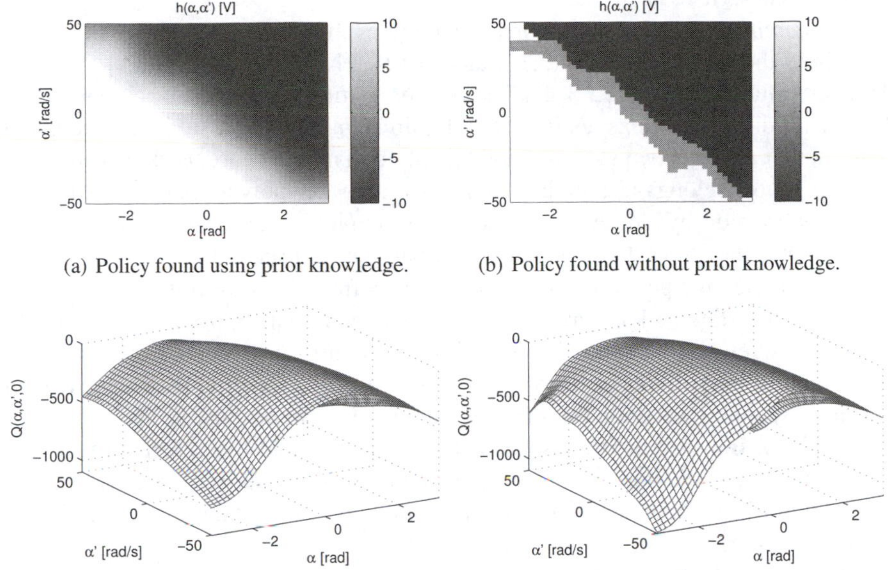

(a) Policy found using prior knowledge.

(b) Policy found without prior knowledge.

(c) Slice through the Q-function found using prior knowledge, for $u = 0$.

(d) Slice through the Q-function found without prior knowledge, for $u = 0$.

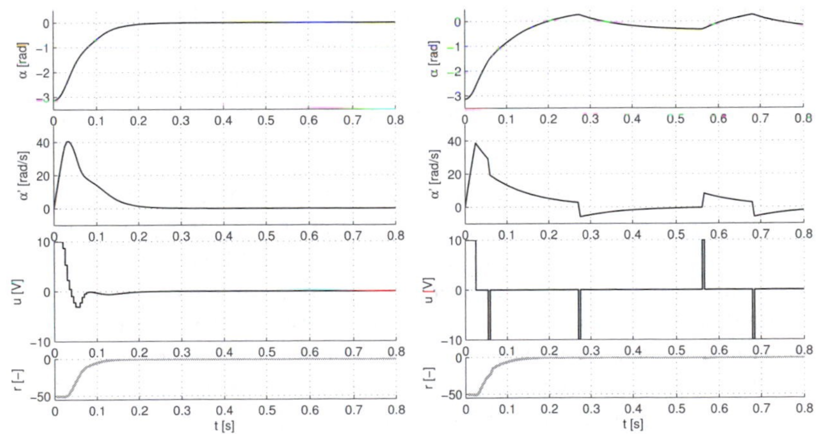

(e) Trajectory controlled by the policy found using prior knowledge.

(f) Trajectory controlled by the policy found without prior knowledge.

FIGURE 5.15
Representative solutions found using prior knowledge (left) and without prior knowledge (right) for the DC motor.

shaped, and distributed on an equidistant 11×11 grid. This state-space approximator is held fixed throughout the experiments, while the action-space approximator is changed as described next.

The *continuous-action* approximator combines the RBFs with Chebyshev polynomials of the first kind, as in (5.12) and (5.13). The degree M_p of the polynomials takes values in the set $\{2,3,4\}$. The *discrete-action* approximator combines the RBFs with discrete actions, as in (5.15). Equidistant discrete actions are used, and their number M takes two values: 3 and 5. Only odd numbers are used to ensure that the zero action belongs to the discrete set. We consider polynomial approximators of degree M_p side by side with discrete-action approximators having $M = M_p + 1$ actions, since they have the same number of parameters (Section 5.5).

The samples for policy evaluation are drawn from a uniform distribution over the continuous state-action space $X \times U$ for the polynomial approximators, and over the state-discrete action space $X \times U_d$ for the discrete-action approximators. For a fair comparison, the number of samples provided per Q-function parameter is held constant, by making the total number of samples n_s proportional to the number of parameters n. Because the state-space approximator is held fixed, n_s is in fact proportional to $M_p + 1$ in the continuous case, and to M in the discrete case. We choose $n_s = 10000$ for the approximators with $M = M_p + 1 = 3$, leading to $n_s = 13334$ for $M_p = 3$ and to $n_s = 16667$ for $M = M_p + 1 = 5$.

The offline LSPI algorithm is considered to have converged when the Euclidean norm of the difference between two consecutive parameter vectors does not exceed $\varepsilon_{\text{LSPI}} = 0.01$, or when limit cycles are detected in the sequence of parameters. The policy resulting from every convergent experiment is evaluated by estimating its average return over the grid (5.16) of initial states.

Results of LSPI with continuous actions, and comparison with the discrete action results

Figure 5.16 shows the performance and execution time of LSPI with continuous-action approximation in comparison with discrete-action approximation. These graphs report mean values and confidence intervals over 20 independent runs,[9] and show experiments with the same number of Q-function parameters at the same horizontal coordinate. In Figure 5.16(a), the performance differences between polynomial and discrete-action approximation are inconclusive for low-degree polynomials ($M_p = 2$). When the degree increases, the performance of the polynomial approximator becomes worse, probably due to overfitting. Polynomial approximation leads to a larger computational cost, which also grows with the degree of the polynomial, as shown in Figure 5.16(b). Among other reasons, this is because the policy improvements (5.14) with polynomial approximation are more computationally demanding than the discrete-action policy improvements.

Figure 5.17 compares a representative continuous-action solution with a representative discrete-action solution. A continuous-action solution found using second-

[9]Note that 2 out of the 20 runs of the experiment with third-degree polynomial approximation were not convergent, and were ignored when computing the means and confidence intervals.

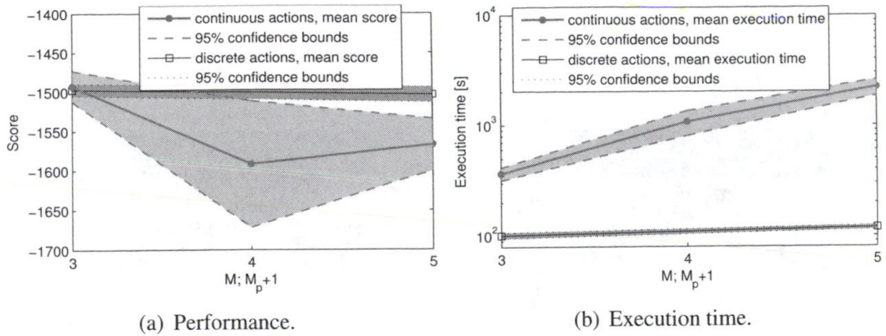

FIGURE 5.16
Comparison between continuous-action and discrete-action approximation in the inverted pendulum problem.

degree polynomial approximation was chosen, because it suffers less from overfitting (see Figure 5.16(a)). A discrete-action solution with $M = 3$ was selected for comparison, because it has the same number of parameters. The two policies have a similar structure, and the Q-functions are also similar. Continuous actions are in fact useful to eliminate chattering in Figure 5.17(e), even though this advantage is not apparent in the numerical scores shown in Figure 5.16(a). This discrepancy can be explained by examining the nature of the swing-up trajectories. Their first part can be well approximated by a "bang-off-bang" control law (Kirk, 2004, Section 5.5), which can be realized using only three discrete actions (maximum action in either direction, or zero action). Any chattering in the final part of the trajectory, although undesirable, will have little influence on the total return, due to the exponential discounting of the rewards. Because swing-ups are necessary from many initial states, in this problem it is difficult to improve upon the returns obtained by the discrete actions.[10] If it is essential to avoid chattering, the problem should be reformulated so that the reward function penalizes chattering more strongly.

5.7 Summary and discussion

This chapter has considered several extensions of LSPI, an originally offline algorithm that represents Q-functions using a linear parametrization, and finds the parameters by LSTD-Q policy evaluation. More specifically, an online variant of LSPI

[10]Note that the return obtained along the continuous-action trajectory of Figure 5.17(e) is slightly worse than for Figure 5.17(f). This is because of the spurious nonmaximal actions in the interval $[0.5, 0.7]$ s. On the other hand, even though the continuous-action trajectory exhibits a steady-state error, the negative contribution of this error to the return is less important than the negative contribution of the chattering in the discrete-action trajectory.

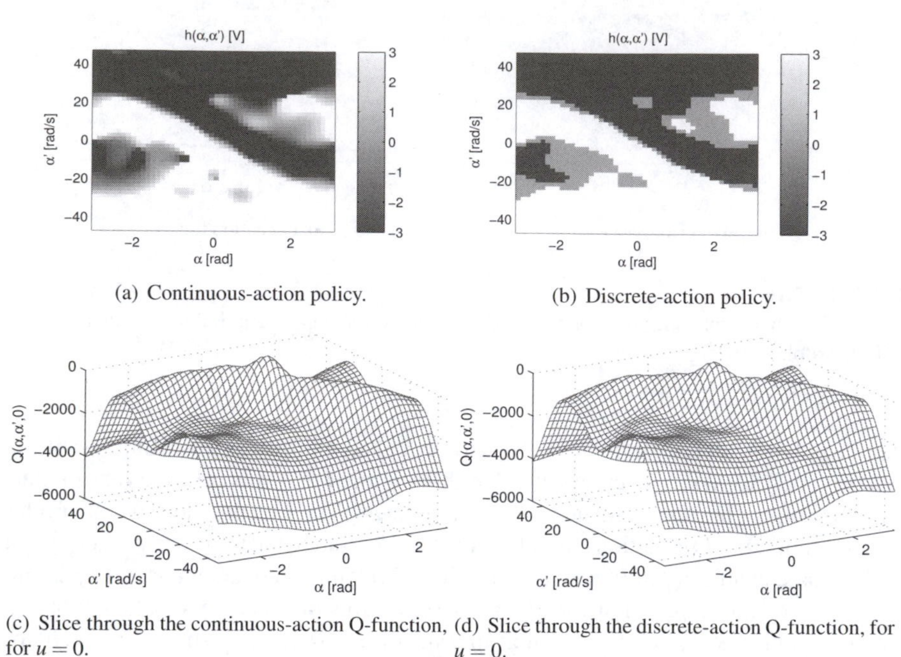

(a) Continuous-action policy.

(b) Discrete-action policy.

(c) Slice through the continuous-action Q-function, for $u = 0$.

(d) Slice through the discrete-action Q-function, for $u = 0$.

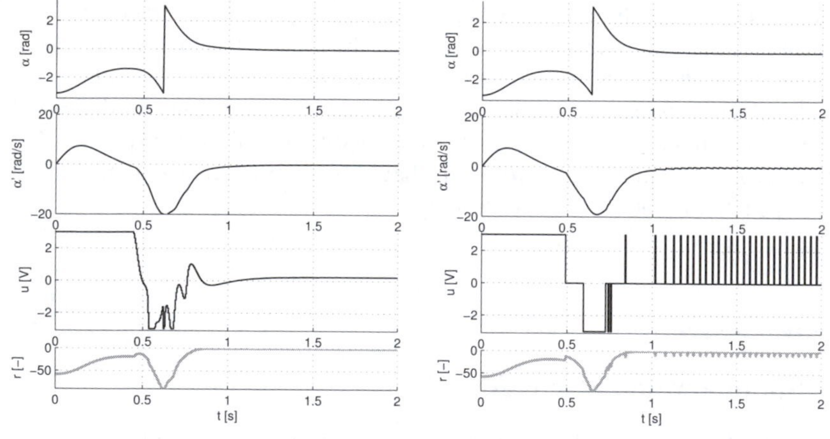

(e) Swing-up using continuous actions.

(f) Swing-up using discrete actions.

FIGURE 5.17

Representative solutions found with continuous actions ($M_\mathrm{p} = 2$; left) and with discrete actions ($M = 3$; right) for the inverted pendulum.

has been introduced, together with an approach to integrate prior knowledge into this variant, and with a continuous-action approximator for LSPI.

Online LSPI provided fast learning in simulation and real-time experiments with an inverted pendulum. In the same problem, online LSPI performed on par with its offline counterpart, and was more stable than online PI with LSPE-Q, probably because LSTD-Q is more resilient than LSPE-Q to the frequent policy improvements necessary in the online setting. Online LSPI also learned to stabilize a two-link manipulator, but the solutions found were suboptimal, due to the limitations of the chosen approximator, which only employed equidistant BFs.

In fact, such equidistant approximators were used in all the examples of this chapter, in order to focus on fairly evaluating the extensions of LSPI that were introduced, in isolation from the difficulties of designing the BFs. Nevertheless, in general the number of equidistant BFs required to achieve a good accuracy may be prohibitive, and it is better to employ a smaller number of well-chosen BFs. This would also help to reduce the computational demands of the algorithms; for instance, some simulation experiments with online LSPI took longer to execute than the interval of time they simulated, and so they cannot be replicated in real-time. While we have not considered the BF design problem in this chapter, in Section 3.6 we have reviewed some approaches to find good BFs automatically in least-squares methods for policy evaluation. These approaches could be adapted to work with online LSPI.

The performance guarantees of offline PI rely on bounded policy evaluation errors. Unfortunately, these guarantees cannot be applied to the online case, because online LSPI improves the policy optimistically, before an accurate policy evaluation can be completed. A different approach is required to theoretically analyze the performance of online, optimistic LSPI. Such an approach may also be useful to analyze online PI with LSPE-Q.

The method presented to integrate prior knowledge into online LSPI considers problems in which the policy is known to be monotonic in the state variables. For an example involving the control of a DC motor, the use of this type of prior knowledge led to much faster (in terms of simulated time) and more reliable learning. The monotonicity requirements can be made less restrictive, e.g., by only requiring that the policy is monotonic in the neighborhood of an equilibrium. More general constraints on the policy can also be considered, but they may be more difficult to enforce in the policy improvement step.

The continuous-action Q-function approximator developed for LSPI combines state-dependent basis functions with orthogonal polynomial approximation in the action space. This approach was evaluated in the inverted pendulum problem, where a second-degree polynomial approximator helped eliminate chattering of the control action, although it did not obtain a better numerical performance (return) than the discrete-action solution. High-degree polynomials can lead to overfitting, so it would be useful to develop continuous-action Q-function approximators that are more resilient to this detrimental effect. However, care must be taken to ensure that greedy actions can be efficiently computed using these approximators.

At this point in the book, we have considered in detail a method for approximate value iteration (Chapter 4), and one for approximate policy iteration (this chapter). In the next and final chapter, we will discuss in depth an algorithm from the third class of DP/RL methods, approximate policy search.

Bibliographical notes

The use of least-squares methods online was proposed, e.g., by Lagoudakis and Parr (2003a) and by Bertsekas (2007), but at the time of this writing, not much is known about how they behave in practice. Online, optimistic PI with LSPE-Q was used by Jung and Polani (2007a). Li et al. (2009) evaluated LSPI with online sample collection, focusing on the issue of exploration. Their method does not perform optimistic policy improvements, but instead fully executes LSPI between consecutive sample-collection episodes. To the best of our knowledge, methods to employ prior knowledge about the policy in LSPI have not yet been studied at the time of this writing. Concerning continuous-action results, Pazis and Lagoudakis (2009) proposed an approach to use continuous actions in LSPI that relies on iteratively refining discrete actions, rather than using polynomial approximation.

6

Approximate policy search with cross-entropy optimization of basis functions

This chapter describes an algorithm for approximate policy search in continuous-state, discrete-action problems. The algorithm looks for the best policy that can be represented using a given number of basis functions associated with discrete actions. The locations and shapes of the basis functions, together with the action assignments, are optimized using the cross-entropy method, so that the empirical return from a representative set of initial states is maximized. The resulting cross-entropy policy search algorithm is evaluated in problems with two to six state variables.

6.1 Introduction

The previous two chapters have considered value iteration and policy iteration techniques for continuous-space problems. While these techniques have many benefits, they also have limitations that can make them unsuitable for certain problems. A central difficulty is that representing value functions accurately becomes very demanding as the dimensionality of the problem increases. This is especially true for uniform-resolution representations of the value function, as seen in Chapters 4 and 5, since the complexity of such a representation grows exponentially with the number of dimensions. Even methods that construct adaptive-resolution approximators are often applied only to relatively low-dimensional problems (Munos and Moore, 2002; Ernst et al., 2005; Mahadevan and Maggioni, 2007).

In this final chapter of the book, we take a different approach, by designing a policy search algorithm that does not require a value function and thereby avoids the difficulty discussed above. Instead, this algorithm parameterizes the policy and searches for optimal parameters that lead to maximal returns (see also Section 3.7). We focus on the case where prior knowledge about the policy is not available, which means that a flexible policy parametrization must be employed. Since this flexible parametrization may lead to a nondifferentiable optimization criterion with many local optima, a gradient-free global optimization technique is required. We thus build on the framework of gradient-free policy search discussed in Section 3.7.2. Together with Chapters 4 and 5, this chapter completes the trio of examples that, at the end of

Chapter 3, we set out to develop for approximate value iteration, approximate policy iteration, and approximate policy search.

To obtain a flexible policy parametrization, we exploit ideas from the optimization of basis functions (BFs) for value function approximation (Section 3.6.2). In particular, we represent the policies using $\mathcal{N}$ state-dependent, *optimized* BFs that are associated with discrete actions in a many-to-one fashion. A discrete (or discretized) action space is therefore required. The type of BFs and their number $\mathcal{N}$ are specified in advance and determine the complexity and the representation power of the parametrization, whereas the locations and shapes of the BFs, together with the action assignments, are optimized by the policy search procedure. The optimization criterion is a weighted sum of the returns from a finite set of representative initial states, in which each return is computed with Monte Carlo simulations. The representative states and the weight function can be used to focus the algorithm on important parts of the state space.

This approach is expected to be efficient in high-dimensional state spaces, because its computational demands are not inherently related to the number of state variables. Instead, they depend on how many representative initial states are chosen (since simulations must be run from every such state), and on how difficult it is to find a policy that obtains near-optimal returns from these initial states (since this dictates the complexity of the optimization problem). Note that such a policy can be significantly easier to find than a globally optimal one, because it only needs to take good actions in the state space subregions reached by near-optimal trajectories from the representative states. In contrast, a globally optimal policy must take good actions over the entire state space, which can be significantly larger than these subregions.

We select the cross-entropy (CE) method to optimize the parameters of the policy (Rubinstein and Kroese, 2004). The resulting algorithm for CE policy search with adaptive BFs is evaluated in three problems, gradually increasing in dimensionality: the optimal control of a double integrator, the balancing of a bicycle, and the control of the treatment for infection by the human immunodefficiency virus (HIV). The two-dimensional double-integrator is used to study CE policy search when it looks for a policy that performs well over the entire state space. In this setting, CE policy search is also compared to value iteration and policy iteration with uniform-resolution approximation, and to a policy search variant that employs a different optimization algorithm called DIRECT (Jones, 2009). In the four-dimensional bicycle balancing problem, we study the effects of the set of representative states and of stochastic transitions. Finally, CE policy search is applied to a realistic, six-dimensional HIV infection control problem.

We next provide a recapitulation of CE optimization in Section 6.2. Section 6.3 then describes the policy parametrization and the CE algorithm to optimize the parameters, and Section 6.4 reports the results of the numerical experiments outlined above. A summary and a discussion are given in Section 6.5.

6.2 Cross-entropy optimization

In this section, the CE method for optimization (Rubinstein and Kroese, 2004) is briefly introduced, specializing the discussion to the context of this chapter. This introduction is mainly intended for the reader who skipped Chapter 4, where the CE method was also described; other readers can safely start reading from only Equation (6.5) onwards, where the discussion is novel. Also note that a more detailed description of the CE method can be found in Appendix B.

Consider the following optimization problem:

$$\max_{a \in \mathscr{A}} s(a) \tag{6.1}$$

where $s : \mathscr{A} \rightarrow \mathbb{R}$ is the score function (optimization criterion) to maximize, and the variable a takes values in the domain $\mathscr{A}$. Denote the maximum of s by s^*. The CE method maintains a density[1] with support $\mathscr{A}$. At each iteration, a number of samples are drawn from this density and the score values of these samples are computed. A smaller number of samples that have the best scores are kept, and the remaining samples are discarded. The density is then updated using the selected samples, such that the probability of drawing better samples is increased at the next iteration. The algorithm stops when the score of the worst selected sample no longer improves significantly.

Formally, a family of probability densities $\{p(\cdot; v)\}$ must be chosen. This family has support $\mathscr{A}$ and is parameterized by v. At each iteration $\tau \geq 1$ of the CE algorithm, a number N_{CE} of samples are drawn from the density $p(\cdot; v_{\tau-1})$, their scores are computed, and the $(1 - \rho_{\text{CE}})$ quantile[2] λ_τ of the sample scores is determined, with $\rho_{\text{CE}} \in (0, 1)$. Then, a so-called associated stochastic problem is defined, which involves estimating the probability that the score of a sample drawn from $p(\cdot; v_{\tau-1})$ is at least λ_τ:

$$\text{P}_{a \sim p(\cdot; v_{\tau-1})}(s(a) \geq \lambda_\tau) = \text{E}_{a \sim p(\cdot; v_{\tau-1})}\{I(s(a) \geq \lambda_\tau)\} \tag{6.2}$$

where I is the indicator function, equal to 1 whenever its argument is true, and 0 otherwise.

The probability (6.2) can be estimated by importance sampling. For the associated stochastic problem, an importance sampling density is one that increases the probability of the interesting event $s(a) \geq \lambda_\tau$. An optimal importance sampling density in the family $\{p(\cdot; v)\}$, in the sense of the smallest cross-entropy (smallest Kullback-Leibler divergence), is given by a parameter that is a solution of:

$$\arg\max_{v} \text{E}_{a \sim p(\cdot; v_{\tau-1})}\{I(s(a) \geq \lambda_\tau) \ln p(a; v)\} \tag{6.3}$$

[1]For simplicity, we will abuse the terminology by using the term "density" to refer to probability density functions (which describe probabilities of continuous random variables), as well as to probability mass functions (which describe probabilities of discrete random variables).

[2]If the score values of the samples are ordered increasingly and indexed such that $s_1 \leq \cdots \leq s_{N_{\text{CE}}}$, then the $(1 - \rho_{\text{CE}})$ quantile is: $\lambda_\tau = s_{\lceil (1-\rho_{\text{CE}}) N_{\text{CE}} \rceil}$.

An approximate solution $\widehat{v}_\tau$ of (6.3) is computed with the so-called stochastic counterpart:

$$\widehat{v}_\tau = v_\tau^\ddagger, \text{ where } v_\tau^\ddagger \in \arg\max_v \frac{1}{N_{\text{CE}}} \sum_{i_s=1}^{N_{\text{CE}}} I(s(a_{i_s}) \geq \lambda_\tau) \ln p(a_{i_s}; v) \tag{6.4}$$

Only the samples that satisfy $s(a_{i_s}) \geq \lambda_\tau$ contribute to this formula, since the contributions of the other samples are made to be zero by the product with the indicator function. In this sense, the updated density parameter only depends on these best samples, and the other samples are discarded.

CE optimization proceeds with the next iteration using the new density parameter $v_\tau = \widehat{v}_\tau$ (note that the probability (6.2) is never actually computed). The updated density aims at generating good samples with a higher probability than the old density, thus bringing $\lambda_{\tau+1}$ closer to the optimum s^*. The goal is to eventually converge to a density that, with very high probability, generates samples close to optimal value(s) of a. The algorithm can be stopped when the $(1-\rho_{\text{CE}})$-quantile of the sample performance improves for $d_{\text{CE}} > 1$ consecutive iterations, but these improvements do not exceed a small positive constant ε_{CE}; alternatively, the algorithm stops when a maximum number of iterations $\tau_{\max}$ is reached. The best score among the samples generated in all the iterations is taken as the approximate solution of the optimization problem (6.1), and the corresponding sample as an approximate location of an optimum.

Instead of setting the new density parameter equal to the solution $\widehat{v}_\tau$ of (6.4), it can also be updated incrementally:

$$v_\tau = \alpha_{\text{CE}} \widehat{v}_\tau + (1-\alpha_{\text{CE}}) v_{\tau-1} \tag{6.5}$$

where $\alpha_{\text{CE}} \in (0,1]$. This so-called smoothing procedure is useful to prevent CE optimization from becoming stuck in local optima (Rubinstein and Kroese, 2004).

Under certain assumptions on $\mathscr{A}$ and $p(\cdot; v)$, the stochastic counterpart (6.4) can be solved analytically. One particularly important case when this happens is when $p(\cdot; v)$ belongs to the natural exponential family (Morris, 1982). For instance, when $\{p(\cdot; v)\}$ is the family of Gaussians parameterized by the mean η and the standard deviation σ (so that $v = [\eta, \sigma]^{\text{T}}$), the solution v_τ of (6.4) consists of the mean and the standard deviation of the best samples, i.e., of the samples a_{i_s} for which $s(a_{i_s}) \geq \lambda_\tau$.

CE optimization has been shown to perform well in many optimization problems, often better than other randomized algorithms (Rubinstein and Kroese, 2004), and has found applications in many areas, among which are biomedicine (Mathenya et al., 2007), power systems (Ernst et al., 2007), vehicle routing (Chepuri and de Mello, 2005), vector quantization (Boubezoul et al., 2008), and clustering (Rubinstein and Kroese, 2004). While the convergence of CE optimization has not yet been proven in general, the algorithm is usually convergent in practice (Rubinstein and Kroese, 2004). For combinatorial (discrete-variable) optimization, the CE method provably converges with probability 1 to a unit mass density, which always generates samples equal to a single point. Furthermore, the probability that this convergence point is in fact an optimal solution can be made arbitrarily close to 1 by using a sufficiently small smoothing parameter α_{CE} (Costa et al., 2007).

6.3 Cross-entropy policy search

In this section, a general approach to policy optimization using the CE method is described, followed by a version of the general algorithm that employs radial basis functions (RBFs) to parameterize the policy.

6.3.1 General approach

Consider a stochastic or deterministic Markov decision process (MDP). In the sequel, we employ the notation for stochastic MDPs, but all the results can easily be specialized to the deterministic case. Denote by D the number of state variables of the MDP (i.e., the dimension of X). We assume that the action space of the MDP is discrete and contains M distinct actions, $U_\text{d} = \{u_1, \dots, u_M\}$. The set U_d can result from the discretization of an originally larger (e.g., continuous) action space U.

The policy parametrization is introduced next, followed by the score function and by the CE procedure to optimize the parameters.

Policy parametrization

The policy is represented using $\mathscr{N}$ basis functions (BFs) defined over the state space and parameterized by a vector $\xi \in \Xi$:

$$\varphi_i(\cdot;\xi) : X \to \mathbb{R}, \quad i = 1, \dots, \mathscr{N}$$

where the dot stands for the state argument x. The parameter vector ξ typically gives the locations and shapes of the BFs. The BFs are associated to discrete actions by a many-to-one mapping, which can be represented as a vector $\vartheta \in \{1, \dots, M\}^{\mathscr{N}}$ that associates each BF φ_i to a discrete action index ϑ_i, or equivalently to a discrete action u_{ϑ_i}. A schematic representation of this parametrization is given in Figure 6.1.

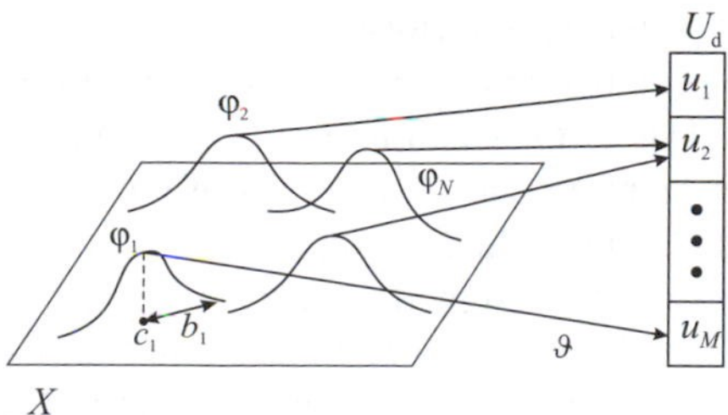

FIGURE 6.1
A schematic representation of the policy parametrization. The vector ϑ associates the BFs to discrete actions. In this example, the BFs are parameterized by their centers c_i and widths b_i, so that $\xi = [c_1^\text{T}, b_1^\text{T}, \dots, c_{\mathscr{N}}^\text{T}, b_{\mathscr{N}}^\text{T}]^\text{T}$. Reproduced with permission from (Buşoniu et al., 2009), © 2009 IEEE.

The complete policy parameter vector is thus $[\xi^{\mathrm{T}}, \vartheta^{\mathrm{T}}]^{\mathrm{T}}$, ranging in the set $\Xi \times \{1,\ldots,M\}^{\mathscr{N}}$. For any x, the policy chooses the action associated to a BF that takes the largest value at x:

$$h(x;\xi,\vartheta) = u_{\vartheta_{i^\ddagger}}, \quad \text{where } i^\ddagger \in \arg\max_i \varphi_i(x;\xi) \tag{6.6}$$

Score function

The goal of CE policy search is to find optimal parameters that maximize the weighted average of the returns obtained from a finite set X_0 of representative initial states. The return from every representative state is estimated using Monte Carlo simulations. This goal was already discussed in our review of gradient-free policy search given in Section 3.7.2, and this explanation is recapitulated here, specialized to CE policy search.

The score function (optimization criterion) can be written (3.63):

$$s(\xi,\vartheta) = \sum_{x_0 \in X_0} w(x_0) \widehat{R}^{h(\cdot;\xi,\vartheta)}(x_0) \tag{6.7}$$

where X_0 is the set of representative states, weighted by $w : X_0 \to (0,1]$.[3] The Monte Carlo estimate of the return for each state $x_0 \in X_0$ is (3.64):

$$\widehat{R}^{h(\cdot;\xi,\vartheta)}(x_0) = \frac{1}{N_{\mathrm{MC}}} \sum_{i_0=1}^{N_{\mathrm{MC}}} \sum_{k=0}^{K} \gamma^k \tilde{\rho}(x_{i_0,k}, h(x_{i_0,k};\xi,\vartheta), x_{i_0,k+1})$$

where $x_{i_0,0} = x_0$, $x_{i_0,k+1} \sim \tilde{f}(x_{i_0,k}, h(x_{i_0,k};\xi,\vartheta),\cdot)$, and N_{MC} is the number of Monte Carlo simulations to carry out. So, each simulation i_0 makes use of a system trajectory that is K steps long and generated using the policy $h(\cdot;\xi,\vartheta)$. The system trajectories are generated independently, so the score computation is unbiased. Given a desired precision $\varepsilon_{\mathrm{MC}} > 0$, the length K can be chosen using (3.65) to guarantee that truncating the trajectory introduces an error of at most $\varepsilon_{\mathrm{MC}}$ in the estimate of the sum along the original, infinitely long trajectory.

The set X_0 of representative initial states, together with the weight function w, determines the performance of the resulting policy. Some problems only require the optimal control of the system from a restricted set of initial states; X_0 should then be equal to this set, or included in it when the set is too large. Also, initial states that are deemed more important can be assigned larger weights. When all the initial states are equally important, the elements of X_0 should be uniformly spread over the state space and identical weights equal to $\frac{1}{|X_0|}$ should be assigned to every element of X_0 (recall that $|\cdot|$ denotes set cardinality). We study the influence of X_0 in Section 6.4.2 for a bicycle balancing problem.

---

[3] More generally, a density $\tilde{w}$ over the initial states can be considered, and the score function is then $\mathrm{E}_{x_0 \sim \tilde{w}(\cdot)} \left\{ R^{h(\cdot;\xi,\vartheta)}(x_0) \right\}$, i.e., the expected value of the return when $x_0 \sim \tilde{w}(\cdot)$. Such a score function can be evaluated by Monte Carlo methods. In this chapter, we only use finite sets X_0 associated with weighting functions w, as in (6.7).

A general algorithm for cross-entropy policy search

A global, gradient-free, mixed-integer optimization problem must be solved to find optimal parameters ξ^*, ϑ^* that maximize the score function (6.7). Several techniques are available to solve this type of problem; in this chapter, we select the CE method as an illustrative example of such a technique. In Section 6.4.1, we compare CE optimization with the DIRECT optimization algorithm (Jones, 2009) in the context of policy search.

In order to define the associated stochastic problem (6.2) for CE optimization, it is necessary to choose a family of densities with support $\Xi \times \{1,\ldots,M\}^{\mathcal{N}}$. In general, Ξ may not be a discrete set, so it is convenient to use separate densities for the two parts ξ and ϑ of the parameter vector. Denote the density for ξ by $p_\xi(\cdot; v_\xi)$, parameterized by v_ξ and with support Ξ, and the density for ϑ by $p_\vartheta(\cdot; v_\vartheta)$, parameterized by v_ϑ and with support $\{1,\ldots,M\}^{\mathcal{N}}$. Let N_{v_ξ} denote the number of elements in the vector v_ξ, and N_{v_ϑ} the number of elements in v_ϑ. Note that densities from which it is easy to sample (see Press et al., 1986, Chapter 7) are usually chosen; e.g., we will later use Gaussian densities for continuous variables and Bernoulli densities for binary variables.

ALGORITHM 6.1

Cross-entropy policy search. Reproduced with permission from (Buşoniu et al., 2009), © 2009 IEEE.

Input: dynamics $\tilde{f}$, reward function $\tilde{\rho}$, discount factor γ,
representative states X_0, weight function w,
density families $\{p_\xi(\cdot; v_\xi)\}, \{p_\vartheta(\cdot; v_\vartheta)\}$, density parameter numbers N_{v_ξ}, N_{v_ϑ},
other parameters $\mathcal{N}$, ρ_{CE}, c_{CE}, α_{CE}, d_{CE}, $\varepsilon_{\mathrm{CE}}$, $\varepsilon_{\mathrm{MC}}$, N_{MC}, $\tau_{\max}$

initialize density parameters $v_{\xi,0}$, $v_{\vartheta,0}$
$N_{\mathrm{CE}} \leftarrow c_{\mathrm{CE}}(N_{v_\xi} + N_{v_\vartheta})$
$\tau \leftarrow 0$
repeat
  $\tau \leftarrow \tau + 1$
  generate samples $\xi_1, \ldots, \xi_{N_{\mathrm{CE}}}$ from $p_\xi(\cdot; v_{\xi,\tau-1})$
  generate samples $\vartheta_1, \ldots, \vartheta_{N_{\mathrm{CE}}}$ from $p_\vartheta(\cdot; v_{\vartheta,\tau-1})$
  compute $s(\xi_{i_s}, \vartheta_{i_s})$ with (6.7), $i_s = 1, \ldots, N_{\mathrm{CE}}$
  reorder and reindex s.t. $s_1 \leq \cdots \leq s_{N_{\mathrm{CE}}}$
  $\lambda_\tau \leftarrow s_{\lceil (1-\rho_{\mathrm{CE}}) N_{\mathrm{CE}} \rceil}$
  $\widehat{v}_{\xi,\tau} \leftarrow v^{\ddagger}_{\xi,\tau}$, where $v^{\ddagger}_{\xi,\tau} \in \arg\max_{v_\xi} \sum_{i_s=\lceil (1-\rho_{\mathrm{CE}}) N_{\mathrm{CE}} \rceil}^{N_{\mathrm{CE}}} \ln p_\xi(\xi_{i_s}; v_\xi)$
  $\widehat{v}_{\vartheta,\tau} \leftarrow v^{\ddagger}_{\vartheta,\tau}$, where $v^{\ddagger}_{\vartheta,\tau} \in \arg\max_{v_\vartheta} \sum_{i_s=\lceil (1-\rho_{\mathrm{CE}}) N_{\mathrm{CE}} \rceil}^{N_{\mathrm{CE}}} \ln p_\vartheta(\vartheta_{i_s}; v_\vartheta)$
  $v_{\xi,\tau} \leftarrow \alpha_{\mathrm{CE}} \widehat{v}_{\xi,\tau} + (1-\alpha_{\mathrm{CE}}) v_{\xi,\tau-1}$
  $v_{\vartheta,\tau} \leftarrow \alpha_{\mathrm{CE}} \widehat{v}_{\vartheta,\tau} + (1-\alpha_{\mathrm{CE}}) v_{\vartheta,\tau-1}$
until ($\tau > d_{\mathrm{CE}}$ **and** $|\lambda_{\tau-\tau'} - \lambda_{\tau-\tau'-1}| \leq \varepsilon_{\mathrm{CE}}$, for $\tau' = 0, \ldots, d_{\mathrm{CE}} - 1$) **or** $\tau = \tau_{\max}$

Output: $\widehat{\xi}^*$, $\widehat{\vartheta}^*$, the best sample; and $\widehat{s}^* = s(\widehat{\xi}^*, \widehat{\vartheta}^*)$

The CE method for policy search is summarized in Algorithm 6.1. For easy reference, Table 6.1 collects the meaning of the parameters and variables playing a role in this algorithm. The stochastic counterparts at lines 11 and 12 of Algorithm 6.1 were simplified, using the fact that the samples are already sorted in the ascending order of their scores. The algorithm terminates after the variation of λ is at most ε_{CE} for d_{CE} consecutive iterations, or when a maximum number $\tau_{\max}$ of iterations has been reached. When $\varepsilon_{\text{CE}} = 0$, the algorithm is stopped only if λ does not change at all for d_{CE} consecutive iterations. The integer $d_{\text{CE}} > 1$ ensures that the decrease of the performance variation below ε_{CE} is not accidental (e.g., due to random effects).

TABLE 6.1
Parameters and variables of cross-entropy policy search. Reproduced with permission from (Buşoniu et al., 2009), © 2009 IEEE.

Symbol	Meaning
$\mathcal{N}$; M	number of BFs; number of discrete actions
ξ; ϑ	BF parameters; assignment of discrete actions to BFs
v_ξ; v_ϑ	parameters of the density for ξ; and for ϑ
N_{CE}	number of samples used at every CE iteration
ρ_{CE}	proportion of samples used in the CE updates
λ	$(1-\rho_{\text{CE}})$ quantile of the sample performance
c_{CE}	how many times the number of samples N_{CE} is larger than the number of density parameters
α_{CE}	smoothing parameter
N_{MC}	number of Monte Carlo simulations for each state
ε_{MC}	precision in estimating the returns
ε_{CE}	convergence threshold
d_{CE}	how many iterations the variation of λ should be at most ε_{CE} to stop the algorithm
τ; $\tau_{\max}$	iteration index; maximum number of iterations

Often it is convenient to use densities with unbounded support (e.g., Gaussians) when the BF parameters are continuous. However, the set Ξ must typically be bounded, e.g., when ξ contains centers of RBFs, which must remain inside a bounded state space. Whenever this situation arises, samples can be generated from the density with a larger (unbounded) support, and those samples that do not belong to Ξ can be rejected. The procedure continues until N_{CE} valid samples are generated, and the rest of the algorithm remains unchanged. The situation is entirely similar for the discrete action assignments ϑ, when it is convenient to use a family of densities $p_\vartheta(\cdot; v_\vartheta)$ with a support larger than $\{1,\ldots,M\}^{\mathcal{N}}$. The theoretical basis of CE optimization remains valid when sample rejection is employed, since an equivalent algorithm that uses all the samples can always be given by making the following two modifications:

- The score function is extended to assign very large negative scores (larger in magnitude than for any valid sample) to samples falling outside the domain.

- At each iteration, the parameters N_{CE} and ρ_{CE} are adapted so that a constant number of valid samples is generated, and a constant number of best samples is used for the parameter updates.

The most important parameters in CE policy search are, like in the general CE optimization, the number of samples, N_{CE}, and the proportion of best samples used to update the density, ρ_{CE}. The parameter c_{CE} is taken greater than or equal to 2, so that the number of samples is a multiple of the number of density parameters. The parameter ρ_{CE} can be taken around 0.01 for large numbers of samples, or it can be larger, around $(\ln N_{CE})/N_{CE}$, if the number of samples is smaller ($N_{CE} < 100$) (Rubinstein and Kroese, 2004). The number $\mathcal{N}$ of BFs determines the representation power of the policy approximator, and a good value for $\mathcal{N}$ depends on the problem at hand. In Section 6.4, we study the effect of varying $\mathcal{N}$ in two example problems. For deterministic MDPs, it suffices to simulate a single trajectory for every initial state in X_0, so $N_{MC} = 1$, whereas in the stochastic case, several trajectories should be simulated, i.e., $N_{MC} > 1$, with a good value of N_{MC} depending on the problem. The parameter $\varepsilon_{MC} > 0$ should be chosen smaller than the difference between the return along good trajectories and the return along undesirable trajectories, so that the optimization algorithm can effectively distinguish between these types of trajectories. This choice can be difficult to make and may require some trial and error. As a default initial value, ε_{MC} can be taken to be several orders of magnitude smaller than the bound $\frac{\|\rho\|_\infty}{1-\gamma}$ on the absolute value of the returns. Since it does not make sense to impose a convergence threshold smaller than the precision of the score function, ε_{CE} should be chosen larger than or equal to ε_{MC}, and a good default value is $\varepsilon_{CE} = \varepsilon_{MC}$.

6.3.2 Cross-entropy policy search with radial basis functions

In this section, we describe a version of CE policy search that uses state-dependent, axis-aligned Gaussian RBFs to represent the policy. Gaussian RBFs are chosen because they are commonly used to represent approximate MDP solutions, see, e.g., Chapter 5 of this book and (Tsitsiklis and Van Roy, 1996; Ormoneit and Sen, 2002; Lagoudakis and Parr, 2003a; Menache et al., 2005). Many other types of BFs could be used instead, including, e.g., splines and polynomials.

We assume that the state space is a D-dimensional hyperbox centered in the origin: $X = \left\{x \in \mathbb{R}^D \,\middle|\, |x| \leq x_{max}\right\}$, where $x_{max} \in (0,\infty)^D$. In this formula, as well as in the sequel, mathematical operations and conditions on vectors, such as the absolute value and relational operators, are applied element-wise. The hyperbox assumption is made here for simplicity and can be relaxed. For example, a simple relaxation is to allow hyperbox state spaces that are not centered in the origin, as will be done for the HIV treatment control problem of Section 6.4.3.

Radial basis functions and their probability density

The Gaussian RBFs are defined by:[4]

$$\varphi_i(x;\xi) = \exp\left[-\sum_{d=1}^{D} \frac{(x_d - c_{i,d})^2}{b_{i,d}^2}\right] \tag{6.8}$$

where D is the number of state variables, $c_i = [c_{i,1},\dots,c_{i,D}]^{\mathrm{T}}$ is the D-dimensional center of the i-th RBF, and $b_i = [b_{i,1},\dots,b_{i,D}]^{\mathrm{T}}$ is its width. Denote the vector of centers by $c = [c_1^{\mathrm{T}},\dots,c_{\mathscr{N}}^{\mathrm{T}}]^{\mathrm{T}}$ and the vector of widths by $b = [b_1^{\mathrm{T}},\dots,b_{\mathscr{N}}^{\mathrm{T}}]^{\mathrm{T}}$. So, $c_{i,d}$ and $b_{i,d}$ are scalars, c_i and b_i are D-dimensional vectors that collect the scalars for all D dimensions, and c and b are $D\mathscr{N}$-dimensional vectors that collect the D-dimensional vectors for all $\mathscr{N}$ RBFs. The RBF parameter vector is $\xi = [c^{\mathrm{T}}, b^{\mathrm{T}}]^{\mathrm{T}}$ and takes values in $\Xi = X^{\mathscr{N}} \times (0,\infty)^{D\mathscr{N}}$, since the centers of the RBFs must lie within the bounded state space, $c \in X^{\mathscr{N}}$, and their widths must be strictly positive, $b \in (0,\infty)^{D\mathscr{N}}$.

To define the associated stochastic problem (6.2) for optimizing the RBF parameters, independent Gaussian densities are selected for each element of the parameter vector ξ. Note that this concatenation of densities can converge to a degenerate distribution that always generates samples equal to a single value, such as a precise optimum location. The density for each center $c_{i,d}$ is parameterized by its mean $\eta_{i,d}^c$ and its standard deviation $\sigma_{i,d}^c$, while the density for a width $b_{i,d}$ is likewise parameterized by $\eta_{i,d}^b$ and $\sigma_{i,d}^b$. Similarly to the centers and widths themselves, we denote the $D\mathscr{N}$-dimensional vectors of means and standard deviations by, respectively, η^c, σ^c for the centers, and by η^b, σ^b for the widths. The parameter of the density for the RBF parameters gathers all these vectors together:

$$v_\xi = [(\eta^c)^{\mathrm{T}}, (\sigma^c)^{\mathrm{T}}, (\eta^b)^{\mathrm{T}}, (\sigma^b)^{\mathrm{T}}]^{\mathrm{T}} \in \mathbb{R}^{4D\mathscr{N}}$$

Note that the support of the density for the RBF parameters is $\mathbb{R}^{2D\mathscr{N}}$, which is larger than the domain of the parameters $\Xi = X^{\mathscr{N}} \times (0,\infty)^{D\mathscr{N}}$, and therefore samples that do not belong to Ξ must be rejected and generated again.

The means and the standard deviations are initialized for all i as follows:

$$\eta_i^c = 0, \quad \sigma_i^c = x_{\max}, \quad \eta_i^b = \frac{x_{\max}}{2(\mathscr{N}+1)}, \quad \sigma_i^b = \eta_i^b$$

where "0" denotes a vector of D zeros. The initial density parameters for the RBF centers ensure a good coverage of the state space, while the parameters for the RBF widths are initialized heuristically to yield a similar overlap between RBFs for different values of $\mathscr{N}$. The Gaussian density belongs to the natural exponential family, so the solution $\widehat{v}_{\xi,\tau}$ of the stochastic counterpart at line 11 of Algorithm 6.1 can be computed explicitly, as the element-wise mean and standard deviation of the best samples (see also Section 6.2). For instance, assuming without loss of generality that

[4]Note that the RBF width parameters in this definition are different from those used in the RBF formula (3.6) of Chapter 3. This new variant makes it easier to formalize the optimization algorithm, but is of course entirely equivalent to the original description of axis-aligned RBFs.

the samples are ordered in the ascending order of their scores, the density parameters for the RBF centers are updated as follows:

$$\widehat{\eta}_{\tau}^{c} = \frac{1}{N_{\text{CE}} - i_{\tau} + 1} \sum_{i_{\text{s}}=i_{\tau}}^{N_{\text{CE}}} c_{i_{\text{s}}}, \qquad \widehat{\sigma}_{\tau}^{c} = \sqrt{\frac{1}{N_{\text{CE}} - i_{\tau} + 1} \sum_{i_{\text{s}}=i_{\tau}}^{N_{\text{CE}}} (c_{i_{\text{s}}} - \widehat{\eta}_{\tau}^{c})^2}$$

where $i_{\tau} = \lceil (1-\rho_{\text{CE}}) N_{\text{CE}} \rceil$ denotes the index of the first of the best samples. Recall also that η_{τ}^{c}, σ_{τ}^{c}, and $c_{i_{\text{s}}}$ are all $D\mathcal{N}$-dimensional vectors, and that mathematical operations are performed element-wise.

Discrete action assignments and their probability density

The vector ϑ containing the assignments of discrete actions to BFs is represented in binary code, using $\mathcal{N}^{\text{bin}} = \lceil \log_2 M \rceil$ bits for each element ϑ_i. Thus, the complete binary representation of ϑ has $\mathcal{N}\mathcal{N}^{\text{bin}}$ bits. A binary representation is convenient because it allows us to work with Bernoulli distributions, as described next.

To define the associated stochastic problem (6.2) for optimizing ϑ, every bit is drawn from a Bernoulli distribution parameterized by its mean $\eta^{\text{bin}} \in [0,1]$ (η^{bin} gives the probability of selecting 1, while the probability of selecting 0 is $1-\eta^{\text{bin}}$). Because every bit has its own Bernoulli parameter, the total number of Bernoulli parameters v_{ϑ} is $\mathcal{N}\mathcal{N}^{\text{bin}}$. Similarly to the Gaussian densities above, this combination of independent Bernoulli distributions can converge to a degenerate distribution concentrated on a single value, such as an optimum. Note that if M is not a power of 2, bit combinations corresponding to invalid indices are rejected and generated again. For instance, if $M = 3$, $\mathcal{N}^{\text{bin}} = 2$ is obtained, the binary value 00 points to the first discrete action u_1 (since the binary representation is zero-based), 01 points to u_2, 10 to u_3, and 11 is invalid and will be rejected.

The mean η^{bin} for every bit is initialized to 0.5, which means that the bits 0 and 1 are initially equiprobable. Since the Bernoulli distribution belongs to the natural exponential family, the solution $\widehat{v}_{\vartheta,\tau}$ of the stochastic counterpart at line 12 of Algorithm 6.1 can be computed explicitly, as the element-wise mean of the best samples in their binary representation.

Computational complexity

We now briefly examine the complexity of this version of CE policy search. The number of density parameters is $N_{v_{\xi}} = 4D\mathcal{N}$ for the RBF centers and widths, and $N_{v_{\vartheta}} = \mathcal{N}\mathcal{N}^{\text{bin}}$ for the action assignments. Therefore, the total number of samples used is $N_{\text{CE}} = c_{\text{CE}}(4D\mathcal{N} + \mathcal{N}\mathcal{N}^{\text{bin}})$. Most of the computational load is generated by the simulations required to estimate the score of each sample. Therefore, neglecting the other computations, the complexity of one CE iteration is:

$$t_{\text{step}} [c_{\text{CE}} \mathcal{N} (4D + \mathcal{N}^{\text{bin}}) \cdot |X_0| \cdot N_{\text{MC}} K] \tag{6.9}$$

where K is the maximum length of each trajectory, and t_{step} is the time needed to compute the policy for a given state and to simulate the controlled system for one time step. Of course, if some trajectories terminate in fewer than K steps, the cost is reduced.

The complexity (6.9) is linear in the number $|X_0|$ of representative states, which suggests one way to control the complexity of CE policy search: by limiting the number of initial states to the minimum necessary. The complexity is linear also in the number of state variables D. However, this does not necessarily mean that the computational cost of CE policy search grows (only) linearly with the problem dimension, since the cost is influenced by the problem also in other ways, such as through the number $\mathscr{N}$ of RBFs required to represent a good policy.

In the upcoming examples, we will use the name "CE policy search" to refer to this RBF-based version of the general procedure given in Algorithm 6.1.

6.4 Experimental study

In the sequel, to assess the performance of CE policy search, extensive numerical experiments are carried out in three problems with a gradually increasing dimensionality: the optimal control of a double integrator (two dimensions), the balancing of a bicycle that rides at a constant speed (four dimensions), and the control of the treatment of an HIV infection (six dimensions).

6.4.1 Discrete-time double integrator

In this section, a double integrator optimal control problem is used to evaluate the effectiveness of CE policy search when looking for a policy that performs well over the entire state space. In this setting, CE policy search is compared with fuzzy Q-iteration, with least-squares policy iteration, and with a policy search variant that employs a different optimization algorithm called DIRECT. The double integrator problem is stated such that (near-)optimal trajectories from any state terminate in a small number of steps, which allows extensive simulation experiments to be run and an optimal solution to be found without excessive computational costs.

Double integrator problem

The double integrator is deterministic, has a continuous state space $X = [-1,1] \times [-0.5,0.5]$, a discrete action space $U_{\mathrm{d}} = \{-0.1, 0.1\}$, and the dynamics:

$$x_{k+1} = f(x_k, u_k) = \operatorname{sat}\left\{[x_{1,k} + x_{2,k}, x_{2,k} + u_k]^{\mathrm{T}}, -x_{\max}, x_{\max}\right\} \tag{6.10}$$

where $x_{\max} = [1, 0.5]^{\mathrm{T}}$ and the saturation is employed to bound the state variables to their domain X. Every state for which $|x_1| = 1$ is terminal, regardless of the value of x_2. (Recall that applying any action in a terminal state brings the process back to the same state, with a zero reward.) The goal is to drive the position x_1 to either boundary of the interval $[-1, 1]$, i.e., to a terminal state, so that when x_1 reaches the boundary, the speed x_2 is as small as possible in magnitude. This goal is expressed by the reward function:

$$r_{k+1} = \rho(x_k, u_k) = -(1 - |x_{1,k+1}|)^2 - x_{2,k+1}^2 x_{1,k+1}^2 \tag{6.11}$$

The product $-x_{2,k+1}^2 x_{1,k+1}^2$ penalizes large values of x_2, but only when x_1 is close to 1, i.e., to a terminal state. The discount factor γ is set to 0.95.

Figure 6.2 presents an optimal solution for this problem. In particular, Figure 6.2(a) shows an accurate representation of an optimal policy, consisting of the optimal actions for a regular grid of 101×101 points covering the state space. These optimal actions were obtained using the following brute-force procedure. All the possible sequences of actions of a sufficient length were generated, and the system was controlled with all these sequences starting from every state on the grid. For every such state, a sequence that produced the best discounted return is by definition optimal, and the first action in this sequence is an optimal action. Note that this brute-force procedure could only be employed because the problem has terminal states, and all the optimal trajectories terminate in a small number of steps. For example, Figure 6.2(b) shows an optimal trajectory from the initial state $x_0 = [0,0]^T$, found by applying an optimal sequence of actions. A terminal state is reached after 8 steps, with a zero final velocity.

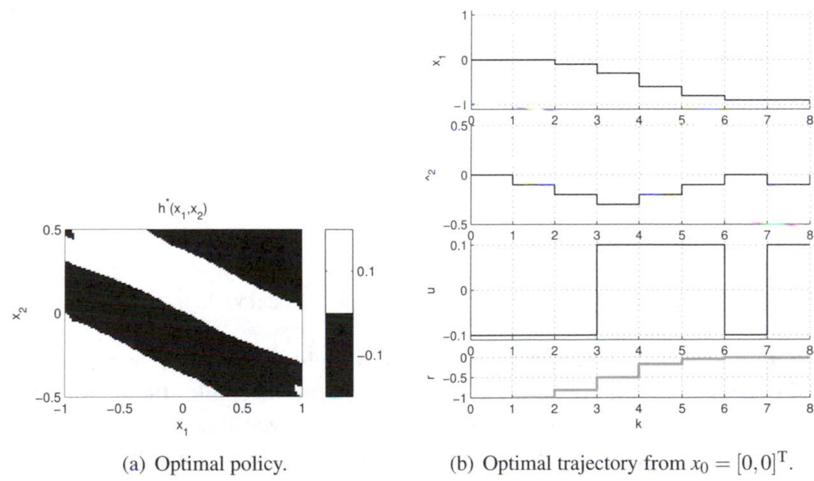

(a) Optimal policy.

(b) Optimal trajectory from $x_0 = [0,0]^T$.

FIGURE 6.2
An optimal solution for the double integrator, found by brute-force search.

Results of CE policy search

To apply CE policy search, we select representative states that are distributed across the entire state space and equally weighted; the algorithm is thus required to achieve a good performance over the entire state space. The set of representative states is:

$$X_0 = \{-1, -0.9, \ldots, 1\} \times \{-0.5, -0.3, -0.1, 0, 0.1, 0.3, 0.5\}$$

and the weight function is $w(x_0) = 1/|X_0|$ for any x_0. This set contains $21 \times 7 = 147$ states, fewer than the grid of Figure 6.2(a). CE policy search is run while gradually increasing the number $\mathscr{N}$ of BFs from 4 to 18. The parameters for the algorithm are set (with little or no tuning) as follows: $c_{\text{CE}} = 10$, $\rho_{\text{CE}} = 0.01$, $\alpha_{\text{CE}} = 0.7$, $\varepsilon_{\text{CE}} = \varepsilon_{\text{MC}} = 0.001$, $d_{\text{CE}} = 5$, and $\tau_{\max} = 100$. Because the system is deterministic, it is sufficient to simulate only one trajectory from every initial state, i.e., $N_{\text{MC}} = 1$. For every value of $\mathscr{N}$, 20 independent runs were performed, in which the algorithm always converged before reaching the maximum number of iterations.

Figure 6.3(a) presents the performance of the policies obtained by CE policy search (mean values across the 20 runs, together with 95% confidence intervals on this mean). For comparison, this figure also shows the exact optimal score for X_0, computed by looking for optimal action sequences with the brute-force procedure explained earlier. CE policy search reliably obtains near-optimal performance for $\mathscr{N} \geq 10$, and sometimes finds good solutions for $\mathscr{N}$ as low as 7. Figure 6.3(b) presents the mean execution time of the algorithm, which is roughly affine in $\mathscr{N}$, as expected from (6.9).[5] The 95% confidence intervals are too small to be visible at the scale of the figure, so they are omitted.

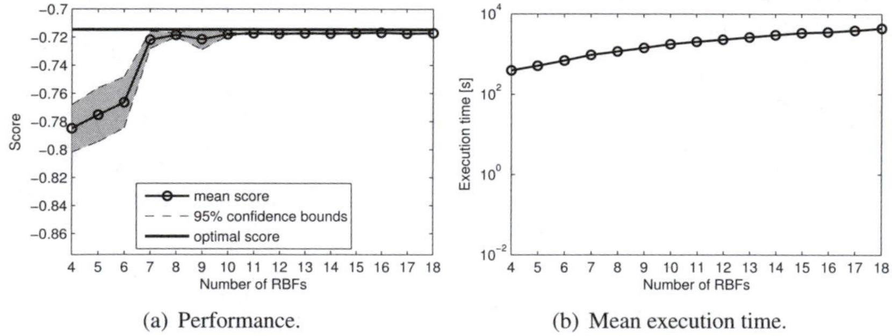

(a) Performance. (b) Mean execution time.

FIGURE 6.3 Results of CE policy search for the double integrator.

Figure 6.4 shows a representative solution found by CE policy search, for $\mathscr{N} = 10$ RBFs (compare with Figure 6.2). The policy found resembles the optimal policy, but the edges where the actions change are more curved due to their dependence on the RBFs. The trajectory starting in $x_0 = [0, 0]^{\text{T}}$ and controlled by this policy is optimal. More specifically, the state and action trajectories are the negatives of those in Figure 6.2(b), and remain optimal despite this "reflection" about the horizontal axis, because the dynamics and the reward function of the double integrator are also symmetric with respect to the origin.

Comparison with value iteration and policy iteration

In this section, CE policy search is compared with representative algorithms for approximate value iteration and policy iteration. From the approximate value iteration

[5]All the computation times reported in this chapter were recorded while running the algorithms in MATLAB® 7 on a PC with an Intel Core 2 Duo E6550 2.33 GHz CPU and with 3 GB RAM.

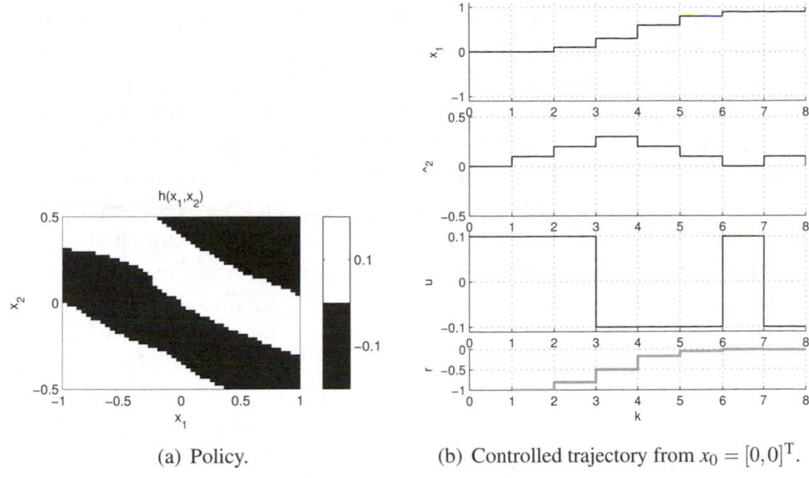

(a) Policy.

(b) Controlled trajectory from $x_0 = [0,0]^{\mathrm{T}}$.

FIGURE 6.4
A representative solution found by CE policy search for the double integrator.

class, fuzzy Q-iteration is selected, which was discussed at length in Chapter 4, and from the policy iteration class, least-squares policy iteration (LSPI) is chosen, which was introduced in Section 3.5.5 and discussed in detail in Chapter 5.

Recall that fuzzy Q-iteration relies on a linearly parameterized Q-function approximator with N state-dependent, normalized membership functions (MFs) $\phi_1, \dots, \phi_N : X \to \mathbb{R}$, and with a discrete set of actions $U_{\mathrm{d}} = \{u_1, \dots, u_M\}$. Approximate Q-values are computed with:

$$\widehat{Q}(x, u_j) = \sum_{i=1}^{N} \phi_i(x) \theta_{[i,j]} \tag{6.12}$$

where $\theta \in \mathbb{R}^{NM}$ is a vector of parameters, and $[i, j] = i + (j-1)N$ denotes the scalar index corresponding to i and j. Fuzzy Q-iteration computes an approximately optimal Q-function of the form (6.12), and then outputs a greedy policy in this Q-function. The Q-function and policy obtained have a bounded suboptimality, as described in Theorem 4.5 of Chapter 4. For the double integrator, the action space is already discrete ($U_{\mathrm{d}} = U = \{-0.1, 0.1\}$ and $M = 2$), so action discretization is not necessary. Triangular MFs are defined (see Example 4.1), distributed on an equidistant grid with N' points along each dimension of the state space; this leads to a total of $N = N'^2$ state-dependent MFs, corresponding to a total of $2N'^2$ parameters. Such a regular placement of MFs provides a uniform resolution over the state space, which is the best option given that prior knowledge about the optimal Q-function is not available. In these experiments, a fuzzy Q-iteration run is considered convergent when the (infinity-norm) difference between consecutive parameter vectors decreases below $\varepsilon_{\mathrm{QI}} = 10^{-5}$.

A similar Q-function approximator is chosen for LSPI, combining state-dependent normalized Gaussian RBFs with the 2 discrete actions. The RBFs are axis-aligned, identically shaped, and their centers are placed on an equidistant grid with N' points along each dimension of X. The widths $b_{i,d}$ of the RBFs along each dimension d were taken identical to the grid spacing along that dimension (using the RBF formula (6.8) from this chapter). This leads to a total of N'^2 RBFs and $2N'^2$ parameters. At every iteration, LSPI approximates the Q-function of the current policy using a batch of transition samples, and then computes an improved, greedy policy in this Q-function. Then, the Q-function of the improved policy is estimated, and so on. The sequence of policies produced in this way eventually converges to a subsequence along which all of the policies have a bounded suboptimality; however, it may not converge to a fixed policy. LSPI is considered convergent when the (two-norm) difference of consecutive parameter vectors decreases below $\varepsilon_{\mathrm{LSPI}} = 10^{-3}$, or when a limit cycle is detected in the sequence of parameters.

The number N' of MFs (in fuzzy Q-iteration) or of BFs (in LSPI) for each state variable is gradually increased from 4 to 18. Fuzzy Q-iteration is a deterministic algorithm, so running it only once for every N' is sufficient. LSPI requires a set of random samples, so each LSPI experiment is run 20 times with independent sets of samples. For $N' = 4$, 1000 samples are used, and for larger N' the number of samples is increased proportionally with the number $2N'^2$ of parameters, which means that $1000N'^2/4^2$ samples are used for each N'. Some side experiments confirmed that this number of samples is sufficient, and that increasing it does not lead to a better performance. Figure 6.5(b) shows the score of the policies computed by fuzzy Q-iteration and LSPI, measured by the average return across the set X_0 of representative states, as in CE policy search (compare with Figure 6.3(a)). The execution time of the algorithms is given in Figure 6.5(b) (compare with Figure 6.3(b)). Some LSPI runs for $N' \leq 8$ did not converge within 100 iterations, and are therefore not taken into account in Figure 6.5.

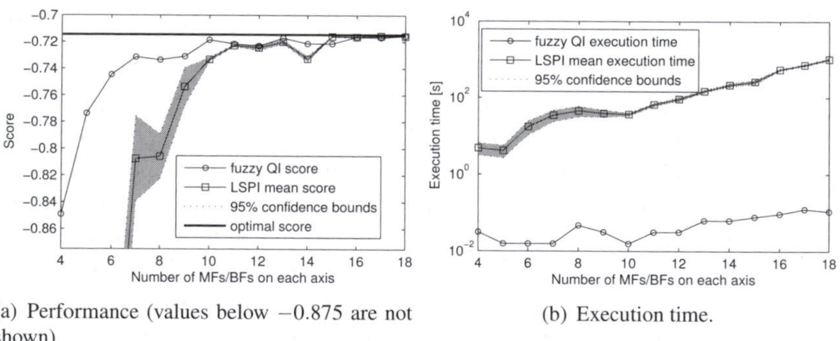

(a) Performance (values below -0.875 are not shown).

(b) Execution time.

FIGURE 6.5 Results of fuzzy Q-iteration and LSPI for the double integrator.

Whereas CE policy search reliably obtained near-optimal performance starting from $\mathcal{N} = 10$ BFs *in total*, fuzzy Q-iteration and LSPI obtain good performance starting from around $N' = 10$ BFs *for each state variable*; the total number of MFs or BFs is N'^2, significantly larger. Furthermore, CE policy search provides a steady performance for larger values of $\mathscr{N}$, whereas fuzzy Q-iteration and LSPI often lead to decreases in performance as the number of MFs or BFs increases. These difference are mainly due to the fact that the MFs of fuzzy Q-iteration and the BFs of LSPI are equidistant and identically shaped, whereas the CE algorithm optimizes parameters encoding the locations and shapes of the BFs. On the other hand, the computational cost of the value function based algorithms is smaller than the cost of CE policy search (for fuzzy Q-iteration, by several orders of magnitude). This indicates that when the performance over the entire state space must be optimized, value or policy iteration may be computationally preferable to CE policy search, at least in low-dimensional problems such as the double integrator. In such problems, CE policy search can be beneficial when the performance must be optimized from only a smaller number of initial states, or when a policy approximator of a fixed complexity is desired and the computational costs to optimize it are not a concern.

Figure 6.6 shows representative solutions found by fuzzy Q-iteration and LSPI for $N' = 10$. While the policies resemble the policy found by the CE algorithm (Fig-

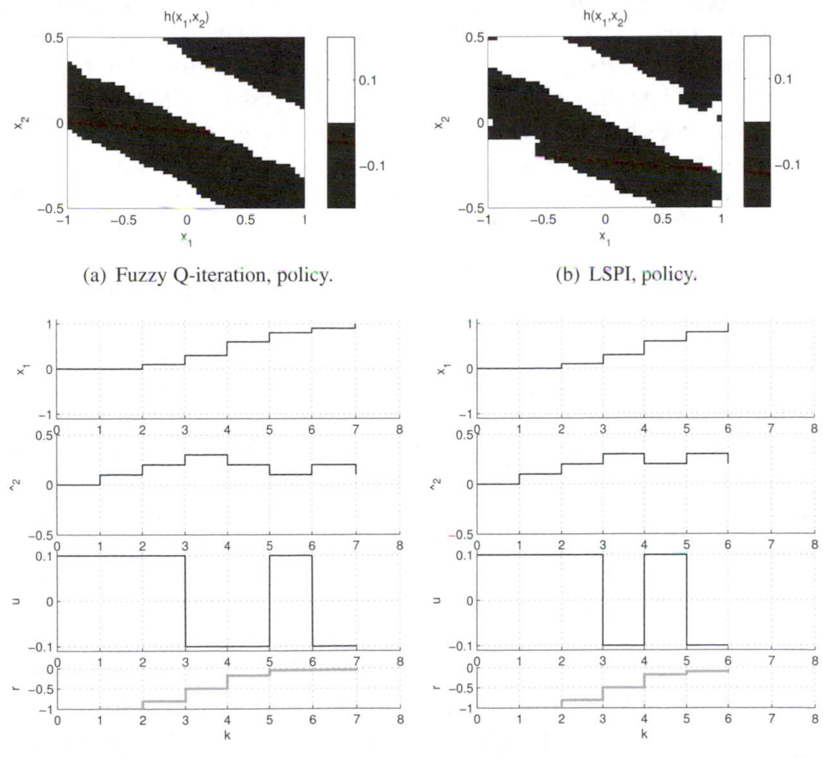

(a) Fuzzy Q-iteration, policy.

(b) LSPI, policy.

(c) Fuzzy Q-iteration, controlled trajectory from $[0,0]^{\mathrm{T}}$.

(d) LSPI, controlled trajectory from $[0,0]^{\mathrm{T}}$.

FIGURE 6.6
Representative solutions found by fuzzy Q-iteration (left) and LSPI (right) for the double integrator.

ure 6.4(a)) and the optimal policy (Figure 6.2(a)), the trajectories from $[0,0]^{\mathrm{T}}$ are suboptimal. For instance, the trajectory obtained by fuzzy Q-iteration (Figure 6.6(c)) reaches the terminal state with a nonzero final velocity after 7 steps. The suboptimality of this trajectory is reflected in the value of its return, which is -2.45. In contrast, the optimal trajectories of Figures 6.2(b) and 6.4(b) reached a zero final velocity, accumulating a better return of -2.43. The LSPI trajectory of Figure 6.6(d) terminates after 6 steps, with an even larger final velocity, leading to a suboptimal return of -2.48. The reason for which the fuzzy Q-iteration solution is better than the LSPI solution is not certain; possibly, the triangular MF approximator used by fuzzy Q-iteration is more appropriate for this problem than the RBF approximator of LSPI.

Comparison of CE and DIRECT optimization

In our policy search approach (Section 6.3), a global, mixed-integer, gradient-free optimization problem must be solved. One algorithm that can address this difficult optimization problem is DIRECT (Jones, 2009), and therefore, in this section, this algorithm is compared with CE optimization in the context of policy search.[6] DIRECT works in hyperbox parameter spaces, by recursively splitting promising hyperboxes in three and sampling the center of each resulting hyperbox. The hyperbox selection procedure leads both to a global exploration of the parameter space, and to a local search in the most promising regions discovered so far. The algorithm is especially suitable for problems in which evaluating the score function is computationally costly (Jones, 2009), as is the case in policy search.

Note that the original parameter space of the RBF policy representation is not a finite hyperbox, because each RBF width can be arbitrarily large, $b_i \in (0,\infty)^D$. However, in practice it is not useful to employ RBFs that are wider than the entire state space, so for the purpose of applying DIRECT, we restrict the width of the RBFs to be at most the width of the state space, i.e., $b_i \leq 2 \cdot x_{\max}$, thereby obtaining a hyperbox parameter space. Another difference with CE policy search is that, for DIRECT, there is no reason to use a binary representation for the action assignments; instead, they are represented directly as integer variables ranging in $1,\ldots,M$. Therefore, the number of parameters to optimize is $\mathscr{N} + 2D\mathscr{N} = 5\mathscr{N}$, consisting of $2D\mathscr{N}$ RBF parameters and $\mathscr{N}$ integer action assignments.

We used DIRECT to optimize the parameters of the policy (6.6) while gradually increasing $\mathscr{N}$ from 4 to 18, as for CE optimization above. DIRECT stops when the score function (6.7) has been evaluated a given number of times; this stopping parameter was set to $2000 \cdot 5\mathscr{N}$ for every $\mathscr{N}$, i.e., 2000 times the number of parameters to optimize. Since DIRECT is a deterministic algorithm, each experiment was run only once. The performance of the policies computed by DIRECT, together with the execution time of the algorithm, are shown in Figure 6.7. For an easy comparison, the CE policy search results from Figure 6.3 are repeated.

DIRECT performs worse than CE optimization for most values of $\mathscr{N}$, while requiring more computations for all values of $\mathscr{N}$. Increasing the allowed number of

[6] We use the DIRECT implementation from the TOMLAB® 7 optimization toolbox for MATLAB.

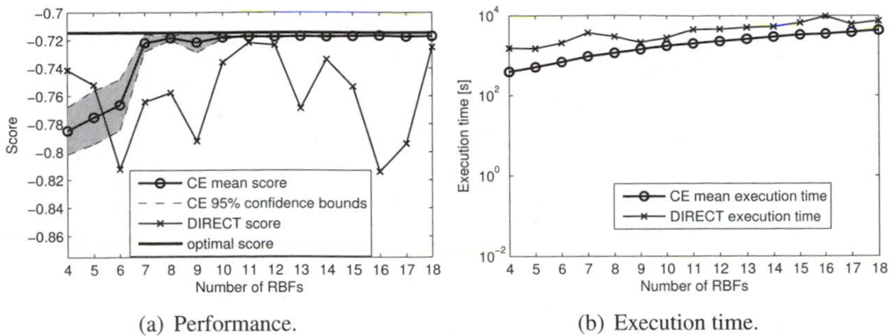

(a) Performance. (b) Execution time.

FIGURE 6.7
Results of DIRECT for the double integrator, compared with the results of CE policy search.

score evaluations may improve the performance of DIRECT, but would also make it more computationally expensive, and therefore less competitive with CE optimization. The poor results of DIRECT may be due to its reliance on splitting the parameter space into hyperboxes: this approach can perform poorly when the parameter space is high-dimensional, as is the case for our policy parametrization.

6.4.2 Bicycle balancing

In this section, CE policy search is applied to a more involved problem than the double integrator, namely the balancing of a bicycle riding at a constant speed on a horizontal surface. The steering column of the bicycle is vertical, which means that the bicycle is not self-stabilizing, but must be actively stabilized to prevent it from falling (a regular bicycle is self-stabilizing under certain conditions, see Åström et al., 2005). This is a variant of a bicycle balancing and riding problem that is widely used as a benchmark for reinforcement learning algorithms (Randløv and Alstrøm, 1998; Lagoudakis and Parr, 2003a; Ernst et al., 2005). We use the bicycle balancing problem to study how CE policy search is affected by changes in the set of representative states and by noise in the transition dynamics.

Bicycle problem

Figure 6.8 provides a schematic representation of the bicycle, which includes the state and control variables. The state variables are the roll angle ω [rad] of the bicycle measured from the vertical axis, the angle α [rad] of the handlebar, equal to 0 when the handlebar is in its neutral position, and the respective angular velocities $\dot{\omega}$, $\dot{\alpha}$ [rad/s]. The control variables are the displacement $\delta \in [-0.02, 0.02]$ m of the bicycle-rider common center of mass perpendicular to the plane of the bicycle, and the torque $\tau \in [-2, 2]$ Nm applied to the handlebar. The state vector is therefore $[\omega, \dot{\omega}, \alpha, \dot{\alpha}]^{\mathrm{T}}$, and the action vector is $u = [\delta, \tau]^{\mathrm{T}}$. The displacement δ can be affected by additive noise z drawn from a uniform density over the interval $[-0.02, 0.02]$ m.

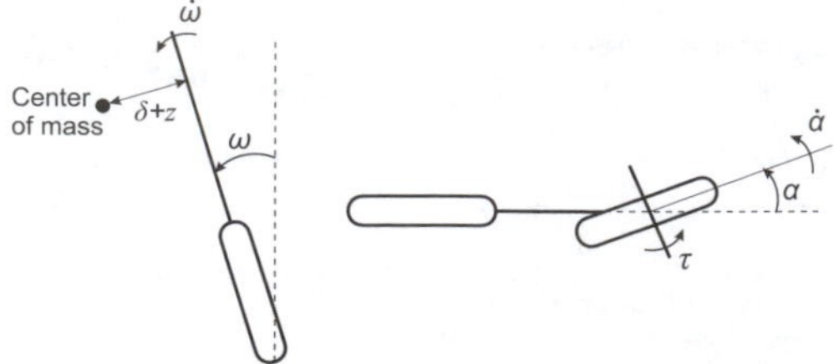

FIGURE 6.8
A schematic representation of the bicycle, as seen from behind (left) and from the top (right).

The continuous-time dynamics of the bicycle are (Ernst et al., 2005):

$$\ddot{\omega} = \frac{1}{J_{\text{bc}}}\Bigg[\sin(\beta)(M_{\text{c}}+M_{\text{r}})gh - \cos(\beta)\Big[\frac{J_{\text{dc}}v}{r}\dot{\alpha} + \operatorname{sign}(\alpha)v^2\Big(\frac{M_{\text{d}}r}{l}(|\sin(\alpha)|+|\tan(\alpha)|) + \frac{(M_{\text{c}}+M_{\text{r}})h}{r_{\text{CM}}}\Big)\Big]\Bigg] \tag{6.13}$$

$$\ddot{\alpha} = \frac{1}{J_{\text{dl}}}\left(\tau - \frac{J_{\text{dv}}v}{r}\dot{\omega}\right) \tag{6.14}$$

where:

$$J_{\text{bc}} = \frac{13}{3}M_{\text{c}}h^2 + M_{\text{r}}(h+d_{\text{CM}})^2 \qquad J_{\text{dc}} = M_{\text{d}}r^2$$

$$J_{\text{dv}} = \frac{3}{2}M_{\text{d}}r^2 \qquad J_{\text{dl}} = \frac{1}{2}M_{\text{d}}r^2$$

$$\beta = \omega + \arctan\frac{\delta+z}{h} \qquad \frac{1}{r_{\text{CM}}} = \begin{cases} \left[(l-c)^2 + \frac{l^2}{\sin^2(\alpha)}\right]^{-1/2} & \text{if } \alpha \neq 0 \\ 0 & \text{otherwise} \end{cases}$$

Note that the noise enters the model in (6.13), via the term β. To obtain the discrete-time transition function, as in (Ernst et al., 2005), the dynamics (6.13)–(6.14) are numerically integrated using the Euler method with a sampling time of $T_{\text{s}} = 0.01$ s (see, e.g., Ascher and Petzold, 1998, Chapter 3). When the magnitude of the roll angle is larger than $\frac{12\pi}{180}$, the bicycle is considered to have fallen, and a terminal, failure state is reached. Additionally, using saturation, the steering angle α is restricted to $[\frac{-80\pi}{180}, \frac{80\pi}{180}]$ to reflect the physical constraints on the handlebar, and the angular velocities $\dot{\omega}$, $\dot{\alpha}$ are restricted to $[-2\pi, 2\pi]$.

Table 6.2 shows the meanings and values of the parameters in the bicycle model.

In the balancing task, the bicycle must be prevented from falling, i.e., the roll angle must be kept within the allowed interval $[\frac{-12\pi}{180}, \frac{12\pi}{180}]$. The following reward function is chosen to express this goal:

$$r_{k+1} = \begin{cases} 0 & \text{if } \omega_{k+1} \in [\frac{-12\pi}{180}, \frac{12\pi}{180}] \\ -1 & \text{otherwise} \end{cases} \tag{6.15}$$

TABLE 6.2 Parameters of the bicycle.

Symbol	Value	Units	Meaning
M_c	15	kg	mass of the bicycle
M_d	1.7	kg	mass of a tire
M_r	60	kg	mass of the rider
g	9.81	m/s^2	gravitational acceleration
v	10/3.6	m/s	velocity of the bicycle
h	0.94	m	height from the ground of the common center of mass (CoM) of the bicycle and the rider
l	1.11	m	distance between the front and back tires at the points where they touch the ground
r	0.34	m	wheel radius
d_{CM}	0.3	m	vertical distance between bicycle CoM and rider CoM
c	0.66	m	horizontal distance between the point where the front wheel touches the ground and the common CoM

Thus, the reward is generally 0, except in the event of reaching a failure state, which is signaled by a (negative) reward of -1. The discount factor is $\gamma = 0.98$.

To apply CE policy search, the rider displacement action is discretized into $\{-0.02, 0, 0.02\}$, and the torque on the handlebar into $\{-2, 0, 2\}$, leading to a discrete action space with 9 elements, which is sufficient to balance the bicycle (as will be seen in the upcoming experiments).

Representative states

We consider the behavior of the bicycle starting from different initial rolls and roll velocities, and the initial steering angle α_0 and velocity $\dot{\alpha}_0$ are always taken to be equal to zero. Two different sets of representative states are employed, in order to study the influence of the representative states on the performance of CE policy search.

The first set of representative states contains a few evenly-spaced values of the roll, while the remaining state variables are zero:

$$X_{0,1} = \left\{\tfrac{-10\pi}{180}, \tfrac{-5\pi}{180}, \ldots, \tfrac{10\pi}{180}\right\} \times \{0\} \times \{0\} \times \{0\}$$

The roll values considered cover the entire acceptable roll domain $[\frac{-12\pi}{180}, \frac{12\pi}{180}]$ except values too close to the boundaries, from which failure is difficult to avoid. The second set is the cross-product of a finer roll grid and a few values of the roll velocity:

$$X_{0,2} = \left\{\tfrac{-10\pi}{180}, \tfrac{-8\pi}{180}, \ldots, \tfrac{10\pi}{180}\right\} \times \left\{\tfrac{-30\pi}{180}, \tfrac{-15\pi}{180}, \ldots, \tfrac{30\pi}{180}\right\} \times \{0\} \times \{0\}$$

For both sets, the representative states are uniformly weighted, i.e., $w(x_0) = 1/|X_0|$ for any $x_0 \in X_0$.

Because a good policy can always prevent the bicycle from falling for any state in $X_{0,1}$, the optimal score (6.7) for this set is 0. This is no longer true for $X_{0,2}$: when ω and $\dot{\omega}$ have the same sign and are too large in magnitude, the bicycle cannot be prevented from falling by any control policy. So, the optimal score for $X_{0,2}$ is strictly negative. To prevent the inclusion of too many such states (from which falling is unavoidable) in $X_{0,2}$, the initial roll velocities are not taken to be too large in magnitude.

Balancing a deterministic bicycle

For the first set of experiments with the bicycle, the noise is eliminated from the simulations by taking $z_k = 0$ at each step k. The CE policy search parameters are the same as for the double-integrator, i.e., $c_{\mathrm{CE}} = 10$, $\rho_{\mathrm{CE}} = 0.01$, $\alpha_{\mathrm{CE}} = 0.7$, $\varepsilon_{\mathrm{CE}} = \varepsilon_{\mathrm{MC}} = 0.001$, $d_{\mathrm{CE}} = 5$, and $\tau_{\max} = 100$. Because the system is deterministic, a single trajectory is simulated from every state in X_0, i.e., $N_{\mathrm{MC}} = 1$. CE policy search is run while gradually increasing the number $\mathscr{N}$ of RBFs from 3 to 8. For each of the two sets of representative states and every value of $\mathscr{N}$, 10 independent runs were performed, during which the algorithm always converged before reaching the maximum number of iterations.

Figure 6.9 presents the performance and execution time of CE policy search (mean values across the 10 runs and 95% confidence intervals). For $X_{0,1}$, in Figure 6.9(a), all the experiments with $\mathscr{N} \geq 4$ reached the optimal score of 0. For $X_{0,2}$, in Figure 6.9(b), the performance is around -0.21 and does not improve as $\mathscr{N}$ grows, which suggests that it is already near optimal. If so, then CE policy search obtains good results with as few as 3 RBFs, which is remarkable. The execution times, shown in Figure 6.9(c), are of course larger for $X_{0,2}$ than for $X_{0,1}$, since $X_{0,2}$ contains more initial states than $X_{0,1}$.[7]

Figure 6.10 illustrates the quality of two representative policies found by CE policy search with $\mathscr{N} = 7$: one for $X_{0,1}$ and the other for $X_{0,2}$. The figure shows how these policies *generalize* to unseen initial states, i.e., how they perform if applied when starting from initial states that do not belong to X_0. These new initial states consist of a grid of values in the $(\omega, \dot{\omega})$ plane; α_0 and $\dot{\alpha}_0$ are always 0. The policy is considered successful from a given initial state if it balances the bicycle for at least 50 s. This duration is chosen to verify whether the bicycle is balanced robustly for a long time; it is roughly 10 times longer than the length of the trajectory used to evaluate the score during the optimization procedure, which was 5.36 s (corresponding to $K = 536$, which is the number of steps necessary to achieve the imposed accuracy of $\varepsilon_{\mathrm{MC}} = 0.001$ in estimating the return). The policy obtained with the smaller set $X_{0,1}$ achieves a reasonable generalization, since it balances the bicycle from a set of initial states larger than $X_{0,1}$. Using the larger set $X_{0,2}$ is more beneficial, because it increases the set of states from which the bicycle is balanced. Recall also that the

[7]The execution times for the bicycle are similar to or larger than those for the double integrator, even though the number of representative states is smaller. This is due to the different nature of the two problems. For the double integrator, the goal requires terminating the task, so better policies lead to earlier termination and to shorter trajectories, which in turn requires less computationally intensive Monte Carlo simulations as the optimization goes on and the policy improves. In contrast, trajectory termination represents a failure for the bicycle, so better policies lead to longer trajectories, which require more computation time to simulate. Overall, this results in a larger computational cost per initial state over the entire optimization process.

bicycle cannot be balanced at all when ω and $\dot{\omega}$ are too large in magnitude and have the same sign, i.e., in the bottom-left and top-right corners of the $(\omega, \dot{\omega})$ plane.

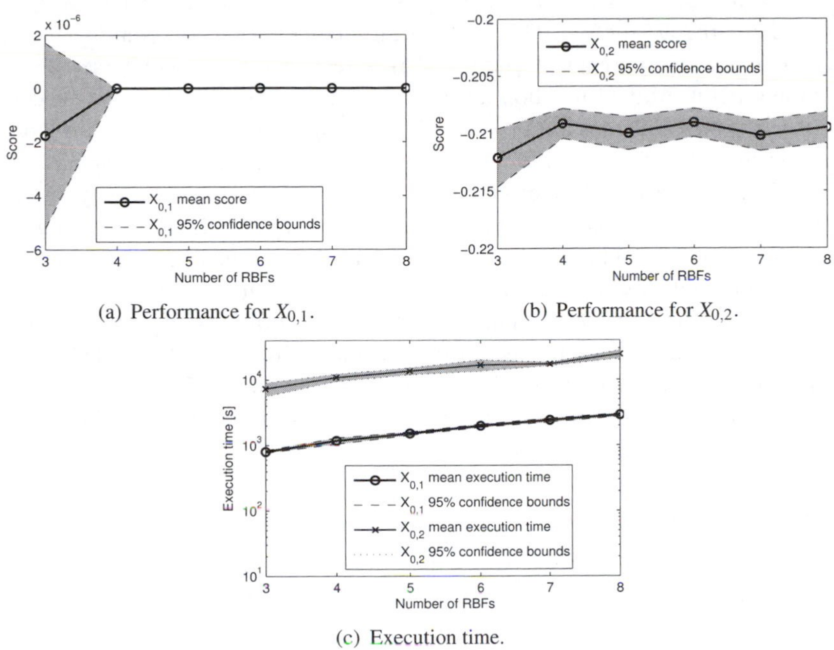

(a) Performance for $X_{0,1}$. (b) Performance for $X_{0,2}$.

(c) Execution time.

FIGURE 6.9 Results of CE policy search for the deterministic bicycle.

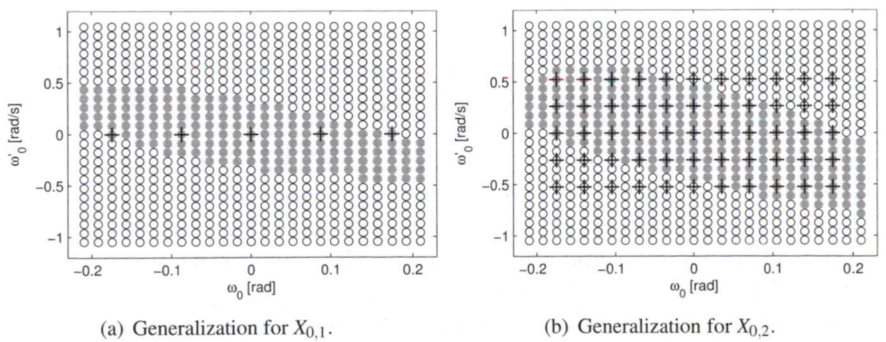

(a) Generalization for $X_{0,1}$. (b) Generalization for $X_{0,2}$.

FIGURE 6.10
Generalization of typical policies found by CE policy search for the deterministic bicycle. White markers indicate the bicycle fell from that initial state, whereas gray markers indicate it was successfully balanced for 50 s. Black crosses mark the representative initial states.

Comparison with fuzzy Q-iteration for the deterministic bicycle

For comparison purposes, fuzzy Q-iteration was run with an equidistant grid of triangular MFs and with the same discrete actions as those employed by CE policy search. The number N' of MFs on each axis was gradually increased, and the first value of N' that gave good performance was 12; the score for this value of N' is 0 for $X_{0,1}$, and -0.2093 for $X_{0,2}$. This leads to a total of $12^4 = 20376$ *equidistant* MFs, vastly more than the number of *optimized* BFs required by CE policy search. The execution time of fuzzy Q-iteration for $N' = 12$ was 1354 s, similar to the execution time of CE policy search with $X_{0,1}$ (see Figure 6.9(c)). In contrast, for the double integrator problem, the execution time of fuzzy Q-iteration was much smaller than that of CE policy search. This discrepancy arises because the complexity of fuzzy Q-iteration grows faster than the complexity of CE policy search, when moving from the two-dimensional double integrator to the four-dimensional bicycle problem. In general, the complexity of fuzzy Q-iteration with triangular MFs grows exponentially with the number of problem dimensions, whereas the complexity (6.9) of CE policy search is linear in the number of dimensions (while crucially depending also on other quantities, such as the number of representative states). Therefore, as the dimension of the problem increases, CE policy search may become preferable to value iteration techniques from a computational point of view.

Balancing a stochastic bicycle

The second set of experiments includes the effects of noise. Recall that the noise z is added to the displacement δ of the bicycle-rider center of mass, and enters the model in the dynamics of $\dot{\omega}$ (6.13), via the term β. The noise z_k is drawn at each step k from a uniform density over $[-0.02, 0.02]$ m. To apply CE policy search, $\mathcal{N} = 7$ RBFs are employed, and $N_{\text{MC}} = 10$ trajectories are simulated from every initial state to compute the score (this value for N_{MC} is not selected to be too large in order to prevent excessive computational costs). The rest of the parameters remain the same as in the deterministic case. For each of the two sets of representative states, 10 independent runs were performed.

The performance of the resulting policies, together with the execution time of the algorithm, are reported in Table 6.3. For easy comparison, the results in the *deterministic* case with $\mathcal{N} = 7$ are also repeated in this table. All the scores for $X_{0,1}$ are optimal, and the scores for $X_{0,2}$ are similar to those obtained in the deterministic case, which illustrates that, in this problem, adding noise does not greatly diminish the potential to accumulate good returns. The execution times are one order of magnitude larger than for the deterministic bicycle, which is expected because $N_{\text{MC}} = 10$, rather than 1 as in the deterministic case.

Figure 6.11 illustrates how the performance of representative policies generalizes to states not belonging to X_0. Whereas in the deterministic case (Figure 6.10) there was little difference between the generalization performance with $X_{0,1}$ and $X_{0,2}$, in the stochastic case this is no longer true. Instead, using the smaller set $X_{0,1}$ of initial states leads to a policy that balances the bicycle for a much smaller portion of the $(\omega, \dot{\omega})$ plane than using $X_{0,2}$. This is because the noise makes the system visit a

TABLE 6.3
Results of CE policy search for the stochastic bicycle, compared with the deterministic case (mean; 95% confidence interval).

Experiment	Score	Execution time [s]
Stochastic, $X_{0,1}$	0; $[0,0]$	22999; $[21716, 24282]$
Deterministic, $X_{0,1}$	0; $[0,0]$	2400; $[2248, 2552]$
Stochastic, $X_{0,2}$	-0.2093; $[-0.2098, -0.2089]$	185205; $[170663, 199748]$
Deterministic, $X_{0,2}$	-0.2102; $[-0.2115, -0.2089]$	17154; $[16119, 18190]$

larger portion of the state space than in the deterministic case, and some of these new states may not have been seen along trajectories starting only in $X_{0,1}$.

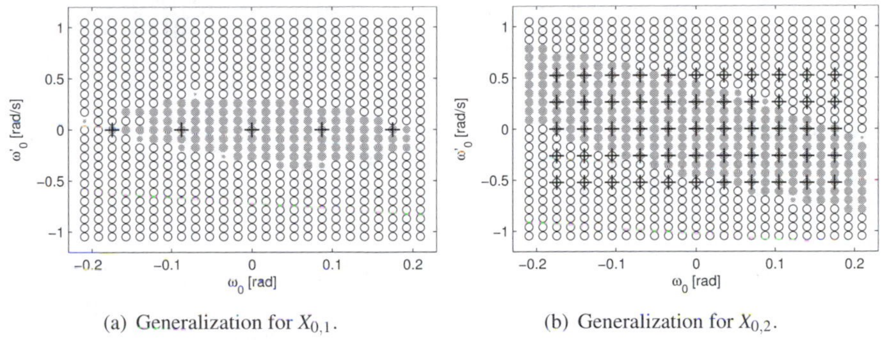

(a) Generalization for $X_{0,1}$. (b) Generalization for $X_{0,2}$.

FIGURE 6.11
Generalization of typical policies found by CE policy search for the stochastic bicycle. White markers indicate the bicycle was never balanced starting from that initial state; the size of the gray markers is proportional to the number of times the bicycle was properly balanced out of 10 experiments. Reproduced with permission from (Buşoniu et al., 2009), © 2009 IEEE.

Figure 6.12 shows how the stochastic bicycle is balanced by a policy found with $X_{0,2}$. The bicycle is successfully prevented from falling, but it is not brought into a vertical position ($\omega = 0$), because the reward function (6.15) makes no difference between zero and nonzero roll angles; it simply indicates that the bicycle should not fall. The control actions exhibit chattering, which is generally necessary when only discrete actions are available to stabilize an unstable system like the bicycle.

6.4.3 Structured treatment interruptions for HIV infection control

In this section, CE policy search is used in a highly challenging simulation problem, involving the optimal control of treatment for an HIV infection. Prevalent HIV treatment strategies involve two types of drugs, called reverse transcriptase inhibitors and protease inhibitors; we will refer to them simply as D1 and D2 in the sequel. The negative side effects of these drugs in the long term motivate the investigation of op-

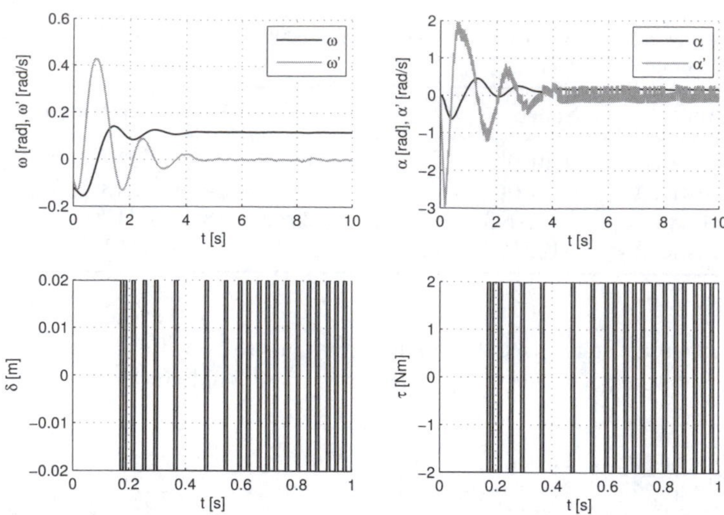

FIGURE 6.12
A controlled trajectory of the bicycle, starting from $\omega_0 = \frac{-7\pi}{180}$, $\dot{\omega}_0 = \frac{-5\pi}{180}$, $\alpha_0 = \dot{\alpha}_0 = 0$. The time axis of the action trajectories is truncated at 1 s to maintain readability.

timal strategies for their use. These strategies might also boost the patient's immune control of the disease (Wodarz and Nowak, 1999). One such strategy involves structured treatment interruptions (STI), where the patient is cycled on and off D1 and D2 therapy (see, e.g., Adams et al., 2004).

HIV infection dynamics and the STI problem

The following six-dimensional, nonlinear model of the HIV infection continuous-time dynamics is considered (Adams et al., 2004):

$$
\begin{aligned}
\dot{T}_1 &= \lambda_1 - d_1 T_1 - (1-\varepsilon_1) k_1 V T_1 \\
\dot{T}_2 &= \lambda_2 - d_2 T_2 - (1-f\varepsilon_1) k_2 V T_2 \\
\dot{T}_1^{\mathrm{t}} &= (1-\varepsilon_1) k_1 V T_1 - \delta T_1^{\mathrm{t}} - m_1 E T_1^{\mathrm{t}} \\
\dot{T}_2^{\mathrm{t}} &= (1-f\varepsilon_1) k_2 V T_2 - \delta T_2^{\mathrm{t}} - m_2 E T_2^{\mathrm{t}} \\
\dot{V} &= (1-\varepsilon_2) N_T \delta (T_1^{\mathrm{t}} + T_2^{\mathrm{t}}) - cV - [(1-\varepsilon_1)\rho_1 k_1 T_1 + (1-f\varepsilon_1)\rho_2 k_2 T_2] V \\
\dot{E} &= \lambda_E + \frac{b_E (T_1^{\mathrm{t}} + T_2^{\mathrm{t}})}{(T_1^{\mathrm{t}} + T_2^{\mathrm{t}}) + K_{\mathrm{b}}} E + \frac{d_E (T_1^{\mathrm{t}} + T_2^{\mathrm{t}})}{(T_1^{\mathrm{t}} + T_2^{\mathrm{t}}) + K_{\mathrm{d}}} E - \delta_E E
\end{aligned}
$$

This model describes two populations of target cells, called type 1 and type 2. The state vector is $x = [T_1, T_2, T_1^{\mathrm{t}}, T_2^{\mathrm{t}}, V, E]^{\mathrm{T}}$, where:

- $T_1 \geq 0$ and $T_2 \geq 0$ are the counts of healthy type 1 and type 2 target cells [cells/ml].

- $T_1^t \geq 0$ and $T_2^t \geq 0$ are the counts of infected type 1 and type 2 target cells [cells/ml].
- $V \geq 0$ is the number of free virus copies [copies/ml].
- $E \geq 0$ is the number of immune response cells [cells/ml].

The positivity of the state variables is ensured during simulations by using saturation. The variables $\varepsilon_1 \in [0, 0.7]$ and $\varepsilon_2 \in [0, 0.3]$ denote the effectiveness of the two drugs D1 and D2.

The values and meanings of the parameters in this model are given in Table 6.4. For a more detailed description of the model and the rationale behind the parameter values, we refer the reader to Adams et al. (2004).

TABLE 6.4 Parameters of the HIV infection model.

Symbol	Value	Units	Meaning
$\lambda_1; \lambda_2$	$10,000; 31.98$	$\frac{\text{cells}}{\text{ml}\cdot\text{day}}$	production rates of target cell types 1 and 2
$d_1; d_2$	$0.01; 0.01$	$\frac{1}{\text{day}}$	death rates of target cell types 1 and 2
$k_1; k_2$	$8 \cdot 10^{-7}; 10^{-4}$	$\frac{\text{ml}}{\text{copies}\cdot\text{day}}$	infection rates of populations 1 and 2
δ	0.7	$\frac{1}{\text{day}}$	infected cell death date
f	0.34	–	treatment effectiveness reduction in population 2
m_1, m_2	$10^{-5}; 10^{-5}$	$\frac{\text{ml}}{\text{cells}\cdot\text{day}}$	immune-induced clearance rates of populations 1 and 2
N_T	100	$\frac{\text{virions}}{\text{cell}}$	virions produced per infected cell
c	13	$\frac{1}{\text{day}}$	virus natural death rate
$\rho_1; \rho_2$	1; 1	$\frac{\text{virions}}{\text{cell}}$	mean number of virions infecting cell types 1 and 2
λ_E	1	$\frac{\text{cells}}{\text{ml}\cdot\text{day}}$	immune effector production rate
b_E	0.3	$\frac{1}{\text{day}}$	maximum birth rate for immune effectors
K_b	100	$\frac{\text{cells}}{\text{ml}}$	saturation constant for immune effector birth
d_E	0.25	$\frac{1}{\text{day}}$	maximum death rate for immune effectors
δ_E	0.1	$\frac{1}{\text{day}}$	natural death rate for immune effectors
K_d	500	$\frac{\text{cells}}{\text{ml}}$	saturation constant for immune effector death

We do not model the relationship between the quantity of administered drugs and their effectiveness, but we assume instead that ε_1 and ε_2 can be directly controlled, which leads to the two-dimensional control vector $u = [\varepsilon_1, \varepsilon_2]^\text{T}$. In STI, drugs are administered either fully (they are "on") or not at all (they are "off"). A fully administered D1 drug corresponds to $\varepsilon_1 = 0.7$, while a fully administered D2 drug corresponds to $\varepsilon_2 = 0.3$. A set of 4 possible discrete actions emerges: $U_\text{d} = \{0, 0.7\} \times \{0, 0.3\}$. Because it is not clinically feasible to change the treatment daily, the state is measured and the drugs are switched on or off once every 5 days (Adams et al., 2004). The system is therefore controlled in discrete time with a sampling time of 5 days. The discrete-time transitions are obtained by numerically integrating the continuous-time dynamics between consecutive time steps.

The HIV dynamics have three uncontrolled equilibria. The *uninfected* equilibrium $x_n = [1000000, 3198, 0, 0, 0, 10]^T$ is unstable: as soon as V becomes nonzero due to the introduction of virus copies, the patient becomes infected and the state drifts away from x_n. The *unhealthy* equilibrium $x_u = [163573, 5, 11945, 46, 63919, 24]^T$ is stable and represents a patient with a very low immune response, for whom the infection has reached dangerous levels. The *healthy* equilibrium $x_h = [967839, 621, 76, 6, 415, 353108]^T$ is stable and represents a patient whose immune system controls the infection without the need for drugs.

We consider the problem of using STI from the unhealthy initial state x_u such that the immune response of the patient is maximized and that the number of virus copies is minimized, while also penalizing the quantity of drugs administered, to account for their side effects. This is represented using the reward function (Adams et al., 2004):

$$\rho(x,u) = -QV - R_1\varepsilon_1^2 - R_2\varepsilon_2^2 + SE \tag{6.16}$$

where $Q = 0.1, R_1 = R_2 = 20000, S = 1000$. The term SE rewards the amount of immune response, $-QV$ penalizes the amount of virus copies, while $-R_1\varepsilon_1^2$ and $-R_2\varepsilon_2^2$ penalize drug use.

Results of CE policy search

In order to apply CE policy search, a discount factor of $\gamma = 0.99$ is used. To compute the score, the number of simulation steps is set to $K = T_f/T_s = 800/5 = 160$, where $T_f = 800$ days is a sufficiently long time horizon for a good policy to control the infection (see Adams et al., 2004; Ernst et al., 2006b). This is different from the examples above, in which a precision ε_{MC} in estimating the return was first imposed, and then the trajectory length was computed accordingly. To limit the effects of the large variation of the state variables, which span several orders of magnitude, a transformed state vector is used, computed as the base 10 logarithm of the original state vector. The policy is represented using $\mathcal{N} = 8$ RBFs, and because we are only interested in applying STI from the unhealthy initial state x_u, only this state is used to compute the score: $X_0 = \{x_u\}$. The other parameters remain unchanged from the earlier examples (Sections 6.4.1 and 6.4.2): $c_{CE} = 10$, $\rho_{CE} = 0.01$, $\alpha_{CE} = 0.7$, $\varepsilon_{CE} = 0.001$, $d_{CE} = 5$, and $\tau_{max} = 100$.

Figure 6.13 shows a trajectory of the HIV system controlled from x_u with a typical policy obtained by CE policy search. The execution time to obtain this policy was 137864 s. For comparison, trajectories obtained with no treatment and with fully administered treatment are also shown. The CE solution switches the D2 drug off after approximately 300 days, but the D1 drug is left on in steady state, which means that the healthy equilibrium x_h is not reached. Nevertheless, the infection is handled better than without STI, and the immune response E in steady state is strong.

Note that because of the high dimensionality of the HIV problem, using a value-function technique with equidistant BF approximation is out of the question.

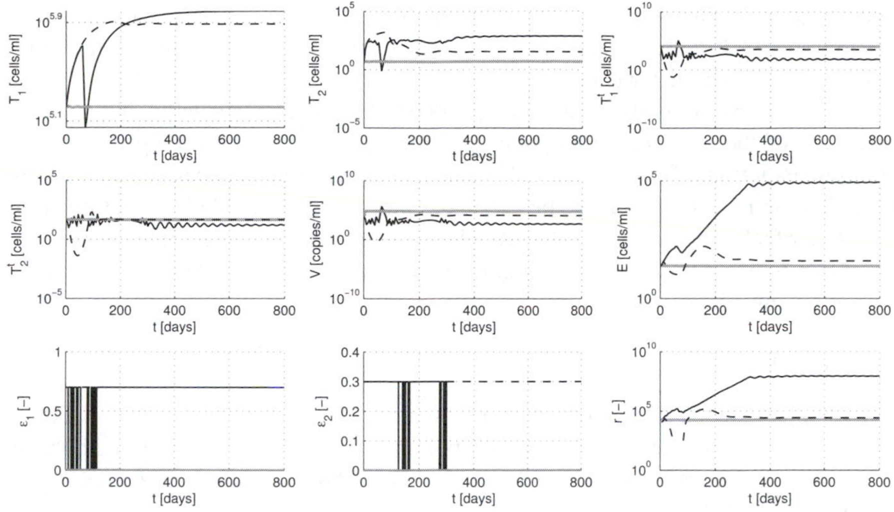

FIGURE 6.13
Trajectories from x_u for HIV infection control. Black, continuous: policy computed with CE policy search. Gray: no treatment. Black, dashed: fully administered treatment. The states and rewards are shown on a logarithmic scale, and negative values of the reward are ignored.

6.5 Summary and discussion

In this final chapter of the book, we have considered a cross-entropy policy search algorithm for continuous-state, discrete-action problems. This algorithm uses a flexible policy parametrization, inspired by the work on BF optimization for value function approximation. The optimization is carried out with the CE method and evaluates the policies by their empirical return from a representative set of initial states. A detailed numerical study of CE policy search has been performed, the more important results of which are the following. The algorithm gives a good performance using only a small number of BFs to represent the policy. When it must optimize the performance over the entire state space, CE policy search is more computationally demanding than value and policy iteration, at least in problems with only a few (e.g., two) dimensions. Nevertheless, given a concise selection of representative states, CE policy search can become computationally preferable to value and policy iteration as the dimensionality increases (e.g., six dimensions or more).

Although CE policy search has been convergent in the experiments of this chapter, it is an open problem whether it provably converges in general. Since CE policy search involves the optimization of both continuous and discrete variables, the CE convergence results of Rubinstein and Kroese (2004); Costa et al. (2007), which only concern the discrete case, cannot be applied directly. The convergence results for the related model-reference adaptive search (Chang et al., 2007) cover more general cases, including stochastic optimization criteria evaluated by Monte Carlo integration (as in CE policy search), but they require the restrictive assumption that the optimal policy parameter is unique.

In this chapter only discrete-action problems have been considered, but a natural extension of the discrete-action policy parametrization can be developed to handle continuous actions, by using the BF values to interpolate between the actions assigned to the BFs. It would also be useful to compare CE and DIRECT optimization with other techniques able to solve the global, mixed-integer, gradient-free problem arising in policy search; such techniques include among others genetic algorithms, simulated annealing, and tabu search.

Bibliographical notes

This chapter extends the authors' earlier work on policy search (Buşoniu et al., 2009). Compared to this earlier work, the policy parametrization has been simplified, the algorithm has been enhanced with a smoothing procedure, and the experimental study has been extended.

Our policy parametrization is inspired by techniques to optimize BFs for value function approximation (e.g., Singh et al., 1995; Menache et al., 2005; Whiteson and Stone, 2006; Bertsekas and Yu, 2009).

Using the CE method to optimize the policy was first proposed by Mannor et al. (2003). Chang et al. (2007, Chapter 4) optimized the policy with the model-reference adaptive search, which is closely related to CE optimization. Both Mannor et al. (2003) and Chang et al. (2007, Chapter 4) focused on solving finite, small MDPs, although they also proposed solving large MDPs using parameterized policies.

Appendix A

Extremely randomized trees

This appendix briefly introduces a nonparametric approximator employing an ensemble of extremely randomized regression trees (extra-trees for short). This approximator was proposed by Geurts et al. (2006) and was combined with the fitted-Q iteration algorithm by Ernst et al. (2005). Our presentation largely follows from material in these two research papers. In this book, fitted Q-iteration with extra-trees approximation was first employed in Section 3.4.5, as an example of approximate value iteration, and then in Section 4.5.2, as a baseline algorithm with which fuzzy Q-iteration was compared. Other successful applications of extra-trees approximation in reinforcement learning can be found in (Ernst et al., 2005, 2006a,b; Jodogne et al., 2006).

A.1 Structure of the approximator

The extra-trees approximator consists of an ensemble of regression trees. Each tree in the ensemble is built from a set of training samples provided in advance, using a procedure that will be detailed in Section A.2. Each tree partitions the input space into a number of disjoint regions, and determines a constant prediction in each region by averaging the output values of the samples that belong to this region.

More formally, assume that a set of N_s training samples is provided:

$$\mathscr{S} = \{(x_{i_\mathrm{s}}, y_{i_\mathrm{s}}) \,|\, i_\mathrm{s} = 1, \ldots, N_\mathrm{s}\}$$

where $x_{i_\mathrm{s}} \in \mathbb{R}^D$ is the i_sth input sample and $y_{i_\mathrm{s}} \in \mathbb{R}$ is the corresponding output sample. A regression problem must be solved, which requires an approximation $\widehat{y}(x)$ of the underlying relationship between the input x and the output y to be inferred from the samples. This relationship may be deterministic or stochastic, and in the latter case, we aim to approximate the expected value of y given x.

Consider the regression tree with index i_tr in the ensemble, and define a function $p_{i_\mathrm{tr}}(x)$ that associates each input x with the region to which it belongs in the partition given by the tree. Then, the prediction (approximate output) $\widehat{y}_{i_\mathrm{tr}}(x)$ of the tree is the average output of the samples from the region $p_{i_\mathrm{tr}}(x)$. We write this as:

$$\sum_{i_\mathrm{s}=1}^{N_\mathrm{s}} \kappa(x, x_{i_\mathrm{s}}) y_{i_\mathrm{s}} \qquad \text{(A.1)}$$

with $\kappa(x, x_{i_s})$ given by:

$$\kappa(x, x_{i_s}) = \frac{I(x_{i_s} \in p_{i_{tr}}(x))}{\sum_{i_s'=1}^{N_s} I(x_{i_s'} \in p_{i_{tr}}(x))} \tag{A.2}$$

Here, the indicator function I is equal to 1 if its argument is true, and 0 otherwise.

The complete, ensemble approximator consists of N_{tr} trees, and averages the predictions of these trees to obtain a final, aggregate prediction:

$$\widehat{y}(x) = \frac{1}{N_{tr}} \sum_{i_{tr}=1}^{N_{tr}} \widehat{y}_{i_{tr}}(x)$$

This final prediction can also be described by an equation of the form (A.1), in which the function $\kappa(x, x_{i_s})$ is now given by:

$$\kappa(x, x_{i_s}) = \frac{1}{N_{tr}} \sum_{i_{tr}=1}^{N_{tr}} \frac{I(x_{i_s} \in p_{i_{tr}}(x))}{\sum_{i_s'=1}^{N_s} I(x_{i_s'} \in p_{i_{tr}}(x))} \tag{A.3}$$

The number of trees N_{tr} is an important parameter of the algorithm. Usually, the more trees, the better the algorithm behaves. However, empirical studies suggest that, often, choosing a number of trees larger than 50 does not significantly improve the accuracy of the approximator (Geurts et al., 2006).

The expression (A.1) was chosen to highlight the relationship between extra-trees and kernel-based approximators. The latter are described by an equation similar to (A.1), see Section 3.3.2. A single tree can be interpreted as a kernel-based approximator with kernel (A.2), whereas for the ensemble of extra-trees the kernel is given by (A.3).

A.2 Building and using a tree

Algorithm A.1 presents the recursive procedure to build one of the trees in the ensemble. Initially, there exists a single root node, which contains the entire set of samples. At every step of the algorithm, each leaf node that contains at least n_{tr}^{min} samples is split, where $n_{tr}^{min} \geq 2$ is an integer parameter. Using a method that will be described shortly, a cut-direction (input dimension) d is selected, together with a scalar cut-point $\bar{x}_d$ (an input value along the selected dimension). The cut-direction and the cut-point constitute a so-called test. Then, the set of samples $\mathscr{S}$ associated with the current node is split into two disjoint sets $\mathscr{S}_{left}$ and $\mathscr{S}_{right}$, respectively containing the samples to the "left" and "right" of the cut-point $\bar{x}_d$:

$$\begin{aligned} \mathscr{S}_{left} &= \left\{ (x_{i_s}, y_{i_s}) \in \mathscr{S} \,\middle|\, x_{i_s,d} < \bar{x}_d \right\} \\ \mathscr{S}_{right} &= \left\{ (x_{i_s}, y_{i_s}) \in \mathscr{S} \,\middle|\, x_{i_s,d} \geq \bar{x}_d \right\} \end{aligned} \tag{A.4}$$

ALGORITHM A.1 Construction of an extremely randomized tree.

Input: set of samples $\mathscr{S}$, parameters N_{tr}, K_{tr}, $n_{\mathrm{tr}}^{\min}$
Output: $\mathscr{T} = \text{BUILDTREE}(\mathscr{S})$

procedure BUILDTREE($\mathscr{S}$)
    if $|\mathscr{S}| < n_{\mathrm{tr}}^{\min}$ **then**
        return a leaf node $\mathscr{T}$ labeled by the value $\frac{1}{|\mathscr{S}|}\sum_{(x,y)\in\mathscr{S}} y$
    else
        $(d, \bar{x}_d) \leftarrow$SELECTTEST($\mathscr{S}$)
        split $\mathscr{S}$ into $\mathscr{S}_{\mathrm{left}}$ and $\mathscr{S}_{\mathrm{right}}$ according to $(d, \bar{x}_d)$; see (A.4)
        $\mathscr{T}_{\mathrm{left}} \leftarrow$BUILDTREE($\mathscr{S}_{\mathrm{left}}$), $\mathscr{T}_{\mathrm{right}} \leftarrow$BUILDTREE($\mathscr{S}_{\mathrm{right}}$)
        $\mathscr{T} \leftarrow$ a node with test $(d, \bar{x}_d)$, left subtree $\mathscr{T}_{\mathrm{left}}$, and right subtree $\mathscr{T}_{\mathrm{right}}$
        return $\mathscr{T}$
    end if
end procedure

procedure SELECTTEST($\mathscr{S}$)
    select K_{tr} cut-directions $\{d_1, \ldots, d_{K_{\mathrm{tr}}}\}$ uniformly random in $\{1, \ldots, D\}$
    for $k = 1, \ldots, K_{\mathrm{tr}}$ **do**
        $x_{d_k,\min} \leftarrow \min_{(x,y)\in\mathscr{S}} x_{d_k}$, $x_{d_k,\max} \leftarrow \max_{(x,y)\in\mathscr{S}} x_{d_k}$
        select a cut-point $\bar{x}_{d_k}$ uniformly random in $(x_{d_k,\min}, x_{d_k,\max}]$
    end for
    return a test $(d_{k'}, \bar{x}_{d_{k'}})$ such that $k' \in \arg\max_k s(d_k, \bar{x}_{d_k}, \mathscr{S})$; see (A.5)
end procedure

A left child node and a right child node are created for the current node, respectively containing these two sets. The selected test is also stored in the node. The procedure continues recursively, until each leaf node contains fewer than $n_{\mathrm{tr}}^{\min}$ samples. Each such leaf node is labeled with the average output of the samples associated with it.

To determine the test at a node, the algorithm generates at random $K_{\mathrm{tr}} \geq 1$ cut-directions and, for each cut-direction, a random cut-point. A score is computed for each of these K_{tr} tests, and a test that maximizes the score is chosen. The score used is the relative variance reduction, which for a test $(d, \bar{x}_d)$ is defined as follows:

$$s(d, \bar{x}_d, \mathscr{S}) = \frac{\mathrm{var}(\mathscr{S}) - \frac{|\mathscr{S}_{\mathrm{left}}|}{N_{\mathrm{s}}}\mathrm{var}(\mathscr{S}_{\mathrm{left}}) - \frac{|\mathscr{S}_{\mathrm{right}}|}{N_{\mathrm{s}}}\mathrm{var}(\mathscr{S}_{\mathrm{right}})}{\mathrm{var}(\mathscr{S})} \tag{A.5}$$

where $\mathscr{S}$ is the set of samples contained by the node considered, $\mathrm{var}(\cdot)$ is the variance of the output y across the argument set, and $|\cdot|$ denotes set cardinality. Note that, if $K_{\mathrm{tr}} = 1$, the cut-direction and the cut-point are chosen completely at random.

Geurts et al. (2006) suggest choosing K_{tr} to be equal to dimensionality D of the input space. As a default value for $n_{\mathrm{tr}}^{\min}$, it is suggested to choose $n_{\mathrm{tr}}^{\min} = 2$, yielding fully developed trees, when the underlying input-output relationship is deterministic; and $n_{\mathrm{tr}}^{\min} = 5$ for stochastic problems. It should be stressed that optimizing these parameters, together with the number of trees N_{tr}, for the problem at hand may improve the approximation accuracy. This optimization could for example be carried out by using a cross-validation technique (see, e.g., Duda et al., 2000).

Algorithm A.2 presents a practical procedure to obtain a prediction (approximate output) from a built tree. To compute this prediction, the algorithm starts from the root node, and applies the test associated with this node. Depending on the result of the test, the algorithm continues along the left subtree or along the right subtree, and so on, until reaching a leaf node. Then, the algorithm returns the label of this leaf node, which is equal to the average output of the associated samples.

ALGORITHM A.2 Prediction using an extremely randomized tree.

Input: tree $\mathscr{T}$, input point x

while $\mathscr{T}$ is not a leaf node **do**
   $(d, \bar{x}_d) \leftarrow$ test associated with root of $\mathscr{T}$
   if $x_d \leq \bar{x}_d$ **then** $\mathscr{T} \leftarrow \mathscr{T}_{\text{left}}$, the left subtree of $\mathscr{T}$
   else $\mathscr{T} \leftarrow \mathscr{T}_{\text{right}}$, the right subtree of $\mathscr{T}$
   end if
end while

Output: the label of $\mathscr{T}$

To clarify the link with the formulas in Section A.1, the input space partition that corresponds to the tree (or, equivalently, the function p) must be defined. This partition contains a number of regions identical to the number of leaf nodes in the tree, and each region consists of all the points in $\mathbb{R}^D$ for which Algorithm A.2 reaches the same leaf node. Equivalently, for any x, $p(x)$ gives the set of points for which Algorithm A.2 would obtain the same leaf node that it reaches when applied to x.

Appendix B

The cross-entropy method

This appendix provides an introduction to the cross-entropy (CE) method. First, the CE algorithm for rare-event simulation is given, followed by the CE algorithm for optimization. The presentation is based on Sections 2.3, 2.4, and 4.2 of the textbook by Rubinstein and Kroese (2004).

B.1 Rare-event simulation using the cross-entropy method

We consider the problem of estimating the probability of a rare event using sampling. Because the event is rare, its probability is small, and straightforward Monte Carlo sampling is impractical because it would require too many samples. Instead, an importance sampling density[1] must be chosen that increases the probability of the interesting event. The CE method for rare-event simulation looks for the best importance sampling density from a given, parameterized class of densities, using an iterative approach. At the first iteration, the algorithm draws a set of samples from an initial density. Using these samples, an easier problem than the original one is defined, in which the probability of the rare event is artificially increased in order to make a good importance sampling density easier to find. This density is then used to obtain better samples in the next iteration, which allow the definition of a more difficult problem, therefore giving a sampling density closer to the optimal one, and so on. When the problems considered at every iteration become at least as difficult as the original problem, the current density can be used for importance sampling in the original problem.

We will next formally describe the CE method for rare-event simulation. Let a be a random vector taking values in the space $\mathscr{A}$. Let $\{p(\cdot;v)\}$ be a family of probability densities on $\mathscr{A}$, parameterized by the vector $v \in \mathbb{R}^{N_v}$m and let a nominal parameter $\bar{v} \in \mathbb{R}^{N_v}$ be given. Given a score function $s : \mathscr{A} \to \mathbb{R}$, the goal is to estimate the probability that $s(a) \geq \lambda$, where the level $\lambda \in \mathbb{R}$ is also given and a is drawn from the density $p(\cdot;\bar{v})$ with the nominal parameter $\bar{v}$. This probability can be written as:

$$\nu = \mathrm{P}_{a\sim p(\cdot;\bar{v})}(s(a) \geq \lambda) = \mathrm{E}_{a\sim p(\cdot;\bar{v})}\{I(s(a) \geq \lambda)\} \tag{B.1}$$

[1] For simplicity, we will abuse the terminology by using the term "density" to refer to probability density functions (which describe probabilities of continuous random variables), as well as to probability mass functions (which describe probabilities of discrete random variables).

where $I(s(a) \geq \lambda)$ is the indicator function, equal to 1 whenever $s(a) \geq \lambda$ and 0 otherwise. When the probability (B.1) is very small (10^{-6} or less), the event $\{s(a) \geq \lambda\}$ is called a rare event.

A straightforward way to estimate v is to use Monte Carlo simulations. A set of random samples $a_1, \ldots, a_{N_{\mathrm{CE}}}$ are drawn from $p(\cdot; \bar{v})$, and the estimated value of v is computed as:

$$\widehat{v} = \frac{1}{N_{\mathrm{CE}}} \sum_{i_s=1}^{N_{\mathrm{CE}}} I(s(a_{i_s}) \geq \lambda) \tag{B.2}$$

However, this procedure is computationally inefficient when $\{s(a) \geq \lambda\}$ is a rare event, since a very large number of samples N_{CE} must be used for an accurate estimation of v. A better way to estimate v is to draw the samples from an importance sampling density $q(\cdot)$ on $\mathscr{A}$, instead of $p(\cdot; \bar{v})$. The density $q(\cdot)$ is chosen to increase the probability of the interesting event $\{s(a) \geq \lambda\}$, thereby requiring fewer samples for an accurate estimation of v. The parameter v can then be estimated using the importance sampling estimator:

$$\widehat{v} = \frac{1}{N_{\mathrm{CE}}} \sum_{i_s=1}^{N_{\mathrm{CE}}} I(s(a_{i_s}) \geq \lambda) \frac{p(a_{i_s}; \bar{v})}{q(a_{i_s})} \tag{B.3}$$

From (B.3), it follows that the importance sampling density:

$$q^*(a) = \frac{I(s(a) \geq \lambda) p(a; \bar{v})}{v} \tag{B.4}$$

makes the argument of the summation equal to v, when substituted into (B.3). Therefore, a single sample a for which $I(s(a) \geq \lambda)$ is nonzero suffices for finding v, and the density $q^*(\cdot)$ is optimal in this sense. It is important to note that the entire procedure is driven by the value of the nominal parameter $\bar{v}$. Hence, among others, q, q^*, and v all depend on $\bar{v}$.

The obvious difficulty is that v is unknown. Moreover, q^* can in general have a complicated shape, which makes it difficult to find. It is often more convenient to choose an importance sampling density from the family of densities $\{p(\cdot; v)\}$. The best importance sampling density in this family can be found by minimizing over the parameter v a measure of the distance between $p(\cdot; v)$ and $q^*(\cdot)$. In the CE method, this measure is the cross-entropy, also known as Kullback-Leibler divergence, which is defined as follows:

$$\begin{aligned} \mathscr{D}(q^*(\cdot), p(\cdot; v)) &= \mathrm{E}_{a \sim q^*(\cdot)} \left\{ \ln \frac{q(a)}{p(a; v)} \right\} \\ &= \int q^*(a) \ln q^*(a) \mathrm{d}a - \int q^*(a) \ln p(a; v) \mathrm{d}a \end{aligned} \tag{B.5}$$

The first term in this distance does not depend on v, and by using (B.4) in the second term, we obtain a parameter that minimizes the cross-entropy as:

$$\begin{aligned} v^* = v^\ddagger, \text{ where } v^\ddagger &\in \arg\max_v \int \frac{I(s(a) \geq \lambda) p(a; \bar{v})}{v} \ln p(a; v) \mathrm{d}a, \\ \text{i.e., } v^\ddagger &\in \arg\max_v \mathrm{E}_{a \sim p(\cdot; \bar{v})} \{ I(s(a) \geq \lambda) \ln p(a; v) \} \end{aligned} \tag{B.6}$$

Unfortunately, the expectation $\mathrm{E}_{a \sim p(\cdot;\bar{v})}\{I(s(a) \geq \lambda) \ln p(a;v)\}$ is difficult to compute by direct Monte Carlo sampling, because the indicators $I(s(a) \geq \lambda)$ will still be 0 for a most of the samples. Therefore, this expectation must also be computed with importance sampling. For an importance sampling density given by the parameter z, the maximization problem (B.6) is rewritten as:

$$v^* = v^\ddagger, \text{ where } v^\ddagger \in \arg\max_v \mathrm{E}_{a \sim p(\cdot;z)}\{I(s(a) \geq \lambda) W(a;\bar{v},z) \ln p(a;v)\} \tag{B.7}$$

in which $W(a;\bar{v},z) = p(a;\bar{v})/p(a;z)$. An approximate solution $\widehat{v}^*$ is computed by drawing a set of random samples $a_1, \ldots, a_{N_{\mathrm{CE}}}$ from the importance density $p(\cdot;z)$ and solving:

$$\widehat{v}^* = v^\ddagger, \text{ where } v^\ddagger \in \arg\max_v \frac{1}{N_{\mathrm{CE}}} \sum_{i_s=1}^{N_{\mathrm{CE}}} I(s(a_{i_s}) \geq \lambda) W(a;\bar{v},z) \ln p(a_{i_s};v) \tag{B.8}$$

The problem (B.8) is called the *stochastic counterpart* of (B.7).

Under certain assumptions on $\mathscr{A}$ and $p(\cdot;v)$, the stochastic counterpart can be solved explicitly. One particularly important case when this happens is when $p(\cdot;v)$ belongs to the natural exponential family (Morris, 1982). For instance, when $\{p(\cdot;v)\}$ is the family of Gaussians parameterized by mean η and standard deviation σ (so, $v = [\eta, \sigma]^{\mathrm{T}}$), the solution of the stochastic counterpart is the mean and the standard deviation of the best samples:

$$\widehat{\eta} = \frac{\sum_{i_s=1}^{N_{\mathrm{CE}}} I(s(a_{i_s}) \geq \lambda) a_{i_s}}{\sum_{i_s=1}^{N_{\mathrm{CE}}} I(s(a_{i_s}) \geq \lambda)} \tag{B.9}$$

$$\widehat{\sigma} = \sqrt{\frac{\sum_{i_s=1}^{N_{\mathrm{CE}}} I(s(a_{i_s}) \geq \lambda)(a_{i_s} - \widehat{\eta})^2}{\sum_{i_s=1}^{N_{\mathrm{CE}}} I(s(a_{i_s}) \geq \lambda)}} \tag{B.10}$$

Choosing directly a good importance sampling parameter z is difficult. If z is poorly chosen, most of the indicators $I(s(a_{i_s}) \geq \lambda)$ in (B.8) will be 0, and the estimated parameter $\widehat{v}^*$ will be a poor approximation of the optimal parameter v^*. To alleviate this difficulty, the CE algorithm uses an iterative approach. Each iteration τ can be viewed as the application of the above methodology using a modified level λ_τ and the importance density parameter $z = v_{\tau-1}$, where:

- The level λ_τ is chosen at each iteration such that the probability of event $\{s(a) \geq \lambda_\tau\}$ under density $p(\cdot;v_{\tau-1})$ is approximately $\rho_{\mathrm{CE}} \in (0,1)$, with ρ_{CE} chosen not too small (e.g., $\rho_{\mathrm{CE}} = 0.05$).
- The parameter $v_{\tau-1}$, $\tau \geq 2$, is the solution of the stochastic counterpart at the previous iteration; v_0 is initialized at $\bar{v}$.

The value λ_τ is computed as the $(1 - \rho_{\mathrm{CE}})$ quantile of the score values of the random sample $a_1, \ldots, a_{N_{\mathrm{CE}}}$, which are drawn from $p(\cdot;v_{\tau-1})$. If these score values are ordered increasingly and indexed such that $s_1 \leq \cdots \leq s_{N_{\text{CE}}}$, then the $(1-\rho_{\text{CE}})$ quantile is:

$$\lambda_\tau = s_{\lceil (1-\rho_{\text{CE}}) N_{\text{CE}} \rceil} \tag{B.11}$$

where $\lceil \cdot \rceil$ rounds the argument to the next greater or equal integer number (ceiling).

When the inequality $\lambda_{\tau^*} \geq \lambda$ is satisfied for some $\tau^* \geq 1$, the rare-event probability is estimated using the density $p(\cdot; v_{\tau^*})$ for importance sampling ($N_1 \in \mathbb{N}^*$):

$$\widehat{v} = \frac{1}{N_1} \sum_{i_s=1}^{N_1} I(s(a_{i_s}) \geq \lambda) W(a_{i_s}; \bar{v}, v_{\tau^*}) \tag{B.12}$$

B.2 Cross-entropy optimization

Consider the following optimization problem:

$$\max_{a \in \mathscr{A}} s(a) \tag{B.13}$$

where $s : \mathscr{A} \to \mathbb{R}$ is the score function (optimization criterion) to maximize, and the variable a takes values in the domain $\mathscr{A}$. Denote the maximum by s^*. The CE method for optimization maintains a density with support $\mathscr{A}$. At each iteration, a number of samples are drawn from this density and the score values of these samples are computed. A (smaller) number of samples that have the best scores are kept, and the remaining samples are discarded. The density is then updated using the selected samples, such that during the next iteration the probability of drawing better samples is increased. The algorithm stops when the score of the worst selected sample no longer improves significantly.

Formally, a family of densities $\{p(\cdot; v)\}$ with support $\mathscr{A}$ and parameterized by v must be chosen. An *associated stochastic problem* to (B.13) is the problem of finding the probability:

$$\nu(\lambda) = \mathrm{P}_{a \sim p(\cdot; v')}(s(a) \geq \lambda) = \mathrm{E}_{a \sim p(\cdot; v')}\{I(s(a) \geq \lambda)\} \tag{B.14}$$

where the random vector a has the density $p(\cdot; v')$ for some parameter vector v'. Consider now the problem of estimating $\nu(\lambda)$ for a λ that is close to s^*. Typically, $\{s(a) \geq \lambda\}$ is a rare event. The CE procedure can therefore be exploited to solve (B.14).

Contrary to the CE method for rare-event simulation, in optimization there is no known nominal λ; its place is taken by s^*, which is unknown. The CE method for optimization circumvents this difficulty by redefining the associated stochastic problem at every iteration τ, using the density with parameter $v_{\tau-1}$, so that λ_τ is expected to converge to s^* as τ increases. Consequently, the stochastic counterpart at iteration τ of CE optimization:

$$\widehat{v}_\tau = v_\tau^\ddagger, \text{ where } v_\tau^\ddagger \in \arg\max_v \frac{1}{N_{\text{CE}}} \sum_{i_s=1}^{N_{\text{CE}}} I(s(a_{i_s}) \geq \lambda_\tau) \ln p(a_{i_s}; v) \tag{B.15}$$

is different from the one used in rare-event simulation (B.8), and corresponds to maximizing over v the expectation $\mathrm{E}_{a\sim p(\cdot;v_{\tau-1})}\{I(s(a_{i_s}) \geq \lambda_\tau) \ln p(a_{i_s};v)\}$. This determines the (approximately) optimal parameter associated with $\mathrm{P}_{a\sim p(\cdot;v_{\tau-1})}(s(a) \geq \lambda_\tau)$, rather than with $\mathrm{P}_{a\sim p(\cdot;\bar{v})}(s(a) \geq \lambda_\tau)$ as in rare-event simulation. So, the term W from (B.7) and (B.8) does not play a role here. The parameter $\bar{v}$, which in rare-event simulation was the fixed nominal parameter under which the rare-event probability has to be estimated, no longer plays a role either. Instead, in CE optimization an initial value v_0 of the density parameter is required, which only serves to define the associated stochastic problem at the first iteration, and which can be chosen in a fairly arbitrary way.

Instead of setting the new density parameter equal to the solution $\widehat{v}_\tau$ of (B.15), it can also be updated incrementally:

$$v_\tau = \alpha_{\mathrm{CE}}\widehat{v}_\tau + (1-\alpha_{\mathrm{CE}})v_{\tau-1} \tag{B.16}$$

where $\alpha_{\mathrm{CE}} \in (0,1]$. This so-called "smoothing procedure" is useful to prevent CE optimization from becoming stuck in local optima (Rubinstein and Kroese, 2004).

The most important parameters in the CE method for optimization are the number N_{CE} of samples and the quantile ρ_{CE} of best samples used to update the density. The number of samples should be at least a multiple of the number of parameters N_v, so $N_{\mathrm{CE}} = c_{\mathrm{CE}} N_v$ with $c_{\mathrm{CE}} \in \mathbb{N}$, $c_{\mathrm{CE}} \geq 2$. The parameter ρ_{CE} can be chosen around 0.01 for large numbers of samples, or it can be larger, around $(\ln N_{\mathrm{CE}})/N_{\mathrm{CE}}$, if there are only a few samples ($N_{\mathrm{CE}} < 100$) (Rubinstein and Kroese, 2004). The smoothing parameter α_{CE} is often chosen around 0.7.

The CE method for optimization is summarized in Algorithm B.1. Note that at line 8, the stochastic counterpart (B.15) was simplified by using the fact that the samples are already sorted in the ascending order of their scores. When $\varepsilon_{\mathrm{CE}} = 0$, the

---

ALGORITHM B.1 Cross-entropy optimization.

Input: family $\{p(\cdot;v)\}$, score function s,
    parameters $\rho_{\mathrm{CE}}, N_{\mathrm{CE}}, d_{\mathrm{CE}}, \varepsilon_{\mathrm{CE}}, \alpha_{\mathrm{CE}}, \tau_{\max}$
1: $\tau \leftarrow 1$
2: initialize density parameters v_0
3: **repeat**
4:     generate sample $a_1, \dots, a_{N_{\mathrm{CE}}}$ from $p(\cdot;v_{\tau-1})$
5:     compute scores $s(a_{i_s})$, $i_s = 1, \dots, N_{\mathrm{CE}}$
6:     reorder and reindex s.t. $s_1 \leq \cdots \leq s_{N_{\mathrm{CE}}}$
7:     $\lambda_\tau \leftarrow s_{\lceil (1-\rho_{\mathrm{CE}})N_{\mathrm{CE}} \rceil}$
8:     $\widehat{v}_\tau \leftarrow v_\tau^\ddagger$, where $v_\tau^\ddagger \in \arg\max_v \sum_{i_s=\lceil (1-\rho_{\mathrm{CE}})N_{\mathrm{CE}} \rceil}^{N_{\mathrm{CE}}} \ln p(a_{i_s};v)$
9:     $v_\tau \leftarrow \alpha_{\mathrm{CE}}\widehat{v}_\tau + (1-\alpha_{\mathrm{CE}})v_{\tau-1}$
10:     $\tau \leftarrow \tau + 1$
11: **until** ($\tau > d_{\mathrm{CE}}$ **and** $|\lambda_{\tau-\tau'} - \lambda_{\tau-\tau'-1}| \leq \varepsilon_{\mathrm{CE}}$, for $\tau' = 0, \dots, d_{\mathrm{CE}}-1$) **or** $\tau = \tau_{\max}$
Output: $\widehat{a}^*$, the best sample encountered at any iteration τ; and $\widehat{s}^* = s(\widehat{a}^*)$

---

algorithm terminates when λ remains constant for d_{CE} consecutive iterations. When $\varepsilon_{\mathrm{CE}} > 0$, the algorithm terminates when λ improves for d_{CE} consecutive iterations, but these improvements do not exceed $\varepsilon_{\mathrm{CE}}$. The integer $d_{\mathrm{CE}} > 1$ accounts for the random nature of the algorithm, by ensuring that the latest performance improvements did not decrease below $\varepsilon_{\mathrm{CE}}$ accidentally (due to random effects), but that instead the decrease remains steady for d_{CE} iterations. A maximum number of iterations τ_{max} is also chosen, to ensure that the algorithm terminates in finite time.

CE optimization has been shown to lead to good performance, often outperforming other randomized algorithms (Rubinstein and Kroese, 2004), and has found many applications in recent years, e.g., in biomedicine (Mathenya et al., 2007), power systems (Ernst et al., 2007), vehicle routing (Chepuri and de Mello, 2005), vector quantization (Boubezoul et al., 2008), and clustering (Rubinstein and Kroese, 2004). While the convergence of CE optimization has not yet been proven in general, the algorithm is usually convergent in practice (Rubinstein and Kroese, 2004). For combinatorial (discrete-variable) optimization, the CE method provably converges with probability 1 to a unit mass density, which always generates samples equal to a single point. Furthermore, the probability that this convergence point is in fact an optimal solution can be made arbitrarily close to 1 by using a sufficiently small smoothing parameter α_{CE} (Costa et al., 2007).

Symbols and abbreviations

List of symbols and notations

The most important mathematical symbols and notations used in this book are listed below, organized by topic.

General notations

$\lvert\cdot\rvert$	absolute value (for numeric arguments); cardinality (for sets)
$\lVert\cdot\rVert_p$	p-norm of the argument
$\lfloor\cdot\rfloor$	the largest integer smaller than or equal to the argument (floor)
$\lceil\cdot\rceil$	the smallest integer larger than or equal to the argument (ceiling)
$g(\cdot;\theta)$	generic function g has argument "$\cdot$" and is parameterized by θ
L_g	Lipschitz constant of generic function g

Probability theory

p	probability density
$a \sim p(\cdot)$	random sample a is drawn from the density p
$\mathrm{P}(\cdot)$	probability of the random variable given as argument
$\mathrm{E}\{\cdot\}$	expectation of the random variable given as argument
η; σ	mean of a Gaussian density; standard deviation of a Gaussian density
η^{bin}	parameter (mean, success probability) of a Bernoulli density

Classical dynamic programming and reinforcement learning

x; X	state; state space
u; U	control action; action space
r	reward
f; $\tilde{f}$	deterministic transition function; stochastic transition function
ρ; $\tilde{\rho}$	reward function for deterministic transitions; reward function for stochastic transitions
h; $\tilde{h}$	deterministic control policy; stochastic control policy
R	return
γ	discount factor
k; K	discrete time index; discrete time horizon
T_s	sampling time
Q; V	Q-function; V-function

Q^h; V^h	Q-function of policy h; V-function of policy h
Q^*; V^*	optimal Q-function; optimal V-function
h^*	optimal policy
$\mathscr{Q}$	set of all Q-functions
T	Q-iteration mapping
T^h	policy evaluation mapping for policy h
ℓ; L	primary iteration index; number of primary iterations
τ	secondary iteration index
α	learning rate (step size)
ε	exploration probability
$\varepsilon_{\mathrm{QI}}, \varepsilon_{\mathrm{PE}}$, etc.	convergence threshold, always subscripted by algorithm type (in these examples, $\varepsilon_{\mathrm{QI}}$ for Q-iteration and $\varepsilon_{\mathrm{PE}}$ for policy evaluation)

Approximate dynamic programming and reinforcement learning

$\widehat{Q}$; $\widehat{V}$	approximate Q-function; approximate V-function
$\widehat{h}$	approximate policy
d; D	index of state dimension (variable); number of state space dimensions
F	approximation mapping
P	projection mapping
U_{d}	set of discrete actions
θ; ϕ	value function parameter vector; basis functions for value function approximation
ϑ; φ	policy parameter vector; basis functions for policy approximation
$\bar{\phi}$	state-dependent basis functions for Q-function approximation
κ	kernel function
n	number of parameters and of state-action basis functions for Q-function approximation
$\mathscr{N}$	number of parameters and of state-dependent basis functions for policy approximation
N	number of state-dependent basis functions for Q-function approximation
M	number of discrete actions
n_{s}	number of state-action samples for Q-function approximation
$\mathscr{N}_{\mathrm{s}}$	number of state samples for policy approximation
l	index of Q-function parameter and of state-action basis function
i	index of state-dependent basis function, of policy parameter, or of discrete state
j	index of discrete action
$[i,j]$	scalar index corresponding to the two-dimensional index (i,j); usually, $[i,j] = i + (j-1)N$
l_{s}	index of state-action sample
i_{s}	index of state sample, as well as generic sample index

ξ; Ξ	parameter vector of basis functions; set of such parameter vectors
c	center of basis function, given as a vector
b; B	width of basis function, given as a vector; and given as a matrix
ς	approximation error
Γ	matrix on the left-hand side of the projected Bellman equation
Λ	matrix on the right-hand side of the projected Bellman equation
z	vector on the right-hand side of the projected Bellman equation
w	weight function of approximation errors, of representative states, etc.
s	score function
X_0	set of representative initial states
N_{MC}	number of Monte Carlo simulations for each representative state
ε_{MC}	admissible error in the estimation of the return along a trajectory

Fuzzy Q-iteration

χ; μ	fuzzy set; membership function
ϕ	in this context, normalized membership function (degree of fulfillment)
x_i	core of the ith fuzzy set
S	asynchronous fuzzy Q-iteration mapping
δ_x; δ_u	state resolution step; action resolution step

Online and continuous-action least-squares policy iteration

K_θ	number of transitions between two consecutive policy improvements
ε_{d}	decay rate of exploration probability
T_{trial}	trial length
δ_{mon}	monotonicity direction
ψ	polynomial
M_{p}	degree of polynomial approximator

Cross-entropy policy search and cross-entropy optimization

v	parameter vector of a parameterized probability density
$I\{\cdot\}$	indicator function, equal to 1 when the argument is true, and 0 otherwise
N_{CE}	number of samples used at every iteration
ρ_{CE}	proportion of samples used in the cross-entropy updates
λ	probability level or $(1-\rho_{\text{CE}})$ quantile of the sample performance
c_{CE}	how many times the number of samples N_{CE} is larger than the number of density parameters
α_{CE}	smoothing parameter
ε_{CE}	convergence threshold of the cross-entropy algorithm
d_{CE}	how many iterations the variation of λ should be at most ε_{CE} to stop the algorithm

Experimental studies

t	continuous time variable
α	in this context, angle
τ	in this context, motor torque
Q_{rew}	weight matrix for the states, used in the reward function
R_{rew}	weight matrix or scalar for the actions, used in the reward function

Furthermore, the following conventions are adopted throughout the book:

- All the vectors used are column vectors. The transpose of a vector is denoted by the superscript T, such that, e.g., the transpose of θ is θ^{T}.
- Boldface notation is used for vector or matrix representations of functions and mappings, e.g., $\boldsymbol{Q}$ is a vector representation of a Q-function Q. However, ordinary vectors and matrices are displayed in a normal font, e.g., θ, Γ.
- Calligraphic notation is used to differentiate variables related to policy approximation, from variables related to value function approximation. For instance, the policy parameter is ϑ, whereas the value function parameter is θ.

List of abbreviations

The following list collects, in alphabetical order, the abbreviations used in this book.

BF	basis function
CE	cross-entropy
DC	direct current
DP	dynamic programming
HIV	human immunodeficiency virus
LSPE (LSPE-Q)	least-squares policy evaluation (for Q-functions)
LSPI	least-squares policy iteration
LSTD (LSTD-Q)	least-squares temporal difference (for Q-functions)
MDP	Markov decision process
MF	membership function
PI	policy iteration
RBF	radial basis function
RL	reinforcement learning
STI	structured treatment interruptions
TD (TD-Q)	temporal difference (for Q-functions)

Bibliography

Åström, K. J., Klein, R. E., and Lennartsson, A. (2005). Bicycle dynamics and control. *IEEE Control Systems Magazine*, 24(4):26–47.

Abonyi, J., Babuška, R., and Szeifert, F. (2001). Fuzzy modeling with multivariate membership functions: Gray-box identification and control design. *IEEE Transactions on Systems, Man, and Cybernetics—Part B: Cybernetics*, 31(5):755–767.

Adams, B., Banks, H., Kwon, H.-D., and Tran, H. (2004). Dynamic multidrug therapies for HIV: Optimal and STI control approaches. *Mathematical Biosciences and Engineering*, 1(2):223–241.

Antos, A., Munos, R., and Szepesvári, Cs. (2008a). Fitted Q-iteration in continuous action-space MDPs. In Platt, J. C., Koller, D., Singer, Y., and Roweis, S. T., editors, *Advances in Neural Information Processing Systems 20*, pages 9–16. MIT Press.

Antos, A., Szepesvári, Cs., and Munos, R. (2008b). Learning near-optimal policies with Bellman-residual minimization based fitted policy iteration and a single sample path. *Machine Learning*, 71(1):89–129.

Ascher, U. and Petzold, L. (1998). *Computer methods for ordinary differential equations and differential-algebraic equations.* Society for Industrial and Applied Mathematics (SIAM).

Audibert, J.-Y., Munos, R., and Szepesvári, Cs. (2007). Tuning bandit algorithms in stochastic environments. In *Proceedings 18th International Conference on Algorithmic Learning Theory (ALT-07)*, pages 150–165, Sendai, Japan.

Auer, P., Cesa-Bianchi, N., and Fischer, P. (2002). Finite time analysis of multiarmed bandit problems. *Machine Learning*, 47(2–3):235–256.

Auer, P., Jaksch, T., and Ortner, R. (2009). Near-optimal regret bounds for reinforcement learning. In Koller, D., Schuurmans, D., Bengio, Y., and Bottou, L., editors, *Advances in Neural Information Processing Systems 21*, pages 89–96. MIT Press.

Baird, L. (1995). Residual algorithms: Reinforcement learning with function approximation. In *Proceedings 12th International Conference on Machine Learning (ICML-95)*, pages 30–37, Tahoe City, US.

Balakrishnan, S., Ding, J., and Lewis, F. (2008). Issues on stability of ADP feedback controllers for dynamical systems. *IEEE Transactions on Systems, Man, and Cybernetics—Part B: Cybernetics*, 4(38):913–917.

Barash, D. (1999). A genetic search in policy space for solving Markov decision processes. In *AAAI Spring Symposium on Search Techniques for Problem Solving under Uncertainty and Incomplete Information*, Palo Alto, US.

Barto, A. and Mahadevan, S. (2003). Recent advances in hierarchical reinforcement learning. *Discrete Event Dynamic Systems: Theory and Applications*, 13(4):341–379.

Barto, A. G., Sutton, R. S., and Anderson, C. W. (1983). Neuronlike adaptive elements that can solve difficult learning control problems. *IEEE Transactions on Systems, Man, and Cybernetics*, 13(5):833–846.

Berenji, H. R. and Khedkar, P. (1992). Learning and tuning fuzzy logic controllers through reinforcements. *IEEE Transactions on Neural Networks*, 3(5):724–740.

Berenji, H. R. and Vengerov, D. (2003). A convergent actor-critic-based FRL algorithm with application to power management of wireless transmitters. *IEEE Transactions on Fuzzy Systems*, 11(4):478–485.

Bertsekas, D. P. (2005a). *Dynamic Programming and Optimal Control*, volume 1. Athena Scientific, 3rd edition.

Bertsekas, D. P. (2005b). Dynamic programming and suboptimal control: A survey from ADP to MPC. *European Journal of Control*, 11(4–5):310–334. Special issue for the CDC-ECC-05 in Seville, Spain.

Bertsekas, D. P. (2007). *Dynamic Programming and Optimal Control*, volume 2. Athena Scientific, 3rd edition.

Bertsekas, D. P., Borkar, V., and Nedić, A. (2004). Improved temporal difference methods with linear function approximation. In Si, J., Barto, A., and Powell, W., editors, *Learning and Approximate Dynamic Programming*. IEEE Press.

Bertsekas, D. P. and Castañon, D. A. (1989). Adaptive aggregation methods for infinite horizon dynamic programming. *IEEE Transactions on Automatic Control*, 34(6):589–598.

Bertsekas, D. P. and Ioffe, S. (1996). Temporal differences-based policy iteration and applications in neuro-dynamic programming. Technical Report LIDS-P-2349, Massachusetts Institute of Technology, Cambridge, US. Available at http://web.mit.edu/dimitrib/www/Tempdif.pdf.

Bertsekas, D. P. and Shreve, S. E. (1978). *Stochastic Optimal Control: The Discrete Time Case*. Academic Press.

Bertsekas, D. P. and Tsitsiklis, J. N. (1996). *Neuro-Dynamic Programming*. Athena Scientific.

Bertsekas, D. P. and Yu, H. (2009). Basis function adaptation methods for cost approximation in MDP. In *Proceedings 2009 IEEE Symposium on Approximate Dynamic Programming and Reinforcement Learning (ADPRL-09)*, pages 74–81, Nashville, US.

Bethke, B., How, J., and Ozdaglar, A. (2008). Approximate dynamic programming using support vector regression. In *Proceedings 47th IEEE Conference on Decision and Control (CDC-08)*, pages 3811–3816, Cancun, Mexico.

Bhatnagar, S., Sutton, R., Ghavamzadeh, M., and Lee, M. (2009). Natural actor-critic algorithms. *Automatica*, 45(11):2471–2482.

Birge, J. R. and Louveaux, F. (1997). *Introduction to Stochastic Programming*. Springer.

Borkar, V. (2005). An actor-critic algorithm for constrained Markov decision processes. *Systems & Control Letters*, 54(3):207–213.

Boubezoul, A., Paris, S., and Ouladsine, M. (2008). Application of the cross entropy method to the GLVQ algorithm. *Pattern Recognition*, 41(10):3173–3178.

Boyan, J. (2002). Technical update: Least-squares temporal difference learning. *Machine Learning*, 49:233–246.

Bradtke, S. J. and Barto, A. G. (1996). Linear least-squares algorithms for temporal difference learning. *Machine Learning*, 22(1–3):33–57.

Breiman, L. (2001). Random forests. *Machine Learning*, 45(1):5–32.

Breiman, L., Friedman, J., Stone, C. J., and Olshen, R. (1984). *Classification and Regression Trees*. Wadsworth International.

Brown, M. and Harris, C. (1994). *Neurofuzzy Adaptive Modeling and Control*. Prentice Hall.

Bubeck, S., Munos, R., Stoltz, G., and Szepesvári, C. (2009). Online optimization in X-armed bandits. In Koller, D., Schuurmans, D., Bengio, Y., and Bottou, L., editors, *Advances in Neural Information Processing Systems 21*, pages 201–208. MIT Press.

Buşoniu, L., Babuška, R., and De Schutter, B. (2008a). A comprehensive survey of multi-agent reinforcement learning. *IEEE Transactions on Systems, Man, and Cybernetics. Part C: Applications and Reviews*, 38(2):156–172.

Buşoniu, L., Ernst, D., De Schutter, B., and Babuška, R. (2007). Fuzzy approximation for convergent model-based reinforcement learning. In *Proceedings 2007 IEEE International Conference on Fuzzy Systems (FUZZ-IEEE-07)*, pages 968–973, London, UK.

Buşoniu, L., Ernst, D., De Schutter, B., and Babuška, R. (2008b). Consistency of fuzzy model-based reinforcement learning. In *Proceedings 2008 IEEE International Conference on Fuzzy Systems (FUZZ-IEEE-08)*, pages 518–524, Hong Kong.

Buşoniu, L., Ernst, D., De Schutter, B., and Babuška, R. (2008c). Continuous-state reinforcement learning with fuzzy approximation. In Tuyls, K., Nowé, A., Guessoum, Z., and Kudenko, D., editors, *Adaptive Agents and Multi-Agent Systems III*, volume 4865 of *Lecture Notes in Computer Science*, pages 27–43. Springer.

Buşoniu, L., Ernst, D., De Schutter, B., and Babuška, R. (2008d). Fuzzy partition optimization for approximate fuzzy Q-iteration. In *Proceedings 17th IFAC World Congress (IFAC-08)*, pages 5629–5634, Seoul, Korea.

Buşoniu, L., Ernst, D., De Schutter, B., and Babuška, R. (2009). Policy search with cross-entropy optimization of basis functions. In *Proceedings 2009 IEEE International Symposium on Adaptive Dynamic Programming and Reinforcement Learning (ADPRL-09)*, pages 153–160, Nashville, US.

Camacho, E. F. and Bordons, C. (2004). *Model Predictive Control*. Springer-Verlag.

Cao, X.-R. (2007). *Stochastic Learning and Optimization: A Sensitivity-Based Approach*. Springer.

Chang, H. S., Fu, M. C., Hu, J., and Marcus, S. I. (2007). *Simulation-Based Algorithms for Markov Decision Processes*. Springer.

Chepuri, K. and de Mello, T. H. (2005). Solving the vehicle routing problem with stochastic demands using the cross-entropy method. *Annals of Operations Research*, 134(1):153–181.

Chin, H. H. and Jafari, A. A. (1998). Genetic algorithm methods for solving the best stationary policy of finite Markov decision processes. In *Proceedings 30th Southeastern Symposium on System Theory*, pages 538–543, Morgantown, US.

Chow, C.-S. and Tsitsiklis, J. N. (1991). An optimal one-way multigrid algorithm for discrete-time stochastic control. *IEEE Transactions on Automatic Control*, 36(8):898–914.

Costa, A., Jones, O. D., and Kroese, D. (2007). Convergence properties of the cross-entropy method for discrete optimization. *Operations Research Letters*, 35(5):573–580.

Cristianini, N. and Shawe-Taylor, J. (2000). *An Introduction to Support Vector Machines and Other Kernel-Based Learning Methods*. Cambridge University Press.

Davies, S. (1997). Multidimensional triangulation and interpolation for reinforcement learning. In Mozer, M. C., Jordan, M. I., and Petsche, T., editors, *Advances in Neural Information Processing Systems 9*, pages 1005–1011. MIT Press.

Defourny, B., Ernst, D., and Wehenkel, L. (2008). Lazy planning under uncertainties by optimizing decisions on an ensemble of incomplete disturbance trees. In Girgin, S., Loth, M., Munos, R., Preux, P., and Ryabko, D., editors, *Recent Advances in Reinforcement Learning*, volume 5323 of *Lecture Notes in Computer Science*, pages 1–14. Springer.

Defourny, B., Ernst, D., and Wehenkel, L. (2009). Planning under uncertainty, ensembles of disturbance trees and kernelized discrete action spaces. In *Proceedings 2009 IEEE International Symposium on Adaptive Dynamic Programming and Reinforcement Learning (ADPRL-09)*, pages 145–152, Nashville, US.

Deisenroth, M. P., Rasmussen, C. E., and Peters, J. (2009). Gaussian process dynamic programming. *Neurocomputing*, 72(7–9):1508–1524.

Dietterich, T. G. (2000). Hierarchical reinforcement learning with the MAXQ value function decomposition. *Journal of Artificial Intelligence Research*, 13:227–303.

Dimitrakakis, C. and Lagoudakis, M. (2008). Rollout sampling approximate policy iteration. *Machine Learning*, 72(3):157–171.

Dorigo, M. and Colombetti, M. (1994). Robot shaping: Developing autonomous agents through learning. *Artificial Intelligence*, 71(2):321–370.

Doya, K. (2000). Reinforcement learning in continuous time and space. *Neural Computation*, 12(1):219–245.

Duda, R. O., Hart, P. E., and Stork, D. G. (2000). *Pattern Classification*. Wiley, 2nd edition.

Dupacová, J., Consigli, G., and Wallace, S. W. (2000). Scenarios for multistage stochastic programs. *Annals of Operations Research*, 100(1–4):25–53.

Edelman, A. and Murakami, H. (1995). Polynomial roots from companion matrix eigenvalues. *Mathematics of Computation*, 64:763–776.

Engel, Y., Mannor, S., and Meir, R. (2003). Bayes meets Bellman: The Gaussian process approach to temporal difference learning. In *Proceedings 20th International Conference on Machine Learning (ICML-03)*, pages 154–161, Washington, US.

Engel, Y., Mannor, S., and Meir, R. (2005). Reinforcement learning with Gaussian processes. In *Proceedings 22nd International Conference on Machine Learning (ICML-05)*, pages 201–208, Bonn, Germany.

Ernst, D. (2005). Selecting concise sets of samples for a reinforcement learning agent. In *Proceedings 3rd International Conference on Computational Intelligence, Robotics and Autonomous Systems (CIRAS-05)*, Singapore.

Ernst, D., Geurts, P., and Wehenkel, L. (2005). Tree-based batch mode reinforcement learning. *Journal of Machine Learning Research*, 6:503–556.

Ernst, D., Glavic, M., Capitanescu, F., and Wehenkel, L. (2009). Reinforcement learning versus model predictive control: A comparison on a power system problem. *IEEE Transactions on Systems, Man, and Cybernetics—Part B: Cybernetics*, 39(2):517–529.

Ernst, D., Glavic, M., Geurts, P., and Wehenkel, L. (2006a). Approximate value iteration in the reinforcement learning context. Application to electrical power system control. *International Journal of Emerging Electric Power Systems*, 3(1). 37 pages.

Ernst, D., Glavic, M., Stan, G.-B., Mannor, S., and Wehenkel, L. (2007). The cross-entropy method for power system combinatorial optimization problems. In *Proceedings of Power Tech 2007*, pages 1290–1295, Lausanne, Switzerland.

Ernst, D., Stan, G.-B., Gonçalves, J., and Wehenkel, L. (2006b). Clinical data based optimal STI strategies for HIV: A reinforcement learning approach. In *Proceedings 45th IEEE Conference on Decision & Control*, pages 667–672, San Diego, US.

Fantuzzi, C. and Rovatti, R. (1996). On the approximation capabilities of the homogeneous Takagi-Sugeno model. In *Proceedings 5th IEEE International Conference on Fuzzy Systems (FUZZ-IEEE'96)*, pages 1067–1072, New Orleans, US.

Farahmand, A. M., Ghavamzadeh, M., Szepesvári, Cs., and Mannor, S. (2009a). Regularized fitted Q-iteration for planning in continuous-space Markovian decision problems. In *Proceedings 2009 American Control Conference (ACC-09)*, pages 725–730, St. Louis, US.

Farahmand, A. M., Ghavamzadeh, M., Szepesvári, Cs., and Mannor, S. (2009b). Regularized policy iteration. In Koller, D., Schuurmans, D., Bengio, Y., and Bottou, L., editors, *Advances in Neural Information Processing Systems 21*, pages 441–448. MIT Press.

Feldbaum, A. (1961). Dual control theory, Parts I and II. *Automation and Remote Control*, 21(9):874–880.

Franklin, G. F., Powell, J. D., and Workman, M. L. (1998). *Digital Control of Dynamic Systems*. Prentice Hall, 3rd edition.

Geramifard, A., Bowling, M., Zinkevich, M., and Sutton, R. S. (2007). iLSTD: Eligibility traces & convergence analysis. In Schölkopf, B., Platt, J., and Hofmann, T., editors, *Advances in Neural Information Processing Systems 19*, pages 440–448. MIT Press.

Geramifard, A., Bowling, M. H., and Sutton, R. S. (2006). Incremental least-squares temporal difference learning. In *Proceedings 21st National Conference on Artificial Intelligence and 18th Innovative Applications of Artificial Intelligence Conference (AAAI-06)*, pages 356–361, Boston, US.

Geurts, P., Ernst, D., and Wehenkel, L. (2006). Extremely randomized trees. *Machine Learning*, 36(1):3–42.

Ghavamzadeh, M. and Mahadevan, S. (2007). Hierarchical average reward reinforcement learning. *Journal of Machine Learning Research*, 8:2629–2669.

Glorennec, P. Y. (2000). Reinforcement learning: An overview. In *Proceedings European Symposium on Intelligent Techniques (ESIT-00)*, pages 17–35, Aachen, Germany.

Glover, F. and Laguna, M. (1997). *Tabu Search*. Kluwer.

Goldberg, D. E. (1989). *Genetic Algorithms in Search, Optimization and Machine Learning*. Addison-Wesley.

Gomez, F. J., Schmidhuber, J., and Miikkulainen, R. (2006). Efficient non-linear control through neuroevolution. In *Proceedings 17th European Conference on Machine Learning (ECML-06)*, volume 4212 of *Lecture Notes in Computer Science*, pages 654–662, Berlin, Germany.

Gonzalez, R. L. and Rofman, E. (1985). On deterministic control problems: An approximation procedure for the optimal cost I. The stationary problem. *SIAM Journal on Control and Optimization*, 23(2):242–266.

Gordon, G. (1995). Stable function approximation in dynamic programming. In *Proceedings 12th International Conference on Machine Learning (ICML-95)*, pages 261–268, Tahoe City, US.

Gordon, G. J. (2001). Reinforcement learning with function approximation converges to a region. In Leen, T. K., Dietterich, T. G., and Tresp, V., editors, *Advances in Neural Information Processing Systems 13*, pages 1040–1046. MIT Press.

Grüne, L. (2004). Error estimation and adaptive discretization for the discrete stochastic Hamilton-Jacobi-Bellman equation. *Numerische Mathematik*, 99(1):85–112.

Hassoun, M. (1995). *Fundamentals of Artificial Neural Networks*. MIT Press.

Hengst, B. (2002). Discovering hierarchy in reinforcement learning with HEXQ. In *Proceedings 19th International Conference on Machine Learning (ICML-02)*, pages 243–250, Sydney, Australia.

Horiuchi, T., Fujino, A., Katai, O., and Sawaragi, T. (1996). Fuzzy interpolation-based Q-learning with continuous states and actions. In *Proceedings 5th IEEE International Conference on Fuzzy Systems (FUZZ-IEEE-96)*, pages 594–600, New Orleans, US.

Hren, J.-F. and Munos, R. (2008). Optimistic planning of deterministic systems. In Girgin, S., Loth, M., Munos, R., Preux, P., and Ryabko, D., editors, *Recent Advances in Reinforcement Learning*, volume 5323 of *Lecture Notes in Computer Science*, pages 151–164. Springer.

Istratescu, V. I. (2002). *Fixed Point Theory: An Introduction*. Springer.

Jaakkola, T., Jordan, M. I., and Singh, S. P. (1994). On the convergence of stochastic iterative dynamic programming algorithms. *Neural Computation*, 6(6):1185–1201.

Jodogne, S., Briquet, C., and Piater, J. H. (2006). Approximate policy iteration for closed-loop learning of visual tasks. In *Proceedings 17th European Conference on Machine Learning (ECML-06)*, volume 4212 of *Lecture Notes in Computer Science*, pages 210–221, Berlin, Germany.

Jones, D. R. (2009). DIRECT global optimization algorithm. In Floudas, C. A. and Pardalos, P. M., editors, *Encyclopedia of Optimization*, pages 725–735. Springer.

Jouffe, L. (1998). Fuzzy inference system learning by reinforcement methods. *IEEE Transactions on Systems, Man, and Cybernetics—Part C: Applications and Reviews*, 28(3):338–355.

Jung, T. and Polani, D. (2007a). Kernelizing LSPE(λ). In *Proceedings 2007 IEEE Symposium on Approximate Dynamic Programming and Reinforcement Learning (ADPRL-07)*, pages 338–345, Honolulu, US.

Jung, T. and Polani, D. (2007b). Learning robocup-keepaway with kernels. In *Gaussian Processes in Practice*, volume 1 of *JMLR Workshop and Conference Proceedings*, pages 33–57.

Jung, T. and Stone, P. (2009). Feature selection for value function approximation using Bayesian model selection. In *Machine Learning and Knowledge Discovery in Databases, European Conference (ECML-PKDD-09)*, volume 5781 of *Lecture Notes in Computer Science*, pages 660–675, Bled, Slovenia.

Jung, T. and Uthmann, T. (2004). Experiments in value function approximation with sparse support vector regression. In *Proceedings 15th European Conference on Machine Learning (ECML-04)*, volume 3201 of *Lecture Notes in Artificial Intelligence*, pages 180–191, Pisa, Italy.

Kaelbling, L. P. (1993). *Learning in Embedded Systems*. MIT Press.

Kaelbling, L. P., Littman, M. L., and Cassandra, A. R. (1998). Planning and acting in partially observable stochastic domains. *Artificial Intelligence*, 101(1–2):99–134.

Kaelbling, L. P., Littman, M. L., and Moore, A. W. (1996). Reinforcement learning: A survey. *Journal of Artificial Intelligence Research*, 4:237–285.

Kakade, S. (2001). A natural policy gradient. In Dietterich, T. G., Becker, S., and Ghahramani, Z., editors, *Advances in Neural Information Processing Systems 14*, pages 1531–1538. MIT Press.

Kalyanakrishnan, S. and Stone, P. (2007). Batch reinforcement learning in a complex domain. In *Proceedings 6th International Conference on Autonomous Agents and Multi-Agent Systems*, pages 650–657, Honolulu, US.

Keller, P. W., Mannor, S., and Precup, D. (2006). Automatic basis function construction for approximate dynamic programming and reinforcement learning. In *Proceedings 23rd International Conference on Machine Learning (ICML-06)*, pages 449–456, Pittsburgh, US.

Khalil, H. K. (2002). *Nonlinear Systems*. Prentice Hall, 3rd edition.

Kirk, D. E. (2004). *Optimal Control Theory: An Introduction*. Dover Publications.

Klir, G. J. and Yuan, B. (1995). *Fuzzy Sets and Fuzzy Logic: Theory and Applications*. Prentice Hall.

Knuth, D. E. (1976). Big Omicron and big Omega and big Theta. *SIGACT News*, 8(2):18–24.

Kolter, J. Z. and Ng, A. (2009). Regularization and feature selection in least-squares temporal difference learning. In *Proceedings 26th International Conference on Machine Learning (ICML-09)*, pages 521–528, Montreal, Canada.

Konda, V. (2002). *Actor-Critic Algorithms*. PhD thesis, Massachusetts Institute of Technology, Cambridge, US.

Konda, V. R. and Tsitsiklis, J. N. (2000). Actor-critic algorithms. In Solla, S. A., Leen, T. K., and Müller, K.-R., editors, *Advances in Neural Information Processing Systems 12*, pages 1008–1014. MIT Press.

Konda, V. R. and Tsitsiklis, J. N. (2003). On actor-critic algorithms. *SIAM Journal on Control and Optimization*, 42(4):1143–1166.

Kruse, R., Gebhardt, J. E., and Klowon, F. (1994). *Foundations of Fuzzy Systems*. Wiley.

Lagoudakis, M., Parr, R., and Littman, M. (2002). Least-squares methods in reinforcement learning for control. In *Methods and Applications of Artificial Intelligence*, volume 2308 of *Lecture Notes in Artificial Intelligence*, pages 249–260. Springer.

Lagoudakis, M. G. and Parr, R. (2003a). Least-squares policy iteration. *Journal of Machine Learning Research*, 4:1107–1149.

Lagoudakis, M. G. and Parr, R. (2003b). Reinforcement learning as classification: Leveraging modern classifiers. In *Proceedings 20th International Conference on Machine Learning (ICML-03)*, pages 424–431. Washington, US.

Levine, W. S., editor (1996). *The Control Handbook*. CRC Press.

Lewis, R. M. and Torczon, V. (2000). Pattern search algorithms for linearly constrained minimization. *SIAM Journal on Optimization*, 10(3):917–941.

Li, L., Littman, M. L., and Mansley, C. R. (2009). Online exploration in least-squares policy iteration. In *Proceedings 8th International Joint Conference on Autonomous Agents and Multiagent Systems (AAMAS-09)*, volume 2, pages 733–739, Budapest, Hungary.

Lin, C.-K. (2003). A reinforcement learning adaptive fuzzy controller for robots. *Fuzzy Sets and Systems*, 137(3):339–352.

Lin, L.-J. (1992). Self-improving reactive agents based on reinforcement learning, planning and teaching. *Machine Learning*, 8(3–4):293–321. Special issue on reinforcement learning.

Liu, D., Javaherian, H., Kovalenko, O., and Huang, T. (2008). Adaptive critic learning techniques for engine torque and air-fuel ratio control. *IEEE Transactions on Systems, Man, and Cybernetics—Part B: Cybernetics*, 38(4):988–993.

Lovejoy, W. S. (1991). Computationally feasible bounds for partially observed Markov decision processes. *Operations Research*, 39(1):162–175.

Maciejowski, J. M. (2002). *Predictive Control with Constraints*. Prentice Hall.

Madani, O. (2002). On policy iteration as a Newton's method and polynomial policy iteration algorithms. In *Proceedings 18th National Conference on Artificial Intelligence and 14th Conference on Innovative Applications of Artificial Intelligence AAAI/IAAI-02*, pages 273–278, Edmonton, Canada.

Mahadevan, S. (2005). Samuel meets Amarel: Automating value function approximation using global state space analysis. In *Proceedings 20th National Conference on Artificial Intelligence and the 17th Innovative Applications of Artificial Intelligence Conference (AAAI-05)*, pages 1000–1005, Pittsburgh, US.

Mahadevan, S. and Maggioni, M. (2007). Proto-value functions: A Laplacian framework for learning representation and control in Markov decision processes. *Journal of Machine Learning Research*, 8:2169–2231.

Mamdani, E. (1977). Application of fuzzy logic to approximate reasoning using linguistic systems. *IEEE Transactions on Computers*, 26:1182–1191.

Mannor, S., Rubinstein, R. Y., and Gat, Y. (2003). The cross-entropy method for fast policy search. In *Proceedings 20th International Conference on Machine Learning (ICML-03)*, pages 512–519, Washington, US.

Marbach, P. and Tsitsiklis, J. N. (2003). Approximate gradient methods in policy-space optimization of Markov reward processes. *Discrete Event Dynamic Systems: Theory and Applications*, 13(1–2):111–148.

Matarić, M. J. (1997). Reinforcement learning in the multi-robot domain. *Autonomous Robots*, 4(1):73–83.

Mathenya, M. E., Resnic, F. S., Arora, N., and Ohno-Machado, L. (2007). Effects of SVM parameter optimization on discrimination and calibration for post-procedural PCI mortality. *Journal of Biomedical Informatics*, 40(6):688–697.

Melo, F. S., Meyn, S. P., and Ribeiro, M. I. (2008). An analysis of reinforcement learning with function approximation. In *Proceedings 25th International Conference on Machine Learning (ICML-08)*, pages 664–671, Helsinki, Finland.

Menache, I., Mannor, S., and Shimkin, N. (2005). Basis function adaptation in temporal difference reinforcement learning. *Annals of Operations Research*, 134(1):215–238.

Millán, J. d. R., Posenato, D., and Dedieu, E. (2002). Continuous-action Q-learning. *Machine Learning*, 49(2–3):247–265.

Moore, A. W. and Atkeson, C. R. (1995). The parti-game algorithm for variable resolution reinforcement learning in multidimensional state-spaces. *Machine Learning*, 21(3):199–233.

Morris, C. (1982). Natural exponential families with quadratic variance functions. *Annals of Statistics*, 10(1):65–80.

Munos, R. (1997). Finite-element methods with local triangulation refinement for continuous reinforcement learning problems. In *Proceedings 9th European Conference on Machine Learning (ECML-97)*, volume 1224 of *Lecture Notes in Artificial Intelligence*, pages 170–182, Prague, Czech Republic.

Munos, R. (2006). Policy gradient in continuous time. *Journal of Machine Learning Research*, 7:771–791.

Munos, R. and Moore, A. (2002). Variable-resolution discretization in optimal control. *Machine Learning*, 49(2–3):291–323.

Munos, R. and Szepesvári, Cs. (2008). Finite time bounds for fitted value iteration. *Journal of Machine Learning Research*, 9:815–857.

Murphy, S. (2005). A generalization error for Q-learning. *Journal of Machine Learning Research*, 6:1073–1097.

Nakamura, Y., Moria, T., Satoc, M., and Ishiia, S. (2007). Reinforcement learning for a biped robot based on a CPG-actor-critic method. *Neural Networks*, 20(6):723–735.

Nedić, A. and Bertsekas, D. P. (2003). Least-squares policy evaluation algorithms with linear function approximation. *Discrete Event Dynamic Systems: Theory and Applications*, 13(1–2):79–110.

Ng, A. Y., Harada, D., and Russell, S. (1999). Policy invariance under reward transformations: Theory and application to reward shaping. In *Proceedings 16th International Conference on Machine Learning (ICML-99)*, pages 278–287, Bled, Slovenia.

Ng, A. Y. and Jordan, M. I. (2000). PEGASUS: A policy search method for large MDPs and POMDPs. In *Proceedings 16th Conference in Uncertainty in Artificial Intelligence (UAI-00)*, pages 406–415, Palo Alto, US.

Nocedal, J. and Wright, S. J. (2006). *Numerical Optimization*. Springer-Verlag, 2nd edition.

Ormoneit, D. and Sen, S. (2002). Kernel-based reinforcement learning. *Machine Learning*, 49(2–3):161–178.

Panait, L. and Luke, S. (2005). Cooperative multi-agent learning: The state of the art. *Autonomous Agents and Multi-Agent Systems*, 11(3):387–434.

Parr, R., Li, L., Taylor, G., Painter-Wakefield, C., and Littman, M. (2008). An analysis of linear models, linear value-function approximation, and feature selection for reinforcement learning. In *Proceedings 25th Annual International Conference on Machine Learning (ICML-08)*, pages 752–759, Helsinki, Finland.

Pazis, J. and Lagoudakis, M. (2009). Binary action search for learning continuous-action control policies. In *Proceedings of the 26th Annual International Conference on Machine Learning (ICML-09)*, pages 793–800, Montreal, Canada.

Pérez-Uribe, A. (2001). Using a time-delay actor-critic neural architecture with dopamine-like reinforcement signal for learning in autonomous robots. In Wermter, S., Austin, J., and Willshaw, D. J., editors, *Emergent Neural Computational Architectures Based on Neuroscience*, volume 2036 of *Lecture Notes in Computer Science*, pages 522–533. Springer.

Perkins, T. and Barto, A. (2002). Lyapunov design for safe reinforcement learning. *Journal of Machine Learning Research*, 3:803–832.

Peters, J. and Schaal, S. (2008). Natural actor-critic. *Neurocomputing*, 71(7–9):1180–1190.

Pineau, J., Gordon, G. J., and Thrun, S. (2006). Anytime point-based approximations for large POMDPs. *Journal of Artificial Intelligence Research (JAIR)*, 27:335–380.

Porta, J. M., Vlassis, N., Spaan, M. T., and Poupart, P. (2006). Point-based value iteration for continuous POMDPs. *Journal of Machine Learning Research*, 7:2329–2367.

Powell, W. B. (2007). *Approximate Dynamic Programming: Solving the Curses of Dimensionality*. Wiley.

Press, W. H., Flannery, B. P., Teukolsky, S. A., and Vetterling, W. T. (1986). *Numerical Recipes: The Art of Scientific Computing*. Cambridge University Press.

Prokhorov, D. and Wunsch, D.C., I. (1997). Adaptive critic designs. *IEEE Transactions on Neural Networks*, 8(5):997–1007.

Puterman, M. L. (1994). *Markov Decision Processes—Discrete Stochastic Dynamic Programming*. Wiley.

Randløv, J. and Alstrøm, P. (1998). Learning to drive a bicycle using reinforcement learning and shaping. In *Proceedings 15th International Conference on Machine Learning (ICML-98)*, pages 463–471, Madison, US.

Rasmussen, C. E. and Kuss, M. (2004). Gaussian processes in reinforcement learning. In Thrun, S., Saul, L. K., and Schölkopf, B., editors, *Advances in Neural Information Processing Systems 16*. MIT Press.

Rasmussen, C. E. and Williams, C. K. I. (2006). *Gaussian Processes for Machine Learning*. MIT Press.

Ratitch, B. and Precup, D. (2004). Sparse distributed memories for on-line value-based reinforcement learning. In *Proceedings 15th European Conference on Machine Learning (ECML-04)*, volume 3201 of *Lecture Notes in Computer Science*, pages 347–358, Pisa, Italy.

Reynolds, S. I. (2000). Adaptive resolution model-free reinforcement learning: Decision boundary partitioning. In *Proceedings Seventeenth International Conference on Machine Learning (ICML-00)*, pages 783–790, Stanford University, US.

Riedmiller, M. (2005). Neural fitted Q-iteration – first experiences with a data efficient neural reinforcement learning method. In *Proceedings 16th European Conference on Machine Learning (ECML-05)*, volume 3720 of *Lecture Notes in Computer Science*, pages 317–328, Porto, Portugal.

Riedmiller, M., Peters, J., and Schaal, S. (2007). Evaluation of policy gradient methods and variants on the cart-pole benchmark. In *Proceedings 2007 IEEE Symposium on Approximate Dynamic Programming and Reinforcement Learning (ADPRL-07)*, pages 254–261, Honolulu, US.

Rubinstein, R. Y. and Kroese, D. P. (2004). *The Cross Entropy Method: A Unified Approach to Combinatorial Optimization, Monte-Carlo Simulation, and Machine Learning*. Springer.

Rummery, G. A. and Niranjan, M. (1994). On-line Q-learning using connectionist systems. Technical Report CUED/F-INFENG/TR166, Engineering Department, Cambridge University, UK. Available at http://mi.eng.cam.ac.uk/reports/svr-ftp/rummery_tr166.ps.Z.

Russell, S. and Norvig, P. (2003). *Artificial Intelligence: A Modern Approach*. Prentice Hall, 2nd edition.

Russell, S. J. and Zimdars, A. (2003). Q-decomposition for reinforcement learning agents. In *Proceedings 20th International Conference of Machine Learning (ICML-03)*, pages 656–663, Washington, US.

Santamaria, J. C., Sutton, R. S., and Ram, A. (1998). Experiments with reinforcement learning in problems with continuous state and action spaces. *Adaptive Behavior*, 6(2):163–218.

Santos, M. S. and Vigo-Aguiar, J. (1998). Analysis of a numerical dynamic programming algorithm applied to economic models. *Econometrica*, 66(2):409–426.

Schervish, M. J. (1995). *Theory of Statistics*. Springer.

Schmidhuber, J. (2000). Sequential decision making based on direct search. In Sun, R. and Giles, C. L., editors, *Sequence Learning*, volume 1828 of *Lecture Notes in Computer Science*, pages 213–240. Springer.

Schölkopf, B., Burges, C., and Smola, A. (1999). *Advances in Kernel Methods: Support Vector Learning*. MIT Press.

Shawe-Taylor, J. and Cristianini, N. (2004). *Kernel Methods for Pattern Analysis*. Cambridge University Press.

Sherstov, A. and Stone, P. (2005). Function approximation via tile coding: Automating parameter choice. In *Proceedings 6th International Symposium on Abstraction, Reformulation and Approximation (SARA-05)*, volume 3607 of *Lecture Notes in Computer Science*, pages 194–205, Airth Castle, UK.

Shoham, Y., Powers, R., and Grenager, T. (2007). If multi-agent learning is the answer, what is the question? *Artificial Intelligence*, 171(7):365–377.

Singh, S., Jaakkola, T., Littman, M. L., and Szepesvári, Cs. (2000). Convergence results for single-step on-policy reinforcement-learning algorithms. *Machine Learning*, 38(3):287–308.

Singh, S. and Sutton, R. (1996). Reinforcement learning with replacing eligibility traces. *Machine Learning*, 22(1–3):123–158.

Singh, S. P., Jaakkola, T., and Jordan, M. I. (1995). Reinforcement learning with soft state aggregation. In Tesauro, G., Touretzky, D. S., and Leen, T. K., editors, *Advances in Neural Information Processing Systems 7*, pages 361–368. MIT Press.

Singh, S. P., James, M. R., and Rudary, M. R. (2004). Predictive state representations: A new theory for modeling dynamical systems. In *Proceedings 20th Conference in Uncertainty in Artificial Intelligence (UAI-04)*, pages 512–518, Banff, Canada.

Smola, A. J. and Schölkopf, B. (2004). A tutorial on support vector regression. *Statistics and Computing*, 14(3):199–222.

Sutton, R., Maei, H., Precup, D., Bhatnagar, S., Silver, D., Szepesvari, Cs., and Wiewiora, E. (2009a). Fast gradient-descent methods for temporal-difference learning with linear function approximation. In *Proceedings 26th International Conference on Machine Learning (ICML-09)*, pages 993–1000, Montreal, Canada.

Sutton, R. S. (1988). Learning to predict by the method of temporal differences. *Machine Learning*, 3:9–44.

Sutton, R. S. (1990). Integrated architectures for learning, planning, and reacting based on approximating dynamic programming. In *Proceedings 7th International Conference on Machine Learning (ICML-90)*, pages 216–224, Austin, US.

Sutton, R. S. (1996). Generalization in reinforcement learning: Successful examples using sparse coarse coding. In Touretzky, D. S., Mozer, M. C., and Hasselmo, M. E., editors, *Advances in Neural Information Processing Systems 8*, pages 1038–1044. MIT Press.

Sutton, R. S. and Barto, A. G. (1998). *Reinforcement Learning: An Introduction*. MIT Press.

Sutton, R. S., Barto, A. G., and Williams, R. J. (1992). Reinforcement learning is adaptive optimal control. *IEEE Control Systems Magazine*, 12(2):19–22.

Sutton, R. S., McAllester, D. A., Singh, S. P., and Mansour, Y. (2000). Policy gradient methods for reinforcement learning with function approximation. In Solla, S. A., Leen, T. K., and Müller, K.-R., editors, *Advances in Neural Information Processing Systems 12*, pages 1057–1063. MIT Press.

Sutton, R. S., Szepesvári, Cs., and Maei, H. R. (2009b). A convergent $O(n)$ temporal-difference algorithm for off-policy learning with linear function approximation. In Koller, D., Schuurmans, D., Bengio, Y., and Bottou, L., editors, *Advances in Neural Information Processing Systems 21*, pages 1609–1616. MIT Press.

Szepesvári, Cs. and Munos, R. (2005). Finite time bounds for sampling based fitted value iteration. In *Proceedings 22nd International Conference on Machine Learning (ICML-05)*, pages 880–887, Bonn, Germany.

Szepesvári, Cs. and Smart, W. D. (2004). Interpolation-based Q-learning. In *Proceedings 21st International Conference on Machine Learning (ICML-04)*, pages 791–798, Banff, Canada.

Takagi, T. and Sugeno, M. (1985). Fuzzy identification of systems and its applications to modeling and control. *IEEE Transactions on Systems, Man, and Cybernetics*, 15(1):116–132.

Taylor, G. and Parr, R. (2009). Kernelized value function approximation for reinforcement learning. In *Proceedings 26th International Conference on Machine Learning (ICML-09)*, pages 1017–1024, Montreal, Canada.

Thrun, S. (1992). The role of exploration in learning control. In White, D. and Sofge, D., editors, *Handbook for Intelligent Control: Neural, Fuzzy and Adaptive Approaches*. Van Nostrand Reinhold.

Torczon, V. (1997). On the convergence of pattern search algorithms. *SIAM Journal on Optimization*, 7(1):1–25.

Touzet, C. F. (1997). Neural reinforcement learning for behaviour synthesis. *Robotics and Autonomous Systems*, 22(3–4):251–281.

Tsitsiklis, J. N. (1994). Asynchronous stochastic approximation and Q-learning. *Machine Learning*, 16(1):185–202.

Tsitsiklis, J. N. (2002). On the convergence of optimistic policy iteration. *Journal of Machine Learning Research*, 3:59–72.

Tsitsiklis, J. N. and Van Roy, B. (1996). Feature-based methods for large scale dynamic programming. *Machine Learning*, 22(1–3):59–94.

Tsitsiklis, J. N. and Van Roy, B. (1997). An analysis of temporal difference learning with function approximation. *IEEE Transactions on Automatic Control*, 42(5):674–690.

Tuyls, K., Maes, S., and Manderick, B. (2002). Q-learning in simulated robotic soccer – large state spaces and incomplete information. In *Proceedings 2002 International Conference on Machine Learning and Applications (ICMLA-02)*, pages 226–232, Las Vegas, US.

Uther, W. T. B. and Veloso, M. M. (1998). Tree based discretization for continuous state space reinforcement learning. In *Proceedings 15th National Conference on Artificial Intelligence and 10th Innovative Applications of Artificial Intelligence Conference (AAAI-98/IAAI-98)*, pages 769–774, Madison, US.

Vrabie, D., Pastravanu, O., Abu-Khalaf, M., and Lewis, F. (2009). Adaptive optimal control for continuous-time linear systems based on policy iteration. *Automatica*, 45(2):477–484.

Waldock, A. and Carse, B. (2008). Fuzzy Q-learning with an adaptive representation. In *Proceedings 2008 IEEE World Congress on Computational Intelligence (WCCI-08)*, pages 720–725, Hong Kong.

Watkins, C. J. C. H. (1989). *Learning from Delayed Rewards*. PhD thesis, King's College, Oxford, UK.

Watkins, C. J. C. H. and Dayan, P. (1992). Q-learning. *Machine Learning*, 8:279–292.

Whiteson, S. and Stone, P. (2006). Evolutionary function approximation for reinforcement learning. *Journal of Machine Learning Research*, 7:877–917.

Wiering, M. (2004). Convergence and divergence in standard and averaging reinforcement learning. In *Proceedings 15th European Conference on Machine Learning (ECML-04)*, volume 3201 of *Lecture Notes in Artificial Intelligence*, pages 477–488, Pisa, Italy.

Williams, R. J. and Baird, L. C. (1994). Tight performance bounds on greedy policies based on imperfect value functions. In *Proceedings 8th Yale Workshop on Adaptive and Learning Systems*, pages 108–113, New Haven, US.

Wodarz, D. and Nowak, M. A. (1999). Specific therapy regimes could lead to long-term immunological control of HIV. *Proceedings of the National Academy of Sciences of the United States of America*, 96(25):14464–14469.

Xu, X., Hu, D., and Lu, X. (2007). Kernel-based least-squares policy iteration for reinforcement learning. *IEEE Transactions on Neural Networks*, 18(4):973–992.

Xu, X., Xie, T., Hu, D., and Lu, X. (2005). Kernel least-squares temporal difference learning. *International Journal of Information Technology*, 11(9):54–63.

Yen, J. and Langari, R. (1999). *Fuzzy Logic: Intelligence, Control, and Information.* Prentice Hall.

Yu, H. and Bertsekas, D. P. (2006). Convergence results for some temporal difference methods based on least-squares. Technical Report LIDS 2697, Massachusetts Institute of Technology, Cambridge, US. Available at http://www.mit.edu/people/dimitrib/lspe_lids_final.pdf.

Yu, H. and Bertsekas, D. P. (2009). Convergence results for some temporal difference methods based on least squares. *IEEE Transactions on Automatic Control*, 54(7):1515–1531.

List of algorithms

List of Algorithms

Index

RECEIVED
JUL -7 2010